Newnes
Radio and RF
Engineer's
Pocket Book

John Davies
Joseph J. Carr

OXFORD   AUCKLAND   BOSTON
JOHANNESBURG   MELBOURNE   NEW DELHI

Newnes
An imprint of Butterworth-Heinemann
Linacre House, Jordan Hill, Oxford OX2 8DP
225 Wildwood Avenue, Woburn, MA 01801-2041
A division of Reed Educational and Professional Publishing Ltd

A member of the Reed Elsevier plc group

First published 1994 as *Newnes Radio Engineer's Pocket Book*

Second edition 2000

British Library Cataloguing in Publication Data
Davies, John
  Newnes Radio Engineer's Pocket Book
  I. Title
  621.384

ISBN 0 7506 46004

Library of Congress Cataloguing in Publication Data
Davies, John.
  Newnes radio engineer's pocket book/John Davies.
    p.   cm.
  Includes bibliographical references and index.
  ISBN 0 7506 46004
  1. Radio – Handbooks, manuals, etc. 2. Mobile communication
systems – Handbooks, manuals, etc. I. Title. II. Title: Radio engineer's
pocket book.
  TK 6550.D243   94 – 12414
  621.384 – dc20   CIP

Typeset by Laser Words, Madras, India
Printed in Great Britain by Antony Rowe Ltd, Reading, Berkshire

Contents

Preface and acknowledgements xi

1 Propagation of radio waves 1
   1.1   Frequency and wavelength 1
   1.2   The radio frequency spectrum 2
   1.3   The isotropic radiator 2
   1.4   Formation of radio waves 4
   1.5   Behaviour of radio waves 7
   1.6   Methods of propagation 14
   1.7   Other propagation topics 20

2 The decibel scale 28
   2.1   Decibels and the logarithmic scale 28
   2.2   Decibels referred to absolute values 28

3 Transmission lines 38
   3.1   General considerations 38
   3.2   Impedance matching 38
   3.3   Base band lines 39
   3.4   Balanced line hybrids 41
   3.5   Radio frequency lines 42
   3.6   Waveguides 54
   3.7   Other transmission line considerations 56

4 Antennas 61
   4.1   Antenna characteristics 61
   4.2   Antenna types 66
   4.3   VHF and UHF antennas 71
   4.4   Microwave antennas 80
   4.5   Loop antennas 84

5 Resonant circuits 91
   5.1   Series and parallel tuned circuits 91
   5.2 Q factor 93
   5.3   Coupled (band-pass) resonant circuits 93

6 Oscillators 97
   6.1   Oscillator requirements 97
   6.2   Tunable oscillators 97
   6.3   Quartz crystal oscillators 99

 6.4 Frequency synthesizers 101
 6.5 Caesium and rubidium frequency
   standards 107

7 Piezo-electric devices 108
 7.1 Piezo-electric effect 108
 7.2 Quartz crystal characteristics 111
 7.3 Specifying quartz crystals 114
 7.4 Filters 116

8 Bandwidth requirements and modulation 121
 8.1 Bandwidth of signals at base band 121
 8.2 Modulation 124
 8.3 Analogue modulation 124
 8.4 Digital modulation 134
 8.5 Spread spectrum transmission 140

9 Frequency planning 144
 9.1 International and regional planning 144
 9.2 National planning 144
 9.3 Designations of radio emissions 146
 9.4 Bandwidth and frequency
   designations 147
 9.5 General frequency allocations 148
 9.6 Classes of radio stations 151
 9.7 Radio wavebands 154

10 Radio equipment 155
 10.1 Transmitters 155
 10.2 Receivers 160
 10.3 Programmable equipment 170

11 Microwave communication 173
 11.1 Microwave usage 173
 11.2 Propagation 173
 11.3 *K* factor 175
 11.4 Fresnel zones, reflections and multi-path
   fading 177
 11.5 Performance criteria for analogue and
   digital links 179
 11.6 Terminology 180
 11.7 Link planning 180
 11.8 Example of microwave link plan 180

12 Information privacy and encryption 182
 12.1 Encryption principles 182
 12.2 Speech encryption 183

12.3 Data encryption 184
12.4 Code division multiple access (CDMA) or spread spectrum 187
12.5 Classification of security 188

13 Multiplexing 189
13.1 Frequency division multiplex (FDM) 189
13.2 Time division multiplex (TDM) 190
13.3 Code division multiple access (CDMA) 192

14 Speech digitization and synthesis 193
14.1 Pulse amplitude modulation (PAM) 193
14.2 Pulse code modulation (PCM) 194
14.3 Delta modulation 195
14.4 Speech synthesis 196

15 VHF and UHF mobile communication 197
15.1 Operating procedures 197
15.2 Control of base stations 200
15.3 Common base station (CBS) operation 201
15.4 Wide area coverage 202

16 Signalling 209
16.1 Sub-audio signalling 209
16.2 In-band tone and digital signalling 210
16.3 Digital signalling 211
16.4 Standard PSTN tones 213

17 Channel occupancy, availability and trunking 215
17.1 Channel occupancy and availability 215
17.2 Trunking 217
17.3 In-band interrupted scan (IBIS) trunking 218
17.4 Trunking to MPT 1327 specification 219

18 Mobile radio systems 221
18.1 Paging 221
18.2 Cordless telephones 222
18.3 Trunked radio 223
18.4 Analogue cellular radio-telephone networks 225
18.5 Global system mobile (GSM) 226
18.6 Personal communication network (PCN) 227

18.7 Private mobile radio (PMR) 227
18.8 UK CB radio 227

19 Base station site management 229
   19.1 Base station objectives 229
   19.2 Site ownership or accommodation rental? 229
   19.3 Choice of site 230
   19.4 Masts and towers 231
   19.5 Installation of electronic equipment 232
   19.6 Earthing and protection against lightning 233
   19.7 Erection of antennas 235
   19.8 Interference 237
   19.9 Antenna multi-coupling 241
   19.10 Emergency power supplies 242
   19.11 Approval and certification 245

20 Instrumentation 247
   20.1 Accuracy, resolution and stability 247
   20.2 Audio instruments 248
   20.3 Radio frequency instruments 251

21 Batteries 262
   21.1 Cell characteristics 262
   21.2 Non-rechargeable, primary batteries 264
   21.3 Rechargeable batteries 269

22 Satellite communications 273
   22.1 Earth orbits 273
   22.2 Communications by satellite link 275
   22.3 Proposed satellite television formats 282
   22.4 Global positioning system (GPS) 282

23 Connectors and interfaces 284
   23.1 Audio and video connectors 284
   23.2 Co-axial connector 286
   23.3 Interfaces 295

24 Broadcasting 307
   24.1 Standard frequency and time transmissions 307
   24.2 Standard frequency formats 308
   24.3 UK broadcasting bands 310
   24.4 BBC AM radio stations 310
   24.5 BBC VHF broadcasting 311

24.6	UK television channels and transmitters	320
24.7	Characteristics of UHF terrestrial television systems	340
24.8	Terrestrial television channels	343
24.9	Terrestrial television aerial dimensions	347
24.10	AM broadcast station classes (USA)	348
24.11	FM broadcast frequencies and channel numbers (USA)	349
24.12	US television channel assignments	352
24.13	Calculating radio antenna great circle bearings	353

25 Abbreviations and symbols 359

25.1	Abbreviations	359
25.2	Letter symbols by unit name	363
25.3	Electric quantities	368
25.4	Transistor letter symbols	369
25.5	Component symbols	375
25.6	Radiocommunications symbols	381
25.7	Block diagram symbols	385
25.8	Frequency spectrum symbols	387
25.9	Equipment marking symbols	388

26 Miscellaneous data 395

26.1	Fundamental constants	395
26.2	Electrical relationships	395
26.3	Dimensions of physical properties	396
26.4	Fundamental units	396
26.5	Greek alphabet	397
26.6	Standard units	397
26.7	Decimal multipliers	399
26.8	Electronic multiple and sub-multiple conversion	400
26.9	Useful formulae	402
26.10	Colour codes	407
26.11	RC time constants	410
26.12	RL time constants	412
26.13	Reactance of capacitors at spot frequencies	414
26.14	Reactance of inductors at spot frequencies	414
26.15	Boundaries of sea areas	415
26.16	The Beaufort scale	416
26.17	Signal rating codes	416
26.18	World time	418

26.19 International allocation of call signs   426
26.20 Amateur radio   431
26.21 Microwave band designation system   441
26.22 International 'Q' code   442
26.23 RST Code   444
26.24 International Morse Code   444
26.25 Phonetic alphabet   445
26.26 Miscellaneous international
       abbreviations   446
26.27 Post WARC-79 radio astronomy frequency
       allocations   447
26.28 Laws   460
26.29 CCITT recommendations   464
26.30 Powers of numbers   465
26.31 Sound   471
26.32 Paper sizes   475
26.33 Fuses   475
26.34 Statistical formulae   476
26.35 Particles of modern physics   477
26.36 Calculus   478
26.37 Mensuration   478
26.38 Trigonometrical relationships   485
26.39 Transistor circuits and characteristics   486
26.40 Astronomical data   487
26.41 Resistivities of selected metals and
       alloys   489
26.42 Electrical properties of elements   492
26.43 Wire data and drill sizes   498

Glossary   517

Index   579

Preface

This edition of the *Newnes Radio and RF Engineer's Pocket Book* is something special. It is a compendium of information of use to engineers and technologists who are engaged in radio and RF engineering. It has been updated to reflect the changing interests of those communities, and reflects a view of the technology like no other. It is packed with information!

This whole series of books is rather amazing with regard to the range and quality of the information they provide, and this book is no different. It covers topics as diverse as circuit symbols and the abbreviations used for transistors, as well as more complex things as satellite communications and television channels for multiple countries in the English speaking world. It is a truly amazing work.

We hope that you will refer to this book frequently, and will enjoy it as much as we did in preparing it.

John Davies
Joseph J. Carr

Acknowledgements

I gratefully acknowledge the ready assistance offered by the following organizations: Andrew Ltd, Aspen Electronics Ltd, BBC, British Telecommunications plc, Farnell Instruments Ltd, Independent Television Authority, International Quartz Devices Ltd, Jaybeam Ltd, MACOM Greenpar Ltd, Marconi Instruments Ltd, Panorama Antennas Ltd, Radiocommunications Agency, the Radio Authority, RTT Systems Ltd. A special thanks goes to my wife Dorothy for once again putting up with my months of seclusion during the book's preparation.

1

PROPAGATION OF RADIO WAVES

1.1 Frequency and wavelength

There is a fixed relationship between the frequency and the wavelength, which is the distance between identical points on two adjacent waves (Figure 1.1), of any type of wave: sound (pressure), electromagnetic (radio) and light. The type of wave and the speed at which the wavefront travels through the medium determines the relationship. The speed of propagation is slower in higher density media.

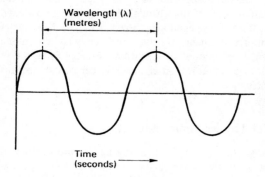

Fig. 1.1 *Frequency and wavelength*

Sound waves travel more slowly than radio and light waves which, in free space, travel at the same speed, approximately 3×10^8 metres per second, and the relationship between the frequency and wavelength of a radio wave is given by:

$$\lambda = \frac{3 \times 10^8}{f} \text{metres}$$

where λ is the wavelength and f is the frequency in hertz (Hz).

1.2 The radio frequency spectrum

The electromagnetic wave spectrum is shown in Figure 1.2: the part usable for radio communication ranges from below 10 kHz to over 100 GHz.

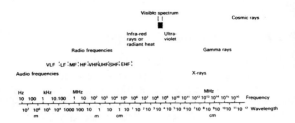

Fig. 1.2 *The electromagnetic wave spectrum*

The radio spectrum is divided into bands and the designation of the bands, their principal use and method of propagation is shown in Table 1.1. Waves of different frequencies behave differently and this, along with the amount of spectrum available in terms of radio communication channels in each band, governs their use.

1.3 The isotropic radiator

A starting point for considering the propagation of radio- or lightwaves is the isotropic radiator, an imaginary point source radiating equally in all directions in free space. Such a radiator placed at the centre of a sphere illuminates equally the complete surface of the sphere. As the surface area of a sphere is given by $4\pi r^2$ where r is the radius of the sphere, the brillance of illumination at any point on the surface varies inversely with the distance from the radiator. In radio terms, the power density at distance from the source is given by:

$$P_d = \frac{P_t}{4\pi r^2}$$

where P_t = transmitted power.

Table 1.1 *Use of radio frequencies*

Frequency band	Designation, use and propagation
3–30 kHz	Very low frequency (VLF). Worldwide and long distance communications. Navigation. Submarine communications. Surface wave.
30–300 kHz	Low frequency (LF). Long distance communications, time and frequency standard stations, long-wave broadcasting. Ground wave.
300–3000 kHz	Medium frequency (MF) or medium wave (MW). Medium-wave local and regional broadcasting. Marine communications. Ground wave.
3–30 MHz	High frequency (HF). 'Short-wave' bands. Long distance communications and short-wave broadcasting. Ionospheric sky wave.
30–300 MHz	Very high frequency (VHF). Short range and mobile communications, television and FM broadcasting. Sound broadcasting. Space wave.
300–3000 MHz	Ultra high frequency (UHF). Short range and mobile communications. Television broadcasting. Point-to-point links. Space wave. Note: The usual practice in the USA is to designate 300–1000 MHz as 'UHF' and above 1000 MHz as 'microwaves'.
3–30 GHz	Microwave or super high frequency (SHF). Point-to-point links, radar, satellite communications. Space wave.
Above 30 GHz	Extra high frequency (EHF). Inter-satellite and micro-cellular radio-telephone. Space wave.

1.4 Formation of radio waves

Radio waves are electromagnetic. They contain both electric and magnetic fields at right angles to each other and also at right angles to the direction of propagation. An alternating current flowing in a conductor produces an alternating magnetic field surrounding it and an alternating voltage gradient – an electric field – along the length of the conductor. The fields combine to radiate from the conductor as in Figure 1.3.

Fig. 1.3 *Formation of electromagnetic wave*

The plane of the electric field is referred to as the E plane and that of the magnetic field as the H plane. The two fields are equivalent to the voltage and current in a wired circuit. They are measured in similar terms, volts per metre and amperes per metre, and the medium through which they propagate possesses an impedance. Where $E = ZI$ in a wired circuit, for an electromagnetic wave:

$$E = ZH$$

where

> E = the RMS value of the electric field strength, V/metre
>
> H = the RMS value of the magnetic field strength, A/metre
>
> Z = characteristic impedance of the medium, ohms

The voltage is that which the wave, passing at the speed of light, would induce in a conductor one metre long.

The characteristic impedance of the medium depends on its permeability (equivalent of inductance) and permittivity (equivalent of capacitance). Taking the accepted figures for free space as:

$$\mu = 4\pi \times 10^{-7} \text{ henrys (H) per metre}$$
$$\text{(permeability) and}$$
$$\varepsilon = 1/36\pi \times 10^9 \text{ farads (F) per metre}$$
$$\text{(permittivity)}$$

then the impedance of free space, Z, is given by:

$$\sqrt{\frac{\mu}{\varepsilon}} = 120\pi = 377 \text{ ohms}$$

The relationship between power, voltage and impedance is also the same for electromagnetic waves as for electrical circuits, $W = E^2/Z$.

The simplest practical radiator is the elementary doublet formed by opening out the ends of a pair of wires. For theoretical considerations the length of the radiating portions of the wires is made very short in relation to the wavelength of the applied current to ensure uniform current distribution throughout their length. For practical applications the length of the radiating elements is one half-wavelength ($\lambda/2$) and the doublet then becomes a dipole antenna (Figure 1.4).

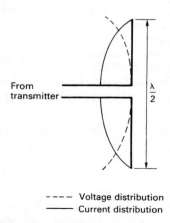

From transmitter

$\frac{\lambda}{2}$

- - - - Voltage distribution
——— Current distribution

Fig. 1.4 *Doublet (dipole) antenna*

When radiation occurs from a doublet the wave is polarized. The electric field lies along the length of the radiator (the E plane) and the magnetic field (the H plane) at right angles to it. If the E plane is vertical, the radiated field is said to vertically polarized. Reference to the E and H planes avoids confusion when discussing the polarization of an antenna.

Unlike the isotropic radiator, the dipole possesses directivity, concentrating the energy in the H plane at the expense of the E plane. It effectively provides a power gain in the direction of the H plane compared with an isotropic radiator. This gain is 1.6 times or 2.15 dBi (dBi means dB relative to an isotropic radiator).

For a direct ray the power transfer between transmitting and receiving isotropic radiators is inversely proportional to the distance between them *in wavelengths*. The free space power loss is given by:

$$\text{Free space loss, dB} = 10 \log_{10} \frac{(4\pi d)^2}{\lambda^2}$$

where d and λ are in metres, or:

$$\text{Free space loss (dB)} = 32.4 + 20 \times \log_{10} d \\ + 20 \times \log_{10} f$$

where d = distance in km and f = frequency in MHz.

The free space power loss, therefore, increases as the square of the distance and the frequency. Examples are shown in Figure 1.5.

With practical antennas, the power gains of the transmitting and receiving antennas, in dBi, must be subtracted from the free space loss calculated as above. Alternatively, the loss may be calculated by:

$$\text{Free space loss (dB)} = 10 \log_{10} \left[\frac{(4\pi d)^2}{\lambda^2} \\ \times \frac{1}{G_t \times G_r} \right]$$

where G_t and G_r are the respective actual gains, not in dB, of the transmitting and receiving antennas.

A major loss in microwave communications and radar systems is atmospheric attenuation (see

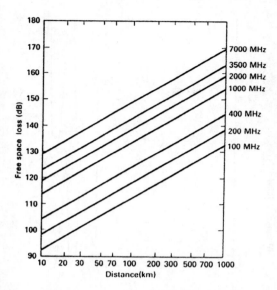

Fig. 1.5 *Free space loss vs. distance and frequency*

Figure 1.6). The attenuation (in decibels per kilometre (dB/km)) is a function of frequency, with especial problems showing up at 22 GHz and 64 GHz. These spikes are caused by *water vapour* and *atmospheric oxygen* absorption of microwave energy, respectively. Operation of any microwave frequency requires consideration of atmospheric losses, but operation near the two principal spike frequencies poses special problems. At 22 GHz, for example, an additional 1 dB/km of loss must be calculated for the system.

1.5 Behaviour of radio waves

1.5.1 Physical effects

The physical properties of the medium through which a wave travels, and objects in or close to its path, affect the wave in various ways.

Absorption

In the atmosphere absorption occurs and energy is lost in heating the air molecules. Absorption caused by

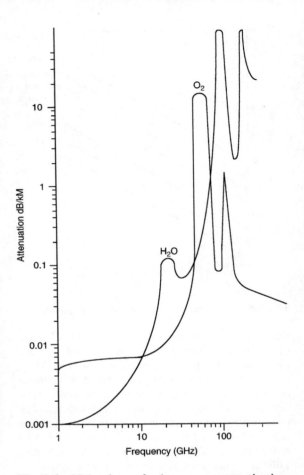

Fig. 1.6 *Major loss of microwave communications and radar systems due to atmospheric attenuation*

this is minimal at frequencies below about 10 GHz but absorption by foliage, particularly when wet, is severe at VHF and above.

Waves travelling along the earth's surface create currents in the earth causing ground absorption which increases with frequency. A horizontally polarized surface wave suffers more ground absorption than a vertically polarized wave because of the 'short-circuiting' by the ground of the electric field. Attenuation at a given frequency is least for propagation over water and greatest over dry ground for a vertically polarized wave.

Refraction and its effect on the radio horizon

As radio waves travel more slowly in dense media and the densest part of the atmosphere is normally the lowest, the upper parts of a wave usually travel faster than the lower. This refraction (Figure 1.7) has the effect of bending the wave to follow the curvature of the earth and progressively tilting the wavefront until eventually the wave becomes horizontally polarized and short-circuited by the earth's conductivity.

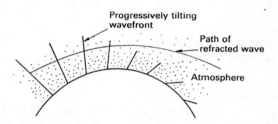

Fig. 1.7 *Effects of refraction*

Waves travelling above the earth's surface (space waves) are usually refracted downwards, effectively increasing the radio horizon to greater than the visual.

The refractive index of the atmosphere is referred to as the K factor; a K factor of 1 indicates zero refraction. Most of the time K is positive at 1.33 and the wave is bent to follow the earth's curvature. The radio horizon is then 4/3 times the visual. However, the density of the atmosphere varies from time to time and in different parts of the world. Density inversions where higher density air is above a region of low density may also occur. Under these conditions the K factor is negative and radio waves are bent away from the earth's surface and are either lost or ducting occurs. A K factor of 0.7 is the worst expected case.

Ducting occurs when a wave becomes trapped between layers of differing density only to be returned at a great distance from its source, possibly creating interference.

Radio horizon distance at VHF/UHF

The radio horizon at VHF/UHF and up is approximately 15% further than the optical horizon. Several equations are used in calculating the distance. If D is the distance to the radio horizon, and H is the antenna height, then:

$$D = k\sqrt{H}$$

- When D is in statute miles (5280 feet) and H in feet, then $k = 1.42$.
- When D is in nautical miles (6000 feet) and H in feet, then $k = 1.23$.
- When D is in kilometres and H is in metres, then $k = 4.12$.

Repeating the calculation for the receiving station and adding the results gives the total path length.

Diffraction

When a wave passes over on the edge of an obstacle some of its energy is bent in the direction of the obstacle to provide a signal in what would otherwise be a shadow. The bending is most severe when the wave passes over a sharp edge (Figure 1.8).

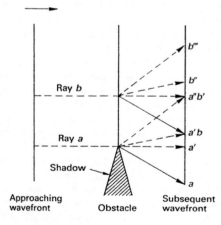

Fig. 1.8 *Effects of diffraction*

As with light waves, the subsequent wavefront consists of wavelets produced from an infinite number of points on the wavefront, rays *a* and *b* in Figure 1.8 (Huygens' principle). This produces a pattern of interfering waves of alternate addition and subtraction.

Reflection

Radio waves are reflected from surfaces lying in and along their path and also, effectively, from ionized layers in the ionosphere – although most of the reflections from the ionized layers are actually the products of refraction. The strength of truly reflected signals increases with frequency, and the conductivity and smoothness of the reflecting surface.

Multi-path propagation

Reflection, refraction and diffraction may provide signals in what would otherwise be areas of no signal, but they also produce interference.

Reflected – or diffracted – signals may arrive at the receiver in any phase relationship with the direct ray and with each other. The relative phasing of the signals depends on the differing lengths of their paths and the nature of the reflection.

When the direct and reflected rays have followed paths differing by an odd number of half-wavelengths they could be expected to arrive at the receiver in anti-phase with a cancelling effect. However, in the reflection process a further phase change normally takes place. If the reflecting surface had infinite conductivity, no losses would occur in the reflection, and the reflected wave would have exactly the same or opposite phase as the incident wave depending on the polarization in relation to the reflecting surface. In practice, the reflected wave is of smaller amplitude than the incident, and the phase relationships are also changed. The factors affecting the phasing are complex but most frequently, in practical situations, approximately 180° phase change occurs on reflection, so that reflected waves travelling an odd number of half-wavelengths arrive in phase with the direct wave while those travelling an even number arrive anti-phase.

As conditions in the path between transmitter and receiver change so does the strength and path length of reflected signals. This means that a receiver may be subjected to signal variations of almost twice the mean level and practically zero, giving rise to severe fading. This type of fading is frequency selective and occurs on troposcatter systems and in the mobile environment where it is more severe at higher frequencies. A mobile receiver travelling through an urban area can receive rapid signal fluctuations caused by additions and cancellations of the direct and reflected signals at half-wavelength intervals. Fading due to the multi-path environment is often referred to as Rayleigh fading and its effect is shown in Figure 1.9. Rayleigh fading, which can cause short signal dropouts, imposes severe restraints on mobile data transmission.

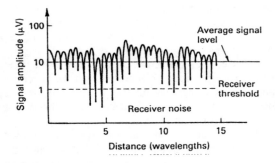

Fig. 1.9 *Multi-path fading*

Noise

The quality of radio signals is not only degraded by the propagation losses: natural or manmade electrical noise is added to them, reducing their intelligibility.

Atmospheric noise includes static from thunderstorms which, unless very close, affects frequencies below about 30 MHz and noise from space is apparent at frequencies between about 8 MHz to 1.5 GHz.

A type of noise with which radio engineers are continually concerned is thermal. Every resistor produces noise spread across the whole frequency spectrum. Its magnitude depends upon the ohmic value of the resistor, its temperature and the bandwidth of the following circuits. The noise voltage produced by a resistor is given by:

$$E_n = \sqrt{4kTBR}$$

where

E_n = noise voltage, V(RMS)

k = Boltzmann's constant

= 1.38×10^{-23} joules/kelvin

T = temperature in degrees K

B = bandwidth of measurement, Hz

R = resistance in ohms

An antenna possesses resistance and its thermal noise, plus that of a receiver input circuit, is a limiting factor to receiver performance.

Noise is produced in every electronic component. Shot noise – it sounds like falling lead shot – caused by the random arrival of electrons at, say, the collector of a transistor, and the random division of electrons at junctions in devices, add to this noise.

Doppler effect

Doppler effect is an apparent shift of the transmitted frequency which occurs when either the receiver or transmitter is moving. It becomes significant in mobile radio applications towards the higher end of the UHF band and on digitally modulated systems.

When a mobile receiver travels directly towards the transmitter each successive cycle of the wave has less distance to travel before reaching the receiving antenna and, effectively, the received frequency is raised. If the mobile travels away from the transmitter, each successive cycle has a greater distance to travel and the frequency is lowered. The variation in frequency depends on the frequency of the wave, its propagation velocity and the velocity of the vehicle containing the receiver. In the situation where the velocity of the vehicle is small compared with the velocity of light, the frequency shift when moving directly towards, or away from, the transmitter is given to sufficient accuracy for most purposes by:

$$f_d = \frac{V}{C} f_t$$

where

f_d = frequency shift, Hz

f_t = transmitted frequency, Hz

$$V = \text{velocity of vehicle, m/s}$$
$$C = \text{velocity of light, m/s}$$

Examples are:

- 100 km/hr at 450 MHz, frequency shift = 41.6 Hz
- 100 km/hr at 1.8 GHz – personal communication network (PCN) frequencies – frequency shift = 166.5 Hz
- Train at 250 km/hr at 900 MHz – a requirement for the GSM pan-European radio-telephone – frequency shift = 208 Hz

When the vehicle is travelling at an angle to the transmitter the frequency shift is reduced. It is calculated as above and the result multiplied by the cosine of the angle of travel from the direct approach (Figure 1.10).

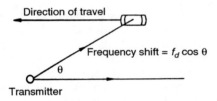

Fig. 1.10 *Doppler frequency shift and angle to transmitter*

In a radar situation Doppler effect occurs on the path to the target and also to the reflected signal so that the above formula is modified to:

$$f_d = \frac{2V}{C} f_t$$

where f_d is now the total frequency shift.

1.6 Methods of propagation

The effects of all of the above phenomena vary with frequency and are used in the selection of frequencies for specific purposes. The behaviour of waves of different frequencies gives rise to the principal types of wave propagation.

Ground wave propagation

Waves in the bands from very low frequencies (VLF, 3–30 kHz), low frequencies (LF, 30–300 kHz) and medium frequencies (MF, 300–3000 kHz) travel close to the earth's surface: the ground wave (Figure 1.11). Transmissions using the ground wave must be vertically polarized to avoid the conductivity of the earth short-circuiting the electric field.

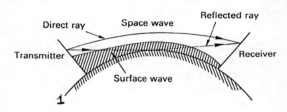

Fig. 1.11 *Components of the ground wave*

The ground wave consists of a surface wave and a space wave. The surface wave travels along the earth's surface, and is attenuated by ground absorption and the tilting of the wavefront due to diffraction. The losses increase with frequency and thus VLF radio stations have a greater range than MF stations. The attenuation is partially offset by the replacement of energy from the upper part of the wave refracted by the atmosphere.

The calculation of the field strength of the surface wave at a distance from a transmitter is complex and affected by several variables. Under plane earth conditions and when the distance is sufficiently short that the earth's curvature can be neglected the field intensity is given by:

$$E_{su} = \frac{2E_0}{d}A$$

where

E_{su} = field intensity in same units as E_0

d = distance in same units of distance as used in E_0

A = a factor calculated from the earth losses, taking frequency, dielectric constant and conductivity into account

E_0 = the free space field produced at unit distance from the transmitter. (With a short (compared with $\lambda/4$) vertical aerial, $2E_0 = 300\sqrt{P}$ mV/m at 1 km where P is the radiated power in kW.) (Terman, 1943)

For a radiated power of 1 kW and ground of average dampness, the distance at which a field of 1 mV/m will exist is given in Table 1.2.

Table 1.2 *Distance at which a field of 1 mV/m will exist for a radiated power of 1 kW and ground of average dampness*

Frequency	Range (km)
100 kHz	200
1 MHz	60
10 MHz	6
100 MHz	1.5

The direct and reflected components of the ground wave produce multi-path propagation and variations in received single strength will arise depending on the different path lengths taken by the two components. When the transmitting and receiving antennas are at ground level the components of the space wave cancel each other and the surface wave is dominant. When the antennas are elevated, the space wave becomes increasingly strong and a height is eventually reached where the surface wave has a negligible effect on the received signal strength.

Sky wave propagation
High frequency (HF) waves between 3 MHz and 30 MHz are effectively reflected by ionized layers in the ionosphere producing the sky wave. Medium frequency waves may also be reflected, but less reliably.

The ionosphere contains several layers of ionized air at varying altitudes (Figure 1.12). The heights and density of the layers vary diurnally, seasonally and with the incidence of sunspot activity. The E and

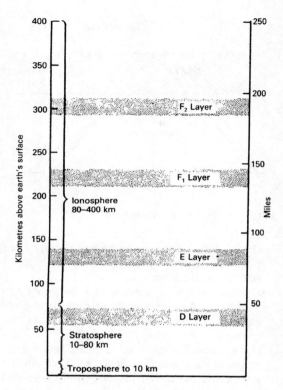

Fig. 1.12 *The ionosphere*

F_2 layers are semi-permanent while the F_1 layer is normally only present during daytime.

Radio waves radiated at a high angle and reflected by these layers return to earth at a distance from the transmitter. The HF reflection process is in reality one of refraction in layers possessing a greater free electron density than at heights above or below them. The speed of propagation is slowed on entering a layer and the wave is bent and, if of a suitable frequency and angle of incidence, returned to earth (Figure 1.13). The terms used are defined as follows:

- *Virtual height*. The height at which a true reflection of the incident wave would have occurred (Figure 1.13).

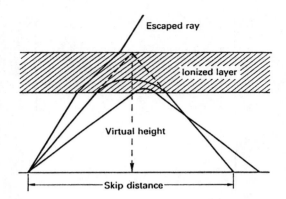

Fig. 1.13 *Sky wave propagation*

- *Critical frequency* (f_c). The highest frequency that would be returned to earth in a wave directed vertically at the layer.
- *Maximum usable frequency (muf)*. The highest frequency that will be returned to earth for a given angle of incidence. If the angle of incidence to the normal is θ, the muf $= f_c / \cos \theta$.
- *Skip distance*. The minimum distance from the transmitter, along the surface of the earth, at which a wave above the critical frequency will be returned to earth (Figure 1.12). Depending on the frequency, the ground wave will exist at a short distance from the transmitter.
- *Sporadic E-layer reflections*. Reflections from the E layer at frequencies higher than those which would normally be returned to earth. They appear to be reflections from electron clouds having sharp boundaries and drifting through the layer. As the name implies the reflections are irregular but occur mostly in summer and at night.

Space wave propagation
The space wave travels through the troposphere (the atmosphere below the ionosphere) between the transmitter and the receiver. It contains both direct and reflected components (see Figure 1.11), and is affected by refraction and diffraction. The importance of these

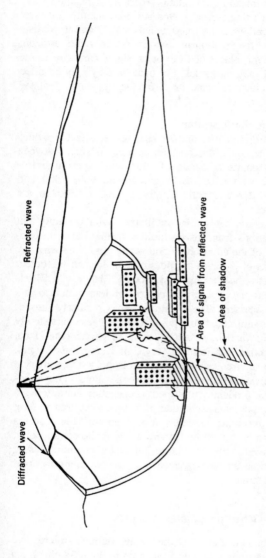

Fig. 1.14 *Pictorial representation of radio coverage from a base station*

Refracted wave

Diffracted wave

Area of signal from reflected wave

Area of shadow

effects varies with frequency, the nature of the terrain and of objects close to the direct path, and the type of communication, e.g. data. Apart from medium-wave broadcasting, space waves are used mainly for communications using frequencies of VHF and upwards.

The range of space waves is the radio horizon. However, places of little or no signal can arise in the lee of radio obstacles. Fortunately, they may be filled with either reflected or diffracted signals as depicted in Figure 1.14.

Tropospheric scatter

The tropospheric, or forward, scatter effect provides reliable, over the horizon, communication between fixed points at bands of ultra and super high frequencies. Usable bands are around 900, 2000 and 5000 MHz and path lengths of 300 to 500 km are typical.

The mechanism is not known with certainty but reflections from discontinuities in the dielectric constant of the atmosphere and scattering of the wave by clouds of particles are possibilities. It is an inefficient process, the scattered power being 60 to 90 dB relative to the incident power, so high transmitter powers are necessary. The phenomenon is regularly present but is subject to two types of fading. One occurs slowly and is due to variations of atmospheric conditions. The other is a form of Rayleigh fading and is rapid, deep and frequency selective. It is due to the scattering occurring at different points over a volume in the atmosphere producing multipath propagation conditions.

Troposcatter technique uses directional transmitting and receiving antennas aimed so that their beams intercept in the troposphere at the mid-distance point. To overcome the fading, diversity reception using multiple antennas spaced over 30 wavelengths apart is common.

1.7 Other propagation topics

Communications in the VHF through microwave regions normally takes place on a 'line-of-sight' basis where the radio horizon defines the limit of sight. In practice, however, the situation is not so neat and simple. There is a transition region between the HF and VHF where long distance ionospheric 'skip' occurs only occasionally. This effect is seen above 25 MHz, and is quite pronounced in the 50 MHz region. Sometimes the region behaves like line-of-sight VHF, and at others like HF shortwave.

1.7.1 Scatter

There are a number of scatter modes of propagation. These modes can extend the radio horizon a considerable amount. Where the radio horizon might be a few tens of kilometres, underscatter modes permit very much longer propagation. For example, a local FM broadcaster at 100 MHz might have a service area of about 40 miles, and might be heard 180 miles away during the summer months when *Sporadic-E* propagation occurs. One summer, a television station in Halifax, Nova Scotia, Canada, was routinely viewable in Washington, DC in the United States during the early morning hours for nearly a week.

Sporadic-E is believed to occur when a small region of the atmosphere becomes differentially ionized, and thereby becomes a species of 'radio mirror.' Ionospheric scatter propagation occurs when clouds of ions exist in the atmosphere. These clouds can exist in both the ionosphere and the troposphere, although the tropospheric model is more reliable for communications. A signal impinging this region may be scattered towards other terrestrial sites which may be a great distance away. The specific distance depends on the geometry of the scenario.

There are at least three different modes of scatter from ionized clouds: *back scatter, side scatter*, and *forward scatter*. The back scatter mode is a bit like radar, in that signal is returned back to the transmitter site, or in regions close to the transmitter. Forward scatter occurs when the reflected signal continues in the same azimuthal direction (with respect to the transmitter), but is redirected toward the Earth's surface. Side scatter is similar to forward scatter, but the azimuthal direction might change.

Unfortunately, there are often multiple reflections from the ionized cloud, and these are shown as 'multiple scatter' in Figure 1.15. When these reflections are

21

able to reach the receiving site, the result is a rapid, fluttery fading that can be of quite profound depths.

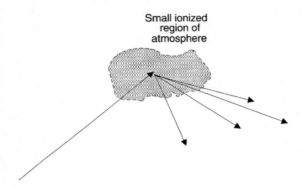

Fig. 1.15 *Multiple scatter*

Meteor scatter is used for communication in high latitude regions. When a meteor enters the Earth's atmosphere it leaves an ionized trail of air behind it. This trail might be thousands of kilometres long, but is very short lived. Radio signals impinging the tubular metre ion trail are reflected back towards Earth. If the density of meteors in the critical region is high, then more or less continuous communications can be achieved. This phenomenon is noted in the low VHF between 50 and about 150 MHz. It can easily be observed on the FM broadcast band if the receiver is tuned to distant stations that are barely audible. If the geometry of the scenario is right, abrupt but short-lived peaks in the signal strength will be noted.

1.7.2 Refraction modes

Refraction is the mechanism for most tropospheric propagation phenomena. The dielectric properties of the air, which are set mostly by the moisture content, are a primary factor in tropospheric refraction. Refraction occurs in both light or radio wave systems when the wave passes between mediums of differing density. Under that situation, the wave path will bend an amount proportional to the difference in density of the two regions.

The general situation is typically found at UHF and microwave frequencies. Because air density normally decreases with altitude, the top of a beam of radio waves typically travels slightly faster than the lower portion of the beam. As a result, those signals refract a small amount. Such propagation provides slightly longer surface distances than are normally expected from calculating the distance to the radio horizon. This phenomenon is called *simple refraction*, and is described by the *K factor*.

Super refraction

A special case of refraction called super refraction occurs in areas of the world where warmed land air flows out over a cooler sea (Figure 1.16). Examples of such areas are deserts that are adjacent to a large body of water: the Gulf of Aden, the southern Mediterranean, and the Pacific Ocean off the coast of Baja, California. Frequent VHF/UHF/microwave communications up to 200 miles are reported in such areas, and up to 600 miles have reportedly been observed.

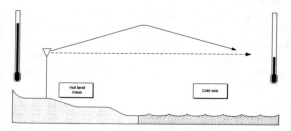

Fig. 1.16 *An example of super refraction*

Ducting

Another form of refraction phenomenon is weather related. Called *ducting*, this form of propagation is actually a special case of super refraction. Evaporation of sea water causes temperature inversion regions to form in the atmosphere. That is, layered air masses in which the air temperature is greater than in the layers below it (note: air temperature normally decreases with altitude, but at the boundary with an inversion region, it begins to increase). The inversion layer forms a 'duct' that acts similarly to a waveguide. Ducting allows long distance communications from lower VHF through microwave frequencies; with 50 MHz being a lower practical limit, and 10 GHz being an ill-defined upper limit. Airborne operators of radio, radar, and other electronic equipment can sometimes note ducting at even higher microwave frequencies.

Antenna placement is critical for ducting propagation. Both the receiving and transmitting antennas must be either: (a) physically inside the duct (as in airborne cases), or (b) able to propagate at an angle such that the signal gets trapped inside the duct. The latter is a function of antenna radiation angle. Distances up to 2500 miles or so are possible through ducting.

Certain paths, where frequent ducting occurs, have been identified: in the United States, the Great Lakes region to the southeastern Atlantic seaboard; Newfoundland to the Canary Islands; across the Gulf of Mexico from Florida to Texas; Newfoundland to the Carolinas; California to Hawaii; and Ascension Island to Brazil.

Subrefraction
Another refractive condition is noted in the polar regions, where colder air from the land mass flows out over warmer seas (Figure 1.17). Called *subrefraction*, this phenomena bends EM waves away from the Earth's surface – thereby reducing the radio horizon by about 30 to 40%.

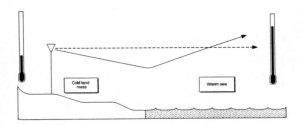

Fig. 1.17 *An example of subrefraction*

All tropospheric propagation that depends upon airmass temperatures and humidity shows diurnal (i.e. over the course of the day) variation caused by the local rising and setting of the sun. Distant signals may vary 20 dB in strength over a 24-hour period. These tropospheric phenomena explain how TV, FM broadcast, and other VHF signals can propagate great distances, especially along seacoast paths, sometimes weak and sometimes nonexistent.

1.7.3 Great circle paths

A great circle is the shortest line between two points on the surface of a sphere, such that it lays on a plane through the Earth's centre and includes the two points. When translated to 'radiospeak', a great circle is the shortest path on the surface of the Earth between two points. Navigators and radio operators use the great circle for similar, but different, reasons. Navigators use it in order to get from here to there, and radio operators use it to get a transmission path from here to there.

The heading of a directional antenna is normally aimed at the receiving station along its great circle path. Unfortunately, many people do not understand the concept well enough, for they typically aim the antenna in the wrong direction. For example, Washington, DC in the USA is on approximately the same latitude as Lisbon, Portugal. If you fly due east, you will have dinner in Lisbon, right? Wrong. If you head due east from Washington, DC, across the Atlantic, the first landfall would be west Africa, somewhere between Ghana and Angola. Why? Because the great circle bearing 90 degrees takes us far south. The geometry of spheres, not flat planes, governs the case.

Long path versus short path

The Earth is a sphere (or, more precisely, an 'oblique spheroid'), so from any given point to any other point there are two great circle paths: the long path (major arc) and the short path (minor arc). In general, the best reception occurs along the short path. In addition, short path propagation is more nearly 'textbook', compared with long path reception. However, there are times when long path is better, or is the only path that will deliver a signal to a specific location from the geographic location in question.

Grey line propagation

The Grey line is the twilight zone between the night and daytime halves of the earth. This zone is also called the *planetary terminator*. It varies up to +23 degrees either side of the north–south longitudinal lines, depending on the season of the year (it runs directly north–south only at the vernal and autumnal equinoxes). The D-layer of the ionosphere absorbs signals in the HF region. This layer disappears almost completely at night, but it builds up during the day. Along the grey line, the D-layer is rapidly decaying west of the line, and has not quite built up east of the line.

Brief periods of abnormal propagation occur along the grey line. Stations on either side of the line can be heard from regions, and at distances, that would otherwise be impossible on any given frequency. For this reason, radio operators often prefer to listen at dawn and dusk for this effect.

1.7.4 Scatter propagation modes

Auroral propagation

The auroral effect produces a luminescence in the upper atmosphere resulting from bursts of particles released from the sun 18 to 48 hours earlier. The light emitted is called the northern lights and the southern lights. The ionized regions of the atmosphere that create the lights form a radio reflection shield, especially at VHF and above, although 15 to 20 MHz effects are known. Auroral propagation effects are normally seen in the higher latitudes, although listeners in the southern tier of states in the USA and Europe are often treated to the reception of signals from the north being reflected from auroral clouds. Similar effects exist in the southern hemisphere.

Non-reciprocal direction

If you listen to the 40 metre (7–7.3 MHz) amateur radio band receiver on the East Coast of the United States, you will sometimes hear European stations – especially in the late afternoon. But when the US amateur tries to work those European stations there is no reply whatsoever. The Europeans are unable to hear the US stations. This propagation anomaly causes the radio wave to travel different paths dependent on which direction it travels, i.e. an east−west signal is not necessarily the reciprocal of a west−east signal. This anomaly can occur when a radio signal travels through a heavily ionized medium in the presence of a magnetic field, which is exactly the situation when the signal travels through the ionosphere in the presence of the Earth's magnetic field.

References

Belcher, R. *et al.* (1989). *Newnes Mobile Radio Servicing Handbook.* Butterworth-Heinemann, Oxford.

Kennedy, G. (1977). *Electronic Communications Systems.* McGraw-Hill Kogashuka, Tokyo.

Terman, F.E. (1943). *Radio Engineers' Handbook.* McGraw-Hill, London.

2

THE DECIBEL SCALE

2.1 Decibels and the logarithmic scale

The range of powers, voltages and currents encountered in radio engineering is too wide to be expressed on linear scale. Consequently, a logarithmic scale based on the decibel (dB, one tenth of a bel) is used. The decibel does not specify a magnitude of a power, voltage or current but a ratio between two values of them. Gains and losses in circuits or radio paths are expressed in decibels.

The ratio between two powers is given by:

$$\text{Gain or loss, dB} = 10 \log_{10} \frac{P_1}{P_2}$$

where P_1 and P_2 are the two powers.

As the power in a circuit varies with the square of the voltage or current, the logarithm of the ratio of these quantities must be multiplied by twenty instead of ten. To be accurate the two quantities under comparison must operate in identical impedances:

$$\text{Gain or loss, dB} = 20 \log_{10} \frac{V_1}{V_2}$$

To avoid misunderstandings, it must be realized that a ratio of 6 dB is 6 dB regardless of whether it is power, voltage or current that is referred to: if it is power, the ratio for 6 dB is four times; if it is voltage or current, the ratio is two times (Table 2.1).

2.2 Decibels referred to absolute values

While the decibel scale expresses ratios only, if a reference value is added to the statement as a suffix it can be used to refer to absolute values. For example, a loss of 10 dB means a reduction in power to a level equal to one tenth of the original and if the statement is −10 dBm the level referred to is 1/10

Table 2.1 *The decibel figures are in the centre column: figures to the left represent decibel loss, and those to the right decibel gain. The voltage and current figures are given on the assumption that there is no change in impedance.*

Voltage or current ratio	Power ratio	dB ← − + →	Voltage or current ratio	Power ratio
1.000	1.000	0	1.000	1.000
0.989	0.977	0.1	1.012	1.023
0.977	0.955	0.2	1.023	1.047
0.966	0.933	0.3	1.035	1.072
0.955	0.912	0.4	1.047	1.096
0.944	0.891	0.5	1.059	1.122
0.933	0.871	0.6	1.072	1.148
0.912	0.832	0.8	1.096	1.202
0.891	0.794	1.0	1.122	1.259
0.841	0.708	1.5	1.189	1.413
0.794	0.631	2.0	1.259	1.585
0.750	0.562	2.5	1.334	1.778
0.708	0.501	3.0	1.413	1.995
0.668	0.447	3.5	1.496	2.239
0.631	0.398	4.0	1.585	2.512
0.596	0.355	4.5	1.679	2.818
0.562	0.316	5.0	1.778	3.162
0.501	0.251	6.0	1.995	3.981
0.447	0.200	7.0	2.239	5.012
0.398	0.159	8.0	2.512	6.310
0.355	0.126	9.0	2.818	7.943
0.316	0.100	10	3.162	10.00
0.282	0.0794	11	3.55	12.6
0.251	0.0631	12	3.98	15.9
0.224	0.0501	13	4.47	20.0
0.200	0.0398	14	5.01	25.1
0.178	0.0316	15	5.62	31.6
0.159	0.0251	16	6.31	39.8
0.126	0.0159	18	7.94	63.1
0.100	0.0100	20	10.00	100.0
3.16×10^{-2}	10^{-3}	30	3.16×10	10^3
10^{-2}	10^{-4}	40	10^2	10^4
3.16×10^{-3}	10^{-5}	50	3.16×10^2	10^5
10^{-3}	10^{-6}	60	10^3	10^6
3.16×10^{-4}	10^{-7}	70	3.16×10^3	10^7
10^{-4}	10^{-8}	80	10^4	10^8
3.16×10^{-5}	10^{-9}	90	3.16×10^4	10^9
10^{-5}	10^{-10}	100	10^5	10^{10}
3.16×10^{-6}	10^{-11}	110	3.16×10^5	10^{11}
10^{-6}	10^{-12}	120	10^6	10^{12}

of a milliwatt. Commonly used suffixes and, where applicable, their absolute reference levels are as follows. Table 2.2 shows the relative levels in decibels at 50 ohms impedance.

Table 2.2 *Relative levels in decibels at 50 ohms impedance*

dBμV	Voltage	dBV	dBm	dBW	Power
−20	100 nV	−140	−127	−157	200 aW
−19	115	−139	−126	−156	250
−18	125	−138	−125	−155	315
−17	140	−137	−124	−154	400
−16	160	−136	−123	−153	500
−15	180	−135	−122	−152	630
−14	200	−134	−121	−151	800
−13	225	−133	−120	−150	1 fW
−12	250	−132	−119	−149	1.25
−11	280	−131	−118	−148	1.6
−10	315	−130	−117	−147	2.0
−9	355	−129	−116	−146	2.5
−8	400	−128	−115	−145	3.15
−7	450	−127	−114	−144	4.0
−6	500	−126	−113	−143	5.0
−5	565	−125	−112	−142	6.3
−4	630	−124	−111	−141	8.0
−3	710	−123	−110	−140	10
−2	800	−122	−109	−139	12.5
−1	900	−121	−108	-138	16
0	1 μV	−120	−107	−137	20
1	1.15	−119	−106	−136	25
2	1.25	−118	−105	−135	31.5
3	1.4	−117	−104	−134	40
4	1.6	−116	−103	−133	50
5	1.8	−115	−102	−132	63
6	2.0	−114	−101	−131	80
7	2.25	−113	−100	−130	100
8	2.5	−112	−99	−129	125
9	2.8	−111	−98	−128	160
10	3.15	−110	−97	−127	200
11	3.55	−109	−96	−126	250
12	4.0	−108	−95	−125	315
13	4.5	−107	−94	−124	400
14	5.0	−106	−93	−123	500
15	5.65	−105	−92	−122	630
16	6.3	−104	−91	−121	800
17	7.1	−103	−90	−120	1 pW
18	8.0	−102	−89	−119	1.25
19	9.0	−101	−88	−118	1.6

Table 2.2 *Continued*

dBμV	Voltage	dBV	dBm	dBW	Power
20	10.0	−100	−87	−117	2.0
21	11.5	−99	−86	−116	2.5
22	12.5	−98	−85	−115	3.15
23	14	−97	−84	−114	4.0
24	16	−96	−83	−113	5.0
25	18	−95	−82	−112	6.3
26	20	−94	−81	−111	8.0
27	22.5	−93	−80	−110	10
28	25	−92	−79	−109	12.5
29	28	−91	−78	−108	16
30	31.5	−90	−77	−107	20
31	35.5	−89	−76	−106	25
32	40	−88	−75	−105	31.5
33	45	−87	−74	−104	40
34	50	−86	−73	−103	50
35	56.5	−85	−72	−102	63
36	63	−84	−71	−101	80
37	71	−83	−70	−100	100
38	80	−82	−69	−99	125
39	90	−81	−68	−98	160
40	100	−80	−67	−97	200
41	115	−79	−66	−96	250
42	125	−78	−65	−95	315
43	140	−77	−64	−94	400
44	160	−76	−63	−93	500
45	180	−75	−62	−92	630
46	200	−74	−61	−91	800
47	225	−73	−60	−90	1 nW
48	250	−72	−59	−89	1.25
49	280	−71	−58	−88	1.6
50	315	−70	−57	−87	2.0
51	355	−69	−56	−86	2.5
52	400	−68	−55	−85	3.15
53	450	−67	−54	−84	4.0
54	500	−66	−53	−83	5.0
55	565	−65	−52	−82	6.3
56	630	−64	−51	−81	8.0
57	710	−63	−50	−80	10
58	800	−62	−49	−79	12.5
59	900	−61	−48	−78	16
60	1 mV	−60	−47	−77	20
61	1.15	−59	−46	−76	25
62	1.25	−58	−45	−75	31.5
63	1.4	−57	−44	−74	40
64	1.6	−56	−43	−73	50
65	1.8	−55	−42	−72	63
66	2.0	−54	−41	−71	80
67	2.25	−53	−40	−70	100

(*continued overleaf*)

Table 2.2 *Continued*

dBμV	Voltage	dBV	dBm	dBW	Power
68	2.5	−52	−39	−69	125
69	2.8	−51	−38	−68	160
70	3.15	−50	−37	−67	200
71	3.55	−49	−36	−66	250
72	4.0	−48	−35	−65	315
73	4.5	−47	−34	−64	400
74	5.0	−46	−33	−63	500
75	5.65	−45	−32	−62	630
76	6.3	−44	−31	−61	800
77	7.1	−43	−30	−60	1 μW
78	8.0	−42	−29	−59	1.25
79	9.0	−41	−28	−58	1.6
80	10 mV	−40	−27	−57	2.0
81	11.5	−39	−26	−56	2.5
82	12.5	−38	−25	−55	3.15
83	14	−37	−24	−54	4.0
84	16	−36	−23	−53	5.0
85	18	−35	−22	−52	6.3
86	20	−34	−21	−51	8.0
87	22.5	−33	−20	−50	10
88	25	−32	−19	−49	12.5
89	28	−31	−18	−48	16
90	31.5	−30	−17	−47	20
91	35.5	−29	−16	−46	25
92	40	−28	−15	−45	31.5
93	45	−27	−14	−44	40
94	50	−26	−13	−43	50
95	56.5	−25	−12	−42	63
96	63	−24	−11	−41	80
97	71	−23	−10	−40	100
98	80	−22	−9	−39	125
99	90	−21	−8	−38	160
100	100	−20	−7	−37	200
101	115	−19	−6	−36	250
102	125	−18	−5	−35	315
103	140	−17	−4	−34	400
104	160	−16	−3	−33	500
105	180	−15	−2	−32	630
106	200	−14	−1	−31	800
107	225	−13	0	−30	1 mW
108	250	−12	1	−29	1.25
109	280	−11	2	−28	1.6
110	315	−10	3	−27	2.0
111	355	−9	4	−26	2.5
112	400	−8	5	−25	3.15
113	450	−7	6	−24	4.0
114	500	−6	7	−23	5.0
115	565	−5	8	−22	6.3
116	630	−4	9	−21	8.0

Table 2.2 *Continued*

dB µV	Voltage	dBV	dBm	dBW	Power
117	710	−3	10	−20	10
118	800	−2	11	−19	12.5
119	900	−1	12	−18	16
120	1 V	0	13	−17	20
121	1.15	1	14	−16	25
122	1.25	2	15	−15	31.5
123	1.4	3	16	−14	40
124	1.6	4	17	−13	50
125	1.8	5	18	−12	63
126	2.0	6	19	−11	80
127	2.25	7	20	−10	100
128	2.5	8	21	−9	125
129	2.8	9	22	−8	160
130	3.15	10	23	−7	200
131	3.55	11	24	−6	250
132	4.0	12	25	−5	315
133	4.5	13	26	−4	400
134	5.0	14	27	−3	500
135	5.65	15	28	−2	630
136	6.3	16	29	−1	800
137	7.1	17	30	0	1 W
138	8.0	18	31	1	1.25
139	9.0	19	32	2	1.6
140	10	20	33	3	2.0
141	11.5	21	34	4	2.5
142	12.5	22	35	5	3.15
143	14	23	36	6	4.0
144	16	24	37	7	5.0
145	18	25	38	8	6.3
146	20	26	39	9	8.0
147	22.5	27	40	10	10
148	25	28	41	11	12.5
149	28	29	42	12	16
150	31.5	30	43	13	20
151	35.5	31	44	14	25
152	40	32	45	15	31.5
153	45	33	46	16	40
154	50	34	47	17	50
155	56.5	35	48	18	63
156	63	36	49	19	80
157	71	37	50	20	100
158	80	38	51	21	125
159	90	39	52	22	160
160	100	40	53	23	200
161	115	41	54	24	250
162	125	42	55	25	315
163	140	43	56	26	400
164	160	44	57	27	500

(*continued overleaf*)

Table 2.2 Continued

dBμV	Voltage	dBV	dBm	dBW	Power
165	180	45	58	28	630
166	200	46	59	29	800
167	225	47	60	30	1 kW
168	250	48	61	31	1.25
169	280	49	62	32	1.6
170	315	50	63	33	2.0
171	355	51	64	34	2.5
172	400	52	65	35	3.15
173	450	53	66	36	4.0
174	500	54	67	37	5.0
175	565	55	68	38	6.3
176	630	56	69	39	8.0
177	710	57	70	40	10
178	800	58	71	41	12.5
179	900	59	72	42	16
180	1kV	60	73	43	20

Decibel Glossary

dBa   stands for dBrn 'adjusted'. This is a weighted circuit noise power referred to -85 dBm, which is 0 dBa. (Historically measured with a noise meter at the receiving end of a line. The meter is calibrated on a 1000 Hz tone such that 1 mW (0 dBm) gives a reading of $+85$ dBm. If the 1 mW is spread over the band 300–3400 Hz as random white noise, the meter will read $+82$ dBa.)

dBa0   circuit noise power in dBa referred to, or measured at, a point of zero relative transmission level (0 dBr). (A point of zero relative transmission level is a point arbitrarily established in a transmission circuit. All other levels are stated relative to this point.) It is preferable to convert circuit noise measurement values from dBa to dBa0 as this makes it unnecessary to know or to state the relative transmission level at the point of measurement.

dBd   used for expressing the gain of an antenna referred to a dipole.

dBi   used for expressing the gain of an antenna referred to an isotropic radiator.

dBμV   decibels relative to 1 microvolt.

Table 2.3 *Binary Decibel Values*

Bits	Max. value	Decibels (dB)
1	2	6.02
2	4	12.04
3	8	18.06
4	16	24.08
5	32	30.10
6	64	36.12
7	128	42.14
8	256	48.16
9	512	54.19
10	1 024	60.21
11	2 048	66.23
12	4 096	72.25
13	8 192	78.27
14	16 384	84.29
15	32 768	90.31
16	65 536	96.33
17	131 072	102.35
18	262 144	108.37
19	524 288	114.39
20	1 048 576	120.41
21	2 097 152	126.43
22	4 194 304	132.45
23	8 388 608	138.47
24	16 777 216	144.49
25	33 554 432	150.51
26	67 108 864	156.54
27	134 217 728	162.56
28	268 435 456	168.58
29	536 870 912	174.60
30	1 073 741 824	180.62
31	2 147 483 648	186.64
32	4 294 967 296	192.66

dbm   decibels relative to 1 milliwatt. The term dBm was originally used for telephone and audio work and, when used in that context, implies an impedance of $600 \, \Omega$, the nominal impedance of a telephone line. When it is desired to define a relative transmission level in a circuit, dBr is preferred.

dBm0   dBm referred to, or measured at, a point of zero transmission level.

dBmp	a unit of noise power in dBm, measured with psophometric weighting. dBmp $= 10 \log_{10}$ pWp $- 90 =$ dBa $- 84 =$ dBm $- 2.5$ (for flat noise 300–3400 Hz).
	pWp $=$ picowatts psophometrically weighted.
dBm0p	the abbreviation for absolute noise power in dBm referred to or measured at a point of zero relative transmission level, psophometrically weighted.
dBr	means dB 'relative level'. Used to define transmission level at various points in a circuit or system referenced to the zero transmission level point.
dBrn	a weighted circuit noise power unit in dB referenced to 1 pW (-90 dBm) which is 0 dBrn.
dBrnc	weighted noise power in dBrn, measured by a noise measuring set with 'C-message' weighting.
dBrnc0	noise measured in dBrnc referred to zero transmission level point.
dBu	decibels relative to 0.775 V, the voltage developed by 1 mW when applied to 600 Ω. dBu is used in audio work when the impedance is not 600 Ω and no specific impedance is implied.
dbV	decibels relative to 1 volt.
dbW	decibels relative to 1 watt.

Note: To convert dBm to dBμV add 107 (e.g. -20 dBm $= -20 + 107 = 87$ dBμV).

The beauty of decibel notation is that system gains and losses can be computed using addition and subtraction rather than multiplication and division. For example, suppose a system consists of an antenna that delivers a -4.7 dBm signal at its terminals (we conveniently neglect the antenna gain by this ploy). The antenna is connected to a 40 dB low-noise amplifier (A1) at the head end, and then through a 370 metre long coaxial cable to a 20 dB gain amplifier (A2), with a loss (L1) of -48 dB. The amplifier is followed by a bandpass filter with a -2.8 dB insertion loss (L2), and a -10 dB attenuator (L3). How does the signal exist at the end of this cascade chain?

S1	−4.7 dBm
A1	40.0 dB
A2	20.0 dB
L1	−48.0 dB
L2	−2.8 dB
L3	−10.0 dB
Total:	−5.5 dBm

Converting dBm to watts

$$P = \frac{10^{\text{dBm}/10}}{1000}$$

Converting any dB to ratio

Power levels: $\dfrac{P1}{P2} = 10^{\text{dB}/10}$

Voltage levels: $\dfrac{V1}{V2} = 10^{\text{dB}/20}$

Current levels: $\dfrac{I1}{I2} = 10^{\text{dB}/20}$

Binary decibel values

Binary numbers are used in computer systems. With the digitization of RF systems it is necessary to understand the decibel values of binary numbers. These binary numbers might be from an analogue-to-digital converter (ADC or A/D) that digitizes the IF amplifier output, or a digital-to-analogue converter (DAC) used to generate the analogue signal in a direct digital synthesis (DDS) signal generator.

3
TRANSMISSION LINES

3.1 General considerations

The purpose of any transmission line is to transfer power between a source and a load with the minimum of loss and distortion in either amplitude, frequency or phase angle.

Electrons travel more slowly in conductors than they do in free space and all transmission lines contain distributed components: resistance, inductance and capacitance. Consequently, lines possess an impedance which varies with frequency, and loss and distortion occur. Because the impedance is not constant over a wide frequency band the insertion loss will not be the same for all frequencies and frequency distortion will arise. A wavefront entering a line from a source takes a finite time to travel its length. This transit time, because of the distributed components, also varies with frequency and creates phase distortion.

3.2 Impedance matching

To transfer the maximum power from a generator into a load the impedance of the load and the internal impedance of the generator – and any intervening transmission line – must be equal.

Figure 3.1 illustrates the simplest case of a generator of internal impedance Z_s equal to 5 ohms and producing an e.m.f. of 20 volts.

When loads of varying impedance, Z_1, are connected the output voltage, V (p.d.) and the power in the load, P_1, varies as follows:

$$Z_1 = 5\,\Omega \quad I = 20/10 = 2\,A \quad V = 10\,V$$
$$Z_1 = 3\,\Omega \quad I = 2.5\,A \quad V = 7.5\,V$$
$$Z_1 = 8.33\,\Omega \quad I = 1.5\,A \quad V = 12.5\,V$$

$$P_1 = V^2/Z_1 = 100/5 = 20\,W$$
$$P_1 = V^2/Z_1 = 56.25/3 = 18.75\,W$$
$$P_1 = V^2/Z_1 = 156.25/8.33 = 18.75\,W$$

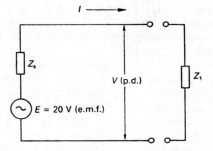

Fig. 3.1 *Impedance matching*

When dealing with alternating current and when transmission lines, particularly at radio frequencies, are interposed between the source and the load, other factors than the power transfer efficiency must also be considered.

3.3 Base band lines

These are the lines which generally operate at comparatively low frequencies carrying information at base band, e.g. speech, music, video or data. Generally provided by the telecommunications or telephone companies, usually on a rental basis, they are no longer likely to be hard wired, solid copper lines, although these may still be obtainable for lengths below about 10 km within one exchange area. Longer lines will probably be multiplexed, and comprised of radio and optical circuits over part of their length.

British Telecom publish specifications for their various grades of private wires (rented lines). The lines are categorized into voice – AccessLine and SpeechLine – and data services – KeyLine, Kilo-Stream and MegaStream – although the data services may also carry voice traffic. AccessLine, SpeechLine and KeyLine are analogue circuits, while the others are digital. KeyLine circuits transmit data in the form of tones and require the use of a modem to transmit data. Within each category the lines are graded by performance. Important parameters with analogue performance targets for the middle grades of line are:

- *Impedance.* May vary between $450\,\Omega$ and $750\,\Omega$. Nominal impedance is $600\,\Omega$. Most line parameters are specified when measured between $600\,\Omega$ non-reactive impedance.
- *Insertion loss.* The difference between power input to and power output from the line expressed in dB usually at a frequency of 800 Hz. Performance target for SpeechLine 6 is 6 dB max.
- *Loss/frequency response.* The variation in loss over the band of frequencies to be transmitted (Figure 3.2). Performance target for SpeechLine 6 is 6 dB to +12 dB between 300–3000 Hz (+ means more loss).
- *Absolute/bulk delay.* The minimum time (μs) for any frequency to travel from the input to the output of the circuit. No figure quoted.
- *Group delay.* The variation in time that the various frequency components will take when passing from the input to the output of the circuit. Figure 3.3 shows the performance limits for a KeyLine 3 (Option 3) circuit.
- *Noise.* Both random and impulsive noise are present on lines. Random noise is measured using a psophometer with a CCITT telephone network. KeyLine 3 (Option 3) performance target is $45\,\mathrm{dBm0_p}$. Impulsive noise is characterized by high amplitude and short duration peaks, such as those produced by switching. Performance target for KeyLine 3 (Option 3) is 21 dBm0 (no more than 18 impulsive noise counts to exceed the threshold in any 15 minute period).

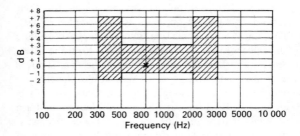

Fig. 3.2 *Target loss/frequency response envelope for medium grade telephone line*

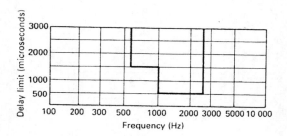

Fig. 3.3 *Delay limits for Keyline 3 (Option 3) circuit*

- *Maximum frequency error.* The frequency error produced on multiplexed circuits when the oscillators used to change the frequencies of the base band signals are not synchronized at the transmitter and receiver. Quoted as not exceeding 2 Hz.

  plus, for digital circuits:

- *Maximum bit rate.* KiloStream, 64 kbit/s, Mega-Stream, 2 or 8 Mbits/s.
- *Bit error rate.* The error parameter (performance targets for both KiloStream and MegaStream in brackets) is quoted in terms of: percentage of error free seconds (99.5), percentage of 1 minute periods with error ratio better than 10⁻⁶ (99.5), percentage of 1 second periods with error ratio better than 10⁻³ (99.95) (BT, 1992).

3.4 Balanced line hybrids

Radio transmitters and receivers are often controlled over a two-wire line. To facilitate this a balanced line hybrid circuit, consisting of two transformers connected back to back as in Figure 3.4, is inserted between the transmitter and receiver, and the line.

A signal from the receiver audio output is fed to winding L_1 of transformer T_1 which induces voltages across L_2 and L_3. The resultant line current also flows through L_4 and produces a voltage across L_6 which would appear as modulation on the transmitter but for the anti-phase voltage appearing across L_5. To ensure that the voltages cancel exactly a variable resistor, and often a capacitor to equalize the frequency response, is connected between L_2 and L_5.

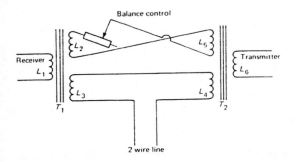

Fig. 3.4 *Balanced line hybrid*

A signal arriving via the line is applied to the transmitter as modulation; that it is also applied to the receiver poses no problem.

3.5 Radio frequency lines

Radio frequency transmission lines possess similar electrical characteristics to base band lines. However, they may be carrying large powers and the effects of a mismatched load are much more serious than a loss of transferred power. Three types of wire RF line are commonly used: a single wire with ground return for MF and LF broadcast transmission, an open pair of wires at HF and co-axial cable at higher frequencies. Waveguides are used at the higher microwave frequencies. RF lines exhibit an impedance characterized by their type and construction.

3.5.1 Characteristic impedance, Z_0

The physical dimensions of an RF transmission line, the spacing between the conductors, their diameters and the dielectric material between them, determine the characteristic impedance of the line, Z_0, which is calculated for the most commonly used types as follows.

Single wire with ground return (Figure 3.5(a)):

$$Z_0 = 138 \log_{10} \frac{4h}{d} \text{ ohms}$$

2-wire balanced, in air (Figure 3.5(b)):

$$Z_0 = \frac{276}{\sqrt{\varepsilon_p}} \log_{10} \frac{2s}{d} \text{ ohms}$$

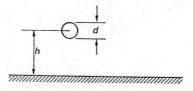

(a) Single wire, ground return line

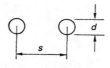

(b) 2-wire balanced air dielectric line

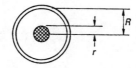

(c) Co-axial cable

Fig. 3.5

Co-axial (Figure 3.5 (c)):

$$Z_0 = \frac{138}{\sqrt{\varepsilon_r}} \log_{10} \frac{R}{r} \text{ ohms}$$

where dimensions are in mm and ε_r = relative permittivity of continuous dielectric.

3.5.2 Insertion loss

The loss in RF cables is quoted in specifications as attenuation in dB per unit length at specific frequencies, the attenuation increasing with frequency. The electrical specifications for cables having a braided outer conductor are given in Tables 3.1 and 3.2. Those for the now commonly used foam dielectric, solid outer conductor cables are provided in Table 3.3, all at the end of this chapter.

Table 3.1 RF cables USA RG series

RG number	Nominal impedance Z_0 (ohms)	Overall diameter (inches)	Velocity factor	Attenuation (dB per 100 ft)					Capacity (pF/ft)	Maximum operating voltage RMS
				1 MHz	10 MHz	100 MHz	1000 MHz	3000 MHz		
RG-5/U	52.5	0.332	0.659	0.21	0.77	2.9	11.5	22.0	28.5	3000
RG-5B/U	50.0	0.332	0.659	0.16	0.66	2.4	8.8	16.7	29.5	3000
RG-6A/U	75.0	0.332	0.659	0.21	0.78	2.9	11.2	21.0	20.0	2700
RG-8A/U	50.0	0.405	0.659	0.16	0.55	2.0	8.0	16.5	30.5	4000
RG-9/U	51.0	0.420	0.659	0.16	0.57	2.0	7.3	15.5	30.0	4000
RG-9B/U	50.0	0.425	0.659	0.175	0.61	2.1	9.0	18.0	30.5	401
RG-10A/U	50.0		0.475	0.659	0.16	0.55	2.0	8.0	16.5	30.5
RG-11A/U	75.0		0.405	0.66	0.18	0.7	2.3	7.8	16.5	20.5
RG-12A/U	75.0		0.475	0.659	0.18	0.66	2.3	8.0	16.5	20.5
RG-13A/U	75.0		0.425	0.659	0.18	0.66	2.3	8.0	16.5	20.5
RG-14A/U	50.0		0.545	0.659	0.12	0.41	1.4	5.5	12.0	30.0

5500	RG-16/U	52.0	0.630	0.670	0.1	0.4	1.2	6.7	16.0	29.5
6000	RG-17A/U	50.0	0.870	0.659	0.066	0.225	0.80	3.4	8.5	30.0
11000	RG-18A/U	50.0	0.945	0.659	0.066	0.225	0.80	3.4	8.5	30.5
11000	RG-19A/U	50.0	1.120	0.659	0.04	0.17	0.68	3.5	7.7	30.5
14000	RG-20A/U	50.0	1.195	0.659	0.04	0.17	0.68	3.5	7.7	30.5
50.0	RG-21/AU	0.332	0.659	1.4	4.4	13.0	43.0	85.0	30.0	2700
53.5	RG-29/U	0.184	0.659	0.33	1.2	4.4	16.0	30.0	28.5	1900
75.0	RG-34A/U	0.630	0.659	0.065	0.29	1.3	6.0	12.5	20.5	5200
75	RG-34B/U	0.630	0.66		0.3	1.4	5.8		21.5	6500

(continued overleaf)

Table 3.1 (continued)

RG number	Nominal impedance Z0 (ohms)	Overall diameter (inches)	Velocity factor	Attenuation (dB per 100 ft)					Capacity (pF/ft)	Maximum operating voltage RMS
				1 MHz	10 MHz	100 MHz	1000 MHz	3000 MHz		
RG-35A/U	75.0	0.945	0.659	0.07	0.235	0.85	3.5	8.60	20.5	10000
RG-54A/U	58.0	0.250	0.659	0.18	0.74	3.1	11.5	21.5	26.5	3000
RG-55/U	53.5	0.206	0.659	0.36	1.3	4.8	17.0	32.0	28.5	1900
RG-55A/U	50.0	0.216	0.659	0.36	1.3	4.8	17.0	32.0	29.5	1900
RG-58/U	53.5	0.195	0.659	0.33	1.25	4.65	17.5	37.5	28.5	1900
RG-58C/U	50.0	0.195	0.659	0.42	1.4	4.9	24.0	45.0	30.0	1900
RG-59A/U	75.0	0.242	0.659	0.34	1.10	3.40	12.0	26.0	20.5	2300
RG-59B/U	75	0.242	0.66		1.1	3.4	12		21	2300
RG-62A/U	93.0	0.242	0.84	0.25	0.85	2.70	8.6	18.5	13.5	750
RG-74A/U	50.0	0.615	0.659	0.10	0.38	1.5	6.0	11.5	30.0	5500

RG-83/U	35.0	0.405	0.66	0.23	0.80	2.8	9.6	24.0	44.0	2000
*RG-213/U	50	0.405	0.66	0.16	0.6	1.9	8.0		29.5	5000
†RG-218/U	50	0.870	0.66	0.066	0.2	1.0	4.4		29.5	11000
‡RG-220/U	50	1.120	0.66	0.04	0.2	0.7	3.6		29.5	14000

*Formerly RG8A/U.
†Formerly RG17A/U.
‡Formerly RG19A/U.

Table 3.2 British UR series

UR number	Nominal impedance Z (ohms)	Overall diameter (inches)	Inner conductor (inches)	Capacity (pF/ft)	Maximum operating voltage RMS	Attenuation (dB per 100 ft)				Nearest RG equivalent
						10 MHz	100 MHz	300 MHz	1000 MHz	
43	52	0.195	0.032	29	2750	1.3	4.3	8.7	18.1	58/U
57	75	0.405	0.044	20.6	5000	0.6	1.9	3.5	7.1	11A/U
63*	75	0.855	0.175	14	4400	0.15	0.5	0.9	1.7	
67	50	0.405	7/0.029	30	4800	0.6	2.0	3.7	7.5	213/U
74	51	0.870	0.188	30.7	15000	0.3	1.0	1.9	4.2	218/U

76	51	0.195	19/0.0066	29	1800	1.6	5.3	9.6	22.0	58C/U
77	75	0.870	0.104	20.5	125000	0.3	1.0	1.9	4.2	164/U
79*		0.855	0.265	21	6000	0.16	0.5	0.9	1.8	
83*	50	0.555	0.168	21	2600	0.25	0.8	1.5	2.8	
85*	75	0.555	0.109	14	2600	0.2	0.7	1.3	2.5	
90	75	0.242	0.022	20	2500	1.1	3.5	6.3	12.3	59B/U

All the above cables have solid dielectric with a velocity factor of 0.66 with the exception of those marked with an asterisk, which are helical membrane and have a velocity factor of 0.96.

Table 3.3 *Foam dielectric, solid outer conductor co-axial RF cables 50 ohms characteristic impedance*

	Superflexible FS-J series		LDF series	
Cable type	FSJ1-50A 1/4″	FSJ4-50B 1/2″	LDF2-50 3/8″	LDF4-50A 1/2″
Min. bending radius, in. (mm)	1.0 (25)	1.25 (32)	3.75 (95)	5.0 (125)
Propagation velocity, %	84	81	88	88
Max. operating frequency, GHz	20.4	10.2	13.5	8.8
Att. dB/100 ft (dB/100 m)				
50 MHz	1.27 (4.17)	0.732 (2.40)	0.75 (2.46)	0.479 (1.57)
100 MHz	1.81 (5.94)	1.05 (3.44)	1.05 (3.44)	0.684 (2.24)
400 MHz	3.70 (12.1)	2.18 (7.14)	2.16 (7.09)	1.42 (4.66)
1000 MHz	6.00 (19.7)	3.58 (11.7)	3.50 (11.5)	2.34 (7.68)
5000 MHz	14.6 (47.9)	9.13 (30.0)	8.80 (28.9)	5.93 (19.5)
10 000 MHz	21.8 (71.5)	15.0 (49.3)	13.5 (44.3)	N/A N/A
Power rating, kW at 25°C (77°F)				
50 MHz	1.33	3.92	2.16	3.63
100 MHz	0.93	2.74	1.48	2.53
400 MHz	0.452	1.32	0.731	1.22
1000 MHz	0.276	0.796	0.445	0.744
5000 MHz	0.11	0.30	0.17	0.29
10 000 MHz	0.072	0.19	0.12	N/A

The claimed advantages of foam dielectric, solid outer conductor cables are:

1. Lower attenuation
2. Improved RF shielding, approximately 30dB improvement
3. High average power ratings because of the improved thermal conductivity of the outer conductor and the lower attenuation.

A disadvantage is that they are not so easy to handle as braided cables.

3.5.3 Voltage standing wave ratio (VSWR)

When an RF cable is mismatched, i.e. connected to a load of a different impedance to that of the cable, not all the power supplied to the cable is absorbed by the load. That which does not enter the load is reflected back down the cable. This reflected power adds to the incident voltage when they are in phase with each other and subtracts from the incident voltage when the two are out of phase. The result is a series of voltage – and current – maxima and minima at half-wavelength intervals along the length of the line (Figure 3.6). The maxima are referred to as antinodes and the minima as nodes.

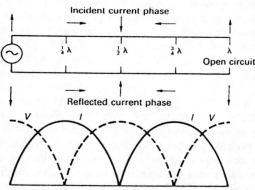

Resultant current and voltage distribution

Fig. 3.6 *Formation of standing waves*

The voltage standing wave ratio is the numerical ratio of the maximum voltage on the line to the minimum voltage: $\text{VSWR} = V_{max}/V_{min}$. It is also given by: $\text{VSWR} = R_L/Z_0$ or Z_0/R_L (depending on which is the larger so that the ratio is always greater than unity) where R_L = the load resistance.

The return loss is the power ratio, in dB, between the incident (forward) power and the reflected (reverse) power.

The reflection coefficient is the numerical ratio of the reflected voltage to the incident voltage.

The VSWR is 1, and there is no reflected power, whenever the load is purely resistive and its value equals the characteristic impedance of the line. When the load resistance does not equal the line impedance, or the load is reactive, the VSWR rises above unity.

A low VSWR is vital to avoid loss of radiated power, heating of the line due to high power loss, breakdown of the line caused by high voltage stress, and excessive radiation from the line. In practice, a VSWR of 1.5:1 is considered acceptable for an antenna system, higher ratios indicating a possible defect.

3.5.4 Transmission line filters, baluns and matching circuits

Use can be made of standing waves on sections of line to provide filters and RF transformers. When a line one-quarter wavelength long (a $\lambda/4$ stub) is open circuit at the load end, i.e. high impedance, an effective short-circuit is presented to the source at the resonant frequency of the section of line, producing an effective band stop filter. The same effect would be produced by a short-circuited $\lambda/2$ section. Unbalanced co-axial cables with an impedance of 50 Ω are commonly used to connect VHF and UHF base stations to their antennas although the antennas are often of a different impedance and balanced about ground. To match the antenna to the feeder and to provide a balance to unbalance transformation (known as a balun), sections of co-axial cable are built into the antenna support boom to act as both a balun and an RF transformer.

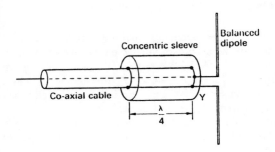

Fig. 3.7 *Sleeve balun*

Balun

The sleeve balun consists of an outer conducting sleeve, one quarter-wavelength long at the operating frequency of the antenna, and connected to the outer conductor of the co-axial cable as in Figure 3.7. When viewed from point Y, the outer conductor of the feeder cable and the sleeve form a short-circuited quarter-wavelength stub at the operating frequency and the impedance between the two is very high. This effectively removes the connection to ground for RF, but not for DC, of the outer conductor of the feeder cable permitting the connection of the balanced antenna to the unbalanced cable without short-circuiting one element of the antenna to ground.

RF transformer

If a transmission line is mismatched to the load variations of voltage and current, and therefore impedance, occur along its length (standing waves). If the line is of the correct length an inversion of the load impedance appears at the input end. When a $\lambda/4$ line is terminated in other than its characteristic impedance an impedance transformation takes place. The impedance at the source is given by:

$$Z_s = \frac{Z_0{}^2}{Z_L}$$

where

Z_s = impedance at input to line
Z_0 = characteristic impedance of line
Z_L = impedance of load

By inserting a quarter-wavelength section of cable having the correct characteristic impedance in a transmission line an antenna of any impedance can be matched to a standard feeder cable for a particular design frequency. A common example is the matching of a folded dipole of $300\,\Omega$ impedance to a $50\,\Omega$ feeder cable.

Let $Z_s = Z_0$ of feeder cable and $Z_0' =$ characteristic impedance of transformer section. Then:

$$Z_0 = \frac{Z_0'{}^2}{Z_L}$$
$$Z_0' = \sqrt{Z_0 Z_L}$$
$$= \sqrt{300 \times 50} = 122\,\Omega$$

53

3.6 Waveguides

At the higher microwave frequencies waveguides which conduct electromagnetic waves, not electric currents, are often used. Waveguides are conductive tubes, either of rectangular, circular or elliptical section which guide the wave along their length by reflections from the tube walls. The walls are not used as conducting elements but merely for containment of the wave. Waveguides are not normally used below about 3 GHz because their cross-sectional dimensions must be comparable to a wavelength at the operating frequency. The advantages of a waveguide over a co-axial cable are lower power loss, low VSWR and a higher operating frequency, but they are more expensive and difficult to install.

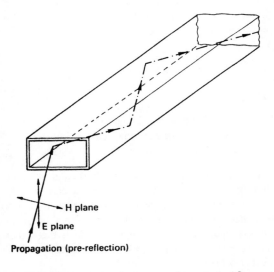

Fig. 3.8 *Propagation in rectangular waveguide*

In a rectangular waveguide an electromagnetic wave is radiated from the source at an angle to the direction of propagation and is bounced off the walls (Figure 3.8). If the wave were transmitted directly along the length of the guide the electric field would be parallel to one of the walls and be short-circuited by it. Radiating the wave at an angle to the walls creates the maximum field at the centre of the guide and zero at the walls, if the dimensions of the guide are correct for the frequency. However, because the wave does not travel directly along the length of the guide, the speed of propagation is less than in space.

In an electro-magnetic wave the electric and magnetic fields, and the direction of propagation, are mutually perpendicular (see Figure 1.3) and such a wave may therefore be thought of as transverse electromagnetic (TEM). In a waveguide though, because of the short-circuiting effect of the walls, a TEM wave cannot exist. A method of making the wave either transverse electric or transverse magnetic is needed.

When a wave is propagated by a reflection either the magnetic or the electric field is changed. The changed field will now contain the normal component perpendicular to the direction of propagation and a component in its direction, i.e. the wave is no longer wholly transverse. It must be either transverse electric or transverse magnetic. The terminology used to distinguish the type of wave differs: the American system uses the field which behaves as it would in free space to describe the type of wave, e.g. when there is no electric field in the direction of propagation the wave is called TE and the mode with no magnetic field in the direction of propagation, TM; the European system uses the field that is modified and an American TE wave is a European TM wave. In the European system H and E may also be used in lieu of (American) TE and TM.

The behaviour of waves in circular waveguides is similar to that in rectangular guides. Circular waveguides minimize feeder attenuation and are particularly suitable for long vertical runs. A single circular waveguide can carry two polarizations with a minimum isolation of 30 dB. Circular waveguides are recommended where attenuation is critical or where multi-band capability is needed.

Elliptical waveguides have the advantages of flexibility, the availability of long continuous runs and reduced cost.

3.7 Other transmission line considerations

3.7.1 Noise factor of coaxial cable transmission line

Any lossy electrical device, including coaxial cable, produces a noise level of its own. The noise factor of coaxial line is:

$$F_{N(COAX)} = 1 + \frac{(L-1)T}{290}$$

where

> $F_{N(COAX)}$ is the noise factor of the coax
> L is the loss expressed as a linear quantity
> T is the physical temperature of the cable in Kelvins

The linear noise factor due to loss can be converted to the noise figure by $10 \log(F_{N(COAX)})$, which can be added to the system noise decibel for decibel.

The attenuation loss figure published in manufacturers' tables is called the *matched line loss* (L_M) because it refers to the situation where the load and characteristic impedance of the line are equal. But we also have to consider the *Total Line Loss* (TLL), which is:

$$TLL = 10 \log \left[\frac{B^2 - C^2}{B(1 - C^2)} \right]$$

where

> $B =$ antilog L_M
> $C = (SWR_{LOAD} - 1)/(SWR_{LOAD} + 1)$
> SWR_{LOAD} is the VSWR at the load end of the line

3.7.2 Types of coaxial cable

Coaxial cable consists of two cylindrical conductors sharing the same axis (hence 'co-axial') and separated by a dielectric. For low frequencies (in flexible cables) the dielectric may be polyethylene or polyethylene foam, but at higher frequencies Teflon® and other materials are used. Also used in some applications, notably high powered broadcasting transmitters, are dry air and dry nitrogen.

Several forms of coaxial line are available. Flexible coaxial cable discussed earlier in this chapter is perhaps the most common form. The outer conductor in such cable is made of either braided wire or foil. Again, television broadcast receiver antennas provide an example of such cable from common experience. Another form of flexible or semi-flexible coaxial line is *helical line* (Figure 3.9) in which the outer conductor is spiral wound. This type of coaxial cable is usually 2.5 or more centimetres in diameter.

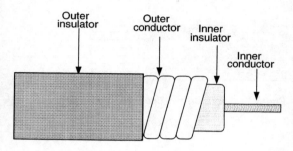

Fig. 3.9 *The helical line*

Hardline is coaxial cable that uses a thin-walled pipe as the outer conductor. Some hardline coax used at microwave frequencies has a rigid outer conductor and a solid dielectric. *Gas-filled line* is a special case of hardline that is hollow (Figure 3.10), the centre conductor being supported by a series of thin ceramic or Teflon insulators. The dielectric is either anhydrous (i.e. dry) nitrogen or some other inert gas.

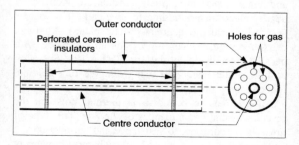

Fig. 3.10 *Gas-filled line*

Some flexible microwave coaxial cable uses a solid 'air-articulated' dielectric (Figure 3.11), in which the inner insulator is not continuous around the centre conductor, but rather is ridged. Reduced dielectric losses increase the usefulness of the cable at higher frequencies. Double shielded coaxial cable (Figure 3.12) provides an extra measure of protection against radiation from the line, and EMI from outside sources from getting into the system.

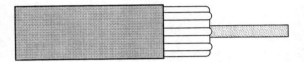

Fig. 3.11 *Solid 'air-articulated' dielectric*

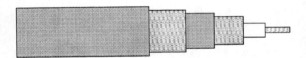

Fig. 3.12 *Double shielded coaxial cable*

3.7.3 Transmission line noise

Transmission lines are capable of generating noise and spurious voltages that are seen by the system as valid signals. Several such sources exist. One source is coupling between noise currents flowing in the outer and inner conductors. Such currents are induced by nearby electromagnetic interference and other sources (e.g. connection to a noisy ground plane). Although coaxial design reduces noise pick-up compared with parallel line, the potential for EMI exists. Selection of high-grade line, with a high degree of shielding, reduces the problem.

Another source of noise is thermal noise in the resistances and conductances of the line. This type of noise is proportional to resistance and temperature.

There is also noise created by mechanical movement of the cable. One species results from movement of the dielectric against the two conductors. This form of noise is caused by electrostatic discharges in much the same manner as the spark created by rubbing a piece of plastic against woollen cloth.

A second species of mechanically generated noise is piezoelectricity in the dielectric. Although more common in cheap cables, one should be aware of it. Mechanical deformation of the dielectric causes electrical potentials to be generated.

Both species of mechanically generated noise can be reduced or eliminated by proper mounting of the cable. Although rarely a problem at lower frequencies, such noise can be significant at microwave frequencies when signals are low.

3.7.4 Coaxial cable capacitance

A coaxial transmission line possesses a certain capacitance per unit of length. This capacitance is defined by:

$$C = \frac{24\varepsilon}{\log(D/d)} \frac{\text{pF}}{\text{Metre}}$$

where

> C is the capacitance
> D is the outside conductor diameter
> d is the inside conductor diameter
> ε is the dielectric constant of the insulator.

A long run of coaxial cable can build up a large capacitance. For example, a common type of coax is rated at 65 pF/metre. A 150 metre roll thus has a capacitance of (65 pF/m) (150 m), or 9750 pF. When charged with a high voltage, as is done in performing breakdown voltage tests at the factory, the cable acts like a charged high voltage capacitor. Although rarely if ever lethal to humans, the stored voltage in new cable can deliver a nasty electrical shock and can irreparably damage electronic components.

3.7.5 Coaxial cable cut-off frequency (F_c)

The normal mode in which a coaxial cable propagates a signal is as a transverse electromagnetic (TEM) wave, but others are possible – and usually undesirable.

There is a maximum frequency above which TEM propagation becomes a problem, and higher modes dominate. Coaxial cable should not be used above a frequency of:

$$F_{\text{CUTOFF}} = \frac{1}{3.76(D+d)\sqrt{\varepsilon}}$$

where

F is the TEM mode cut-off frequency
D is the diameter of the outer conductor in mm
d is the diameter of the inner conductor in mm
ε is the dielectric constant

When maximum operating frequencies for cable are listed it is the TEM mode that is cited. Beware of attenuation, however, when making selections for microwave frequencies. A particular cable may have a sufficiently high TEM mode frequency, but still exhibit a high attenuation per unit length at X or Ku-bands.

References

Andrew Antennas, (1991). *Catalogue 35*, Illinois.
British Telecommunications (1992). *Connect Direct, private circuits*. BT, London.
Edis, E.A. and Varrall, J.E. (1992). *Newnes Telecommunications Pocket Book*. Butterworth-Heinemann, Oxford.
Kennedy, G. (1977). *Electronic Communications Systems*. McGraw-Hill Kogashuka, Tokyo
Terman, F.E. (1943), *Radio Engineers' Handbook*, McGraw-Hill, London.

4

ANTENNAS

4.1 Antenna characteristics

4.1.1 Bandwidth

Stated as a percentage of the nominal design frequency, the bandwidth of an antenna is the band of frequencies over which it is considered to perform acceptably. The limits of the bandwidth are characterized by unacceptable variations in the impedance which changes from resistive at resonance to reactive, the radiation pattern, and an increasing VSWR.

4.1.2 Beamwidth

In directional antennas the beamwidth, sometimes called half-power beamwidth (HPBW), is normally specified as the total width, in degrees, of the main radiation lobe at the angle where the radiated power has fallen by 3 dB below that on the centre line of the lobe (Figure 4.1A).

4.1.3 Directivity and forward gain

All practical antennas concentrate the radiated energy in some directions at the expense of others. They possess directivity but are completely passive; they cannot increase the power applied to them. Nevertheless, it is convenient to express the enhanced radiation in some directions as a power gain.

Antenna gain may be quoted with reference to either an isotropic radiator or the simplest of practical antennas, the dipole. There is a difference of 2.15 dB between the two figures. A gain quoted in dBi is with reference to an isotropic radiator and a gain quoted in dBd is with reference to a dipole. When gain is quoted in dBi, 2.15 dB must be subtracted to relate the gain to that of a dipole.

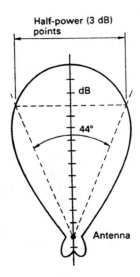

Fig. 4.1A *Half-power beamwidth*

4.1.4 Effective height or length

The current flowing in an antenna varies along its length (see Figure 1.4). If the current were uniform along the length of an antenna it would produce a field appropriate to its physical length, and the effective height or length of the antenna would be its physical length. In practice, because the current is not uniform, the effective length is less than the physical length and is given by:

$$l_{\text{eff}} = \frac{l_{\text{phys}} \times I_{\text{mean}}}{I}$$

where

$$l_{\text{eff}} = \text{effective length}$$
$$l_{\text{phys}} = \text{physical length}$$
$$I = \text{current at feed point}$$

With an antenna which is short in comparison with a wavelength the current can be considered to vary linearly over its length and $I_{\text{mean}} = I/2$. Because the apparent length of a vertical radiator is twice that of its physical dimension due to the mirror image formed below the ground, the effective length of an electrically short vertical antenna may be approximated to be its physical length.

4.1.5 Effective radiated power (erp)

This is the power effectively radiated along the centre line of the main lobe. It is the power supplied to the antenna multiplied by the antenna gain with reference to a dipole.

4.1.6 Radiation resistance and efficiency

The power radiated by an antenna can conveniently be expressed in terms of the value of a resistor which would dissipate the same power that the antenna radiates. This value is referred to as the radiation resistance and is defined as the ratio of the power radiated to the square of the current at the feed point. The efficiency is the ratio of the power radiated to that lost in the antenna. It is given by:

$$\text{eff.} = \frac{R_r}{R_r + R_L} \times 100\%$$

where R_r is the radiation resistance and R_L represents the total loss resistance of the antenna. The sum of the two resistances is the total resistance of the antenna and, for a resonant antenna, is also the impedance.

4.1.7 Front-to-back ratio

The ratio, in dB, of the strength of the radiation (or received signal) in the forward (desired) direction to that in the reverse (unwanted) direction. The front-to-back ratio of the antenna shown in Figure 4.1A is 13 dB.

4.1.8 Impedance

The impedance of an antenna is that presented to the feeder cable connecting it to the transmitter or receiver. It is the result of the vectorial addition of the inductive, capacitive and resistive elements of the antenna. Each resonant antenna possesses an impedance characteristic of the type, and when an antenna operates at its resonant frequency the reactive elements cancel out and the impedance becomes resistive. The radiation resistance plus the losses in the antenna, i.e. the series resistance of the conductors, the shunt resistance of the base material and losses in nearby objects, form the resistive portion of the impedance.

4.1.9 Polarization

The radiated field from an antenna is considered to be polarized in the plane of the length of the conductors which is the plane of the electric field, the E plane. Confusion arises when reference is made to vertical or horizontal polarization and it is preferable when referring to polar diagrams to use the E and H plane references.

Circular polarization, produced by crossed dipoles or helical wound antennas, is occasionally used for point-to-point work at VHF and above to reduce multi-path propagation losses.

Cross polarization discrimination defines how effectively an antenna discriminates between a signal with the correct polarization, i.e. mounted with the elements in the same plane, and one operating at the same frequency with the opposite polarization. 20 dB is typical.

4.1.10 Radiation pattern

A plot of the directivity of an antenna showing a comparison of the power radiated over 360°. Two polar diagrams are required to show the radiation in the E and H planes. The polar diagrams may be calibrated in either linear (voltage) or logarithmic (decibel) forms.

4.1.11 Voltage standing wave ratio (VSWR)

Most VHF and UHF antennas contain an impedance matching device made up of lengths of co-axial cable. Thus the VSWR (see Chapter 3) of these types of antenna varies with the operating frequency, more so than the bandwidth of the antenna alone would produce. At the centre design frequency, the VSWR should, theoretically, be 1:1 but in practice a VSWR less than 1.5:1 is considered acceptable.

4.1.12 Receive aperture

Receiving antennas also possess a property called aperture, or capture area. This concept relates the amount of power that is delivered to a matched receiver to the power density (watts per square metre). The aperture is often larger than the physical area of the antenna, as in the case of the half-wavelength dipole (where the wire fronts a very small physical area), or less as in the case of a parabolic reflector used in microwave reception. Figure 4.1B shows the capture area of a half-wavelength (0.5λ) dipole. It consists of an ellipse with major axes of 0.51λ and 0.34λ. The relationship between gain and aperture is:

$$A_e = \frac{G\lambda^2}{4\pi n}$$

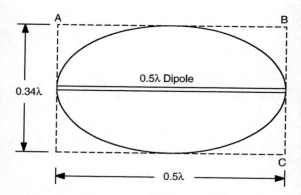

Fig. 4.1B *Capture area of half-wavelength (0.5λ) dipole*

where

A_e is the effective aperture
G is the gain
λ is the wavelength of the signal
n is the aperture effectiveness ($n = 1$ for a perfect no-loss antenna, real values are typically 0.3 to 0.55

4.2 Antenna types

4.2.1 The dipole

The half-wavelength ($\lambda/2$) dipole as described in Section 1.4 is the antenna on which many others are based. Figure 4.2 shows the relative radiation in the E and H planes of a dipole in free space.

The impedance of a half-wavelength dipole is $72\,\Omega$; that of a full wavelength or folded dipole is $300\,\Omega$.

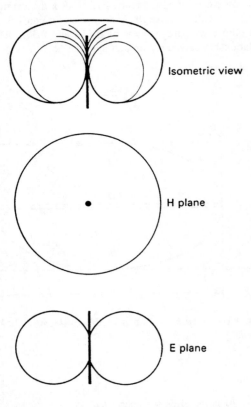

Isometric view

H plane

E plane

Fig. 4.2 *Radiation patterns of a half-wave dipole*

4.2.2 The quarter-wavelength vertical radiator

The quarter-wavelength ($\lambda/4$) vertical radiator is a commonly used antenna for MF broadcasting and for VHF and UHF mobile radio applications. It is derived from the $\lambda/2$ dipole and it is assumed that a mirror image of the radiator is formed below the ground to complete the $\lambda/2$ structure of the dipole as in Figure 4.3. The radiation pattern of a $\lambda/4$ vertical radiator mounted close to a perfect earth shows a strong similarity to that of a dipole. The effect of reducing the size and conductivity of the ground plane raises the angle of radiation.

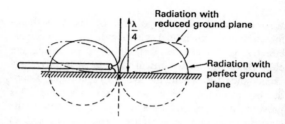

Fig. 4.3 *Quarter-wavelength vertical radiator*

The impedance of a perfect $\lambda/4$ vertical radiator is $36\,\Omega$ but reducing the effectiveness of the ground plane raises the impedance.

4.2.3 LF, MF and HF antennas

Because of the physical lengths involved, LF and MF antennas are usually non-resonant and their impedances do not conform to the resistive $70\,\Omega$ or $36\,\Omega$ of the basic resonant types. The impedance of a non-resonant antenna is usually higher and reactive so an antenna tuning or matching unit is used to couple the antenna efficiently to the transmission line and also act as filter to reduce out-of-band radiations. The matching unit comprises a tuned circuit with either a tap on the coil at the correct impedance point or a separate coupling coil to feed the antenna.

Obtaining an adequate length is always the problem with low frequency antennas and various methods have been used based mainly on the $\lambda/4$ radiator. The horizontal section of the inverted L (Figure 4.4) extends the effective length but, as the ground wave is much used at the lower frequencies, these antennas

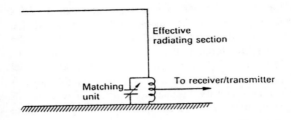

Fig. 4.4 *Inverted L antenna*

are intended for vertical polarization and it is therefore only the down-lead which radiates, or receives, effectively. An alternative method of increasing the effective height of a vertical radiator is to provide a capacitance top where the system of horizontal conductors provides a high capacitance to ground. This prevents the current falling to zero at the top of the antenna, maintaining a higher mean current and so increasing the antenna's effective length.

Dipoles used at HF are mounted horizontally because of their length and have directivity in the horizontal (E) plane. Propagation is mainly by the sky wave and the omni-directional properties in the vertical (H) plane, modified by ground reflections, produce wide angle upwards radiation.

4.2.4 Directional arrays

Broadside array

A broadside array consists of several radiators spaced uniformly along a line, each carrying currents of the same phase. When each radiator has an omni-directional pattern, and the spacing between radiators is less than $3\lambda/4$, maximum radiation occurs at right angles to the line of the array. The power gain is proportional to the length of the array, provided that the length is greater than two wavelengths; this means, effectively, the number of radiators. Figure 4.5 shows a typical H plane polar diagram for an array with vertically mounted radiators and a spacing of $\lambda/2$.

End-fire array

Physically an end-fire array is identical to a broadside except for the feeding arrangements and the spacing

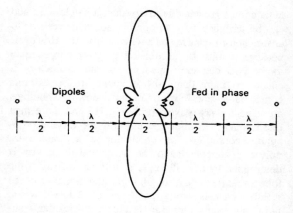

Fig. 4.5 *Broadside array*

of the elements. In an end-fire array the radiators are fed with a phase difference between adjacent radiators equal in radians to the spacing between them in wavelengths. A spacing of λ/4 requires a phase shift of 90° between adjacent radiators. Figure 4.6 shows a typical radiation pattern. An end-fire array concentrates the power in both the E and H planes and the maximum radiation is in the direction of the end of the array with the lagging phase.

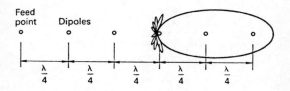

Fig. 4.6 *End-fire array*

Rhombic antenna

A rhombic is a wide band, directional antenna comprised of four non-resonant wire antennas, each several wavelengths long, arranged as shown in Figure 4.7(a) which also shows the radiation pattern for each leg of the rhombus. The lobe angle θ can be varied by adjusting the length, in wavelengths, of each radiator. The antenna has greater directivity than a single wire and can be terminated by an appropriate value resistor to ensure non-resonance and a wide bandwidth. However, because it must be terminated in a resistance equal to the characteristic impedance of the conductors, it cannot be more than 50% efficient. It also exhibits considerable side lobes of radiated power. Rhombics are used for sky wave working at HF and more than one frequency is allocated to allow for varying propagation conditions. The conductors of a rhombic are normally horizontal and the horizontal directivity is determined by the tilt angle, β in Figure 4.7(a). If the lobe angle θ is equal to $(90 - \beta)°$ the radiation in the A lobes cancels, while that from the B lobes, which point in the same direction, is added. The resultant pattern in the horizontal plane is shown in Figure 4.7(b). The vertical directivity is controlled by the height of the conductors above the ground.

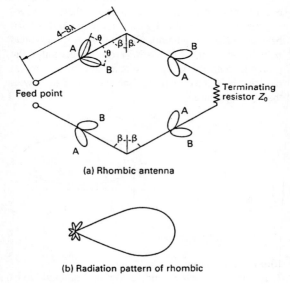

(a) Rhombic antenna

(b) Radiation pattern of rhombic

Fig. 4.7

Log-periodic antenna

An alternative, usable from HF through UHF, to the rhombic for wide band operation is the log-periodic antenna. It is comprised of several dipoles of progressive lengths and spacings as in Figure 4.8, and is resonant over a wide frequency range and may be mounted with either polarization. The dipoles are fed via the support booms and this construction ensures that the resultant phasing of the dipoles is additive in the forward direction producing an end-fire effect. However, because at any one frequency only a few of the dipoles are close to resonance, the forward gain of the antenna is low considering the number of elements it contains.

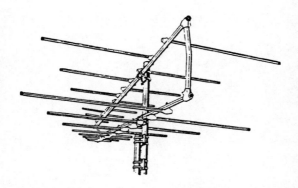

Fig. 4.8 *Log-periodic antenna at VHF frequency*

4.3 VHF and UHF antennas

4.3.1 Base station antennas

Apart from entertainment broadcasting most VHF and UHF systems use vertical polarization and a dipole – or to prevent noise from rain static, the folded dipole – with the conductors mounted vertically is a frequently used antenna for VHF and UHF base station installations. Unfortunately it is often mounted on the side of the support structure in a manner which seriously affects its omnidirectional radiation pattern. Where practical, there should be a minimum spacing of one wavelength between the structure and the rearmost element of the antenna.

To obtain a good omni-directional pattern either an end-fed dipole (Figure 4.9) or a unipole antenna

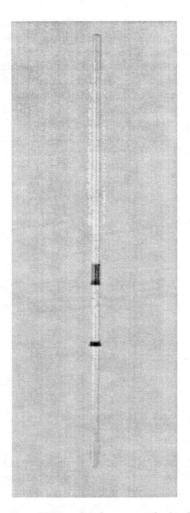

Fig. 4.9 *Type EDV end-fed antenna (by kind permission of C and S Antennas Ltd)*

(Figure 4.10) protruding from the top of the mast or tower is the best option. A unipole is a variation of the vertical quarter-wave radiator and provides a low angle of radiation.

To reduce the likelihood of co-channel interference directional antennas are often necessary. The

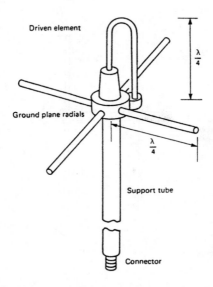

Fig. 4.10 *Folded unipole antenna*

simplest of these is the combination of a $\lambda/2$ dipole and reflector shown in Figure 4.11. The reflector is slightly longer than the dipole and spaced one quarter-wavelength from it. The portion of the signal radiated by the dipole in the direction of the reflector is received and re-transmitted by the reflector, with a 180° phase change occurring in the process. The signal re-transmitted to the rear of the antenna – the direction of the reflector – cancels the signal from the dipole, that towards the front of the antenna adds to the signal from the dipole giving the radiation pattern shown. The power gain of a dipole and reflector, a two-element array, is 3 dBd.

Directivity can be increased by adding directors forward of the dipole, the result is a Yagi–Uda array. The limit to the number of radiators is set by physical constraints and the reduction of bandwidth produced by their addition. At low VHF, a 3-element array is about the practical maximum, while at 1500 MHz, 12-element arrays are commonplace. As a rule of thumb, doubling the number of elements in an array increases

73

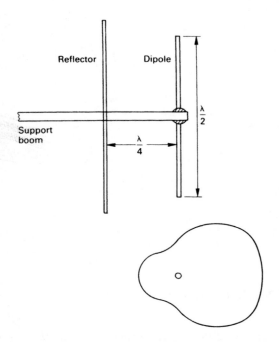

Fig. 4.11 *Dipole and reflector*

the forward gain by 3 dB. Where the maximum front-to-back ratio is essential the single rod reflector can be replaced by a corner reflector screen.

Co-linear antennas provide omni-directional characteristics and power gain in the H plane. A co-linear consists of a number of dipoles stacked vertically and, in the normal configuration, fed so that they radiate in phase and the maximum power is radiated horizontally. Figure 4.12 shows alternative feeding arrangements. One advantage of the co-linear is that the horizontal angle of radiation can be tilted to about 15° downwards by changing the phasing of the elements. The gain of a co-linear is limited, because of the physical lengths involved and losses in the feeding arrangements to 3 dBd at VHF and 6 dBd at UHF.

Figure 4.13 shows a VHF slot antenna. It is made up of three dipoles in parallel with the tips connected together so that an electric field exists across the whole

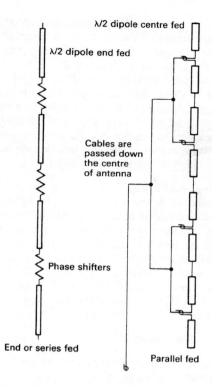

Fig. 4.12 *General construction of co-linears*

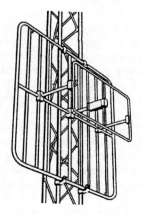

Fig. 4.13 *VHF slot antenna*

75

width of the antenna. It is an expensive antenna, but four such antennas mounted around a tower are often used in high quality installations where a very good omni-directional pattern is essential.

A wide-band alternative to the log-periodic is the conical (discone) antenna (Figure 4.14). It provides unity gain, is omni-directional and has a bandwidth of approximately 3:1, depending on the designed frequency range. In practice there has been a tendency to expect these antennas to perform outside their specified bandwidths with unsatisfactory results.

Stacking and baying

A method of increasing an antenna's directivity is to mount two or more antennas vertically above one another (stacking) or side-by-side (baying), and to feed them so that they radiate in phase. Stacking two dipoles vertically increases the directivity in the E plane and baying them increases the directivity in the H plane, approximately halving the beamwidth in each case.

An array of two stacked plus two bayed antennas approximately halves the beamwidth in both planes.

4.3.2 Mobile antennas

The aerial is the least expensive, and most abused, component of a mobile radio installation. Frequently installed in a manner which does not produce optimum performance it can have a profound effect on the performance of the whole installation.

Most mobile antennas consist of a metal rod forming a quarter-wavelength radiator. The ideal mounting position is the centre of a metallic roof, and as the area of the ground plane is reduced the radiation pattern changes and more of the energy is radiated upwards (not always a bad thing in inner city areas); also, the impedance rises.

The effect of the mounting position on the H plane radiation can be dramatic, resulting in ragged radiation patterns and, in some directions, negligible radiation. Advice on the installation of mobile antennas and the polar diagrams produced by typical installations are illustrated in MPT 1362, *Code of Practice For Installation of Mobile Radio Equipment in Land Based Vehicles*.

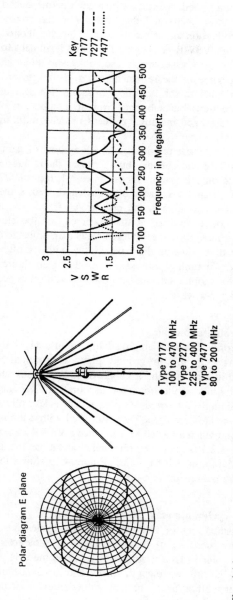

Fig. 4.14 *Discone wide band antenna (by kind permission of Jaybeam Ltd)*

Polar diagram E plane

Key
7177 ⎯⎯⎯
7277 ⎯ ⎯ ⎯
7477 ·········

Frequency in Megahertz

VSWR

- Type 7177
  100 to 470 MHz
- Type 7277
  225 to 400 MHz
- Type 7477
  80 to 200 MHz

As the installation moves away from the ideal and the antenna impedance rises a mismatch is introduced between the antenna and the feeder with the consequent production of standing waves on the feeder. Under high VSWR conditions the cable is subject to higher voltage stresses and it also behaves as an aerial radiating some of the reflected power. This spurious radiation adds to the radiation from the antenna in some directions but subtracts from it in others giving rise to jagged radiation patterns or deep nulls in radiated signal.

Mobile antennas providing a small amount of gain, typically 3 dB and obtained by narrowing the radiation lobes, are on the market. These have a length of 5/8 wavelength and, because the extra length makes the impedance capacitive at the operational frequency, a loading coil is inserted at the lower end of the element to cancel the capacitive reactance. An adjustable metallic disk towards the base of the whip is often provided for tuning purposes. Note that gain figures quoted for mobile antennas are usually with reference to a quarter-wave whip.

Low profile antennas
Low profile antennas are available for use at UHF. They have a built-in ground plane approximately 150 mm in diameter and a height of some 30 mm and have obvious applications for use on high vehicles and, although not strictly covert, where a less obtrusive antenna is required. They are fitted with a tuning screw and when adjusted to resonance a VSWR of better than 1.2:1 is quoted by one maker and a bandwidth of 10 MHz at a VSWR of 2:1. Figure 4.15 shows the radiation pattern for one type.

Motor-cycle antennas
The installations of antennas on motor cycles poses problems because of the absence of a satisfactory ground plane. One frequently used method is to employ a 5/8 wavelength whip and loading coil. Another method uses a pair of grounded downwards-pointing rods to form the lower half of a dipole.

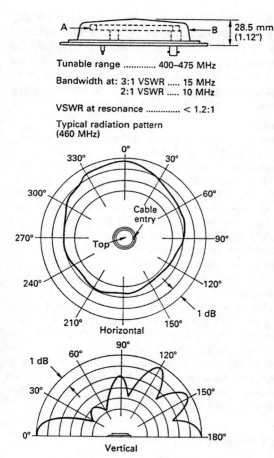

Tunable range .... 400–475 MHz

Bandwidth at: 3:1 VSWR ..... 15 MHz
2:1 VSWR ..... 10 MHz

VSWR at resonance .... < 1.2:1

Typical radiation pattern
(460 MHz)

Antenna mounted on a 40 cm × 40 cm aluminium plate

Fig. 4.15 *Low profile UHF antenna (by kind permission of Panorama Antennas Ltd)*

Hand-portable antennas

Again, because of the lack of a ground plane high performance antennas are difficult to provide for hand-portables, particularly at VHF. Body-worn sets may have an antenna incorporated in the microphone lead but the high current portion of the antenna must then be at a low height and in some directions the radiation must pass through the body, which is highly absorbent, to reach the base station. Helical antennas are frequently used on hand-held sets to reduce the physical length. Useful operating tips are to face the base station when using the radio in low signal areas, while placing the set on a nearby car roof effectively increases the performance of the antenna.

Safety

There are two safety aspects to consider when installing mobile antennas: physical and electrical. The physical considerations are that the antenna must be incapable of inflicting injury when it is in its correct position, and also when it has been bent or damaged. Rear wing mounted antennas need particular care in their positioning; a Band 111 aerial tip is just about eye height when bending over an open boot lid. The same considerations apply to hand-portables, helical antennas being perhaps safer than whips because they are thicker and thus more easily seen. They also have rounded tips.

The electrical dangers are from radiation affecting the body either directly – radiation from a hand-portable helical into the eye is a possible example – or indirectly by affecting electronic equipment. The danger of radiation affecting equipment in the vehicle is increased when the VSWR is high because of increased radiation from the feeder. Advice should be sought from the Radiological Protection Board.

4.4 Microwave antennas

The small antenna elements at microwaves facilitate the construction of highly directive, high gain antennas with high front-to-back ratios.

At frequencies below about 2 GHz, 12- to 24-element Yagi arrays, enclosed in plastic shrouds for weather protection, may be used. At higher frequencies, antennas with dish reflectors are the norm.

The aperture ratio (diameter/wavelength) of a dish governs both its power gain and beamwidth. The power gain of a parabolic dish is given to a close approximation by:

$$\text{Gain} = 10 \log_{10} 6(D/\lambda)^2 \times N, \text{dBi}$$

where D = dish diameter and N = efficiency. Dimensions are in metres. The half-power beam width (HPBW) in degrees is approximately equal to $70\lambda/D$.

A microwave antenna with its dish reflector, or parasitic elements in the case of a Yagi type, is a large structure. Because of the very narrow beamwidths – typically 5° for a 1.8 m dish at 2 GHz – both the antenna mounting and its supporting structure must be rigid and able to withstand high twisting forces to avoid deflection of the beam in high winds. Smooth covers, radomes, fitted to dishes and the fibreglass shrouds which are normally integral with Yagis designed for these applications considerably reduce the wind loading and, for some antenna types, increase the survival wind speed.

The electrical performance of a selection of microwave antennas is given in Table 4.1 and the wind survival and deflection characteristics in Table 4.2 (Andrew Antennas, 1991).

With shrouded Yagis and some dishes low loss foam-filled cables are generally used up to about 2 GHz although special connectors may be required. At higher frequencies, air-spaced or pressurized nitrogen-filled cables are frequently used with waveguides as an alternative.

4.4.1 Omnidirectional normal mode helix

The normal mode helix antenna shown in Figure 4.16 produces an omnidirectional pattern when the antenna is mounted vertically. The diameter (D) of the helical coil should be one-tenth wavelength ($\lambda/10$), while the pitch (i.e. S, the distance between helix loops) is one-twentieth wavelength ($\lambda/20$). An example of the normal mode helix is the 'rubber ducky' antenna used on VHF/UHF two-way radios and scanners.

4.4.2 Axial mode helical antenna

An axial mode helical antenna is shown in Figure 4.17. This antenna fires 'off-the-end' in the direction shown by the arrow. The helix is mounted in the centre of a ground plane that is at least 0.8λ across. For some UHF frequencies some manufacturers have used aluminum pie pans for this purpose. The helix itself is made from

Table 4.1 *2.1–2.2 GHz antennas – electrical characteristics*

Type number	Dia. (m)	Gain (dBi) Bottom	Gain (dBi) Mid-band	Gain (dBi) Top	Beam width (deg.)	Cross pol. disc (dB)	F/B ratio (dB)	VSWR max.
Ultra High Performance Antenna, F-Series Unpressurized – Radome Inc.								
Single polarized								
UHP8F-21	2.4	31.9	32.1	32.3	4.2	32	61	1.10
UHP10F-21	3.0	33.7	33.9	34.0	3.6	33	64	1.10
UHP12F-21	3.7	35.4	35.6	35.8	2.9	32	65	1.10
Dual polarized								
UHX8F-21	2.4	31.9	32.1	32.3	4.2	30	58	1.20
UHX10F-21A	3.0	33.8	34.0	34.2	3.6	32	62	1.20
UHX12F-21A	3.7	35.4	35.6	35.8	2.8	32	67	1.20
High Performance Antenna, F-Series Unpressurized – Radome Inc.								
Single polarized								
HP6F-21B	1.8	29.4	29.6	29.8	5.5	30	46	1.12
HP8F-21A	2.4	31.9	32.1	32.3	4.1	30	53	1.12
HP10F-21A	3.0	33.8	34.0	34.2	3.4	32	55	1.12
HP12F-21A	3.7	35.4	35.6	35.8	2.9	32	56	1.12
Standard Antenna, F-Series Unpressurized								
Single polarized								
P4F-21C	1.2	26.4	26.6	26.8	7.6	30	36	1.15
P6F-21C	1.8	29.8	30.0	30.2	4.9	30	39	1.12
P8F-21C	2.4	32.3	32.5	32.7	3.8	30	40	1.12
P10F-21C	3.0	34.0	34.2	34.4	3.4	30	44	1.12
Grid Antenna, F-Series Unpressurized								
Single polarized								
GP6F-21A	1.8	29.6	29.8	30.0	5.4	31	36	1.15
GP8F-21A	2.4	32.0	32.2	32.4	4.0	35	39	1.15
GP10F-21A	3.0	34.0	34.2	34.4	3.3	40	42	1.15
GP12F-21	3.7	35.5	35.7	35.9	2.8	40	44	1.15

Table 4.2 *Wind survival and deflection characteristics*

Antenna types	Survival ratings Wind velocity (km/h)	Survival ratings Radial ice (mm)	Max. deflection in 110 km wind (degrees)
P4F Series			
Without Radome	160	12	0.1
With Radome	185	12	0.1
Standard Antennas (except P4F Series)			
Without Radome	200	25	0.1
With Standard Radome	200	25	0.1
UHX, UMX, UGX Antennas	200	25	0.1

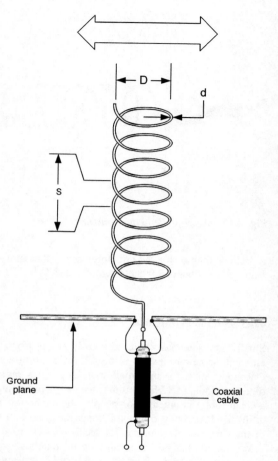

Fig. 4.16 *Normal mode helix antenna*

either heavy copper wire (solid, not stranded) or copper or brass tubing. The copper tubing is a bit easier to work. The dimensions are:

$$D \approx \lambda/3$$
$$S \approx \lambda/4$$
$$\text{Length} \approx 1.44\lambda$$

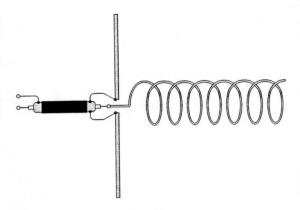

Fig. 4.17 *Axial mode helical antenna*

A 'rule of thumb' for the circumference is that maximum gain is obtained when circumference C is:

$$C = 1.066 + [(N - 5) \times 0.003]$$

4.4.3 Small loop antennas

Small loop antennas are used mostly for receiving, although some designs are also used for transmitting. One application for the small loop antenna is radio direction finding (RDF). Another use is for providing a small footprint antenna for people who cannot erect a full sized receiving antenna. Perhaps the greatest use of the small loop antenna is for receiving stations on crowded radio bands. The small loop antenna has very deep nulls that make it easy to null out interfering co-channel and adjacent channel signals.

4.5 Loop antennas

4.5.1 Small loop antennas defined

Large loop antennas are those with overall wire lengths of 0.5λ to more than 2λ. Small loop antennas, on the other hand, have an overall wire length that is much less than one wavelength (1λ). According to a Second World War US Navy training manual such antennas are those with an overall length of $\leq 0.22\lambda$. Jasik's classic

1961 text on radio antennas uses the figure $\leq 0.17\lambda$, while John Kraus (1950) used the figure $\leq 0.10\lambda$. An amateur radio source, *The ARRL Antenna Book*, recommends $\leq 0.085\lambda$ for small loop antennas. For the purposes of our discussion we will use Kraus's figure of $\leq 0.10\lambda$.

A defining characteristic of small loops versus large loops is seen in the current distribution. In the small loop antenna the current flowing in the loop is uniform in all portions of the loop. In the large loop, however, the current varies along the length of the conductor, i.e. there are current nodes and antinodes.

The small loop antenna also differs from the large loop in the manner of its response to the radio signal. A radio signal is a transverse electromagnetic (TEM) wave, in which magnetic and electrical fields alternate with each other along the direction of travel. The large loop, like most large wire antennas, respond primarily to the *electrical field* component of the TEM, while small loops respond mostly to the *magnetic field* component. The importance of this fact is that it means the small loop antenna is less sensitive to local electromagnetic interference sources such as power lines and appliances. Local EMI consists largely of electrical fields, while radio signals have both magnetic and electrical fields. With proper shielding, the electrical response can be reduced even further.

4.5.2 Small loop geometry

Small loop antennas can be built in any of several different shapes (Figure 4.18). Popular shapes include hexagonal (Figure 4.18A), octagonal (Figure 4.18B), triangular (Figure 4.18C), circular (Figure 4.18D) and square (Figure 4.18E).

The far-field performance of small loop antennas is approximately equal provided that $A^2 \leq \lambda/100$, where A is the loop area and λ is the wavelength of the desired frequency.

The 'standard' loop used in this discussion is the square loop depicted in Figure 4.19. There are two forms of winding used: *depth wound* and *spiral wound*. The difference is that the depth wound has its turns in different parallel planes, while in the spiral wound version all of the turns are in the same plane. The

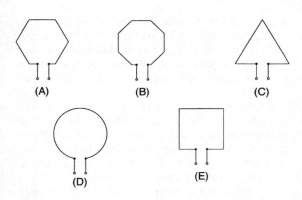

Fig. 4.18 *Small loop antennas: (A) hexagonal, (B) octagonal, (C) triangular, (D) circular and (E) square*

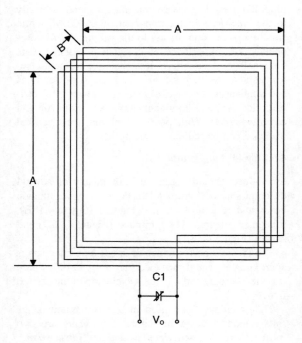

Fig. 4.19 *Square loop*

spiral wound loop theoretically has a deeper null than depth wound, but in practical terms there is usually little difference.

The length of each side of the loop is designated A, while the depth or width is designated B. These dimensions will be used in equations shortly. The constraint on the dimensions is that A should be $\leq 0.10\lambda/4$ and $B \leq A/5$.

The loop may be either tuned or untuned. The differences between tuned and untuned will be discussed shortly.

4.5.3 Small loop antenna patterns

Small loop antennas have patterns opposite those of large loops. The minima, or 'nulls', are perpendicular to the plane of the loop, while the maxima are off the ends. Figure 4.20 shows the directions of maximum and minimum response. The loop antenna is viewed from above. The nulls are orthogonal to the loop axis, while the maxima are along the loop axis.

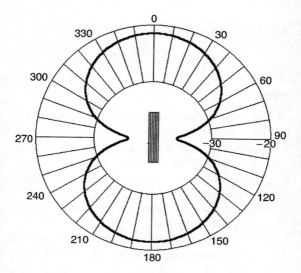

Fig. 4.20 *Small loop azimuth*

The fact that the small loop pattern has nulls perpendicular to the loop axis, i.e. perpendicular to the plane of the loop, is counterintuitive to many people. The advancing radio wave produces alternating regions of high and low amplitude. A *potential difference* exists between any two points. When the loop is aligned such that its axis is parallel to the isopotential lines low signal levels are induced into the loop. If the turns of the loop cut several isopotential lines, a larger signal is induced from this direction.

4.5.4 Signal voltage (V_o) developed by the loop

The actual signal voltage (V_o) at the output of the terminals of an untuned loop is a function of the direction of arrival of the signal (α), as well as the strength of the arriving signal and the design of the loop. The angle α is formed between the loop axis and the advancing isopotential lines of the radio signal.

The output voltage of an untuned loop is:

$$V_o = \frac{2\pi A N E_f \cos \alpha}{\lambda}$$

While the output voltage of a tuned loop is:

$$V_o = \frac{2\pi A N Q E_f \cos \alpha}{\lambda}$$

where

> V_o is the output voltage of an untuned loop in volts (V)
>
> A is the length of one side of the loop
>
> N is the number of turns in the loop
>
> E_f is the strength of the incoming signal in volts per metre (V/m)
>
> α is the angle between the advancing wavefront and the loop axis
>
> λ is the wavelength of the radio signal in metres (m), i.e. the reciprocal of the frequency ($\lambda = 1/F$)
>
> Q is the loaded Q (figure of merit) of the tuned circuit formed by C1 and the loop inductance (typically 10 to 100)

4.5.5 Effective height (H_{eff})

Loop antennas are sometimes described in terms of the *effective height* (H_{eff}) of the antenna. This number is a theoretical construct that compares the output voltage of a small loop with a vertical piece of the same kind of wire that has a height of:

$$H_{eff} = \frac{2\pi NA}{\lambda}$$

where H_{eff} is the effective height in metres, and all other terms are as defined above.

Loop inductance

A loop antenna essentially forms a coil of wire (or other conductor), so will have inductance. There are several methods for calculating loop inductance, but the most common are the *Grover equation* and the *Patterson equation*.

Grover equation:

$$L_{\mu H} = \left[K_1 N^2 A \text{Ln} \left[\frac{K_2 AN}{(N+1)B} \right] \right] + K_3 + \left[\frac{K_4(N+1)B}{AN} \right]$$

where all terms are as previously defined, except $K_1 - K_4$, which are defined in Table 4.3.

Table 4.3

Shape	K_1	K_2	K_3	K_4
Triangle	0.006	1.1547	0.65533	0.1348
Square	0.008	1.4142	0.37942	0.3333
Hexagon	0.012	2.00	0.65533	0.1348
Octagon	0.016	2.613	0.75143	0.07153

Patterson equation:

$$L\mu H = (0.00508A) \times \left[2.303 \log \left(\frac{4A}{d} \right) - \phi \right]$$

where

d is the conductor diameter
ϕ is a factor found in Table 4.4

Table 4.4

Shape	Factor (ϕ)
Circle	2.451
Octagon	2.561
Hexagon	2.66
Pentagon	2.712
Square	2.853
Triangle	3.197

Of these equations, most people will find that the Grover equation most accurately calculates the actual inductance realized when a practical loop is built.

References

Andrew Antennas (1991). *Catalog 35*. Illinois.

Belcher, R. *et al.* (1989). *Newnes Mobile Radio Servicing Handbook*. Butterworth-Heinemann, Oxford.

Davies, J. (1987). *Private Mobile Radio – A Practical Guide*. Heinemann, London.

Green, D. C. (1979). *Radio systems, Vols. II and III*. Pitman, London.

Kennedy, G. (1977). *Electronic Communication Systems*. McGraw-Hill Kogashuka, Tokyo.

Terman, F. E. (1943). *Radio Engineers' Handbook*. McGraw-Hill, London.

5

RESONANT CIRCUITS

5.1 Series and parallel tuned circuits

Tuned resonant circuits composed of inductance and
capacitance are used to generate alternating voltages of
a specific frequency and to select a wanted frequency
or band of frequencies from the spectrum. Figure 5.1
contains the diagrams of series and parallel resonant
circuits including the resistances which account for the
losses present in all circuits. In practice the greatest
loss is in the resistance of the inductor, R_L, with
a smaller loss, r_c, occurring in the dielectric of the
capacitor.

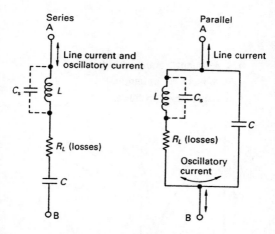

Fig. 5.1 *Series and parallel resonant circuits*

5.1.1 Series resonance

Off resonance, the series circuit exhibits a high impedance to a voltage applied across A and B. This
impedance is formed by the vectorial addition of the
reactances of the inductance and capacitance at the
applied frequency plus the resistances. Ignoring the

dielectric losses and the very small stray shunt capacitance C_s, the resonant impedance is given by:

$$Z = \sqrt{R^2 + \left(\omega L - \frac{1}{\omega C}\right)^2}$$

where

R = resistance of components in ohms
$\omega = 2\pi \times$ frequency in hertz
L = inductance in henries
C = capacitance in farads

At resonance the reactances cancel out and the impedance falls to approximately the value of the resistance, R, and a maximum line current will flow. Figure 5.2 shows the response curves of series and parallel circuits near resonance. The resonant frequency is given by:

$$f = \frac{1}{2\pi\sqrt{LC}}$$

where

f = resonant frequency in hertz
L = inductance in henries
C = capacitance in farads

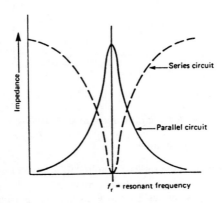

Fig. 5.2 *Variation of impedance around resonance with series and parallel tuned circuits*

5.1.2 Parallel resonance

At resonance, calculated by using the same formula as for a series circuit, the impedance of a parallel tuned circuit is also resistive, and the circulatory current in the circuit is high producing the maximum voltage across the inductance and capacitance. Consequently, at resonance the minimum line current flows. The impedance at resonance or dynamic resistance of a parallel tuned circuit of moderate to high Q is given by:

$$R_\mathrm{d} = \frac{L}{CR}$$

where

R_d = dynamic resistance of circuit

R = resistance of components in ohms

L = inductance in henries

C = capacitance in farads

5.2 Q factor

The voltages across the inductor and capacitor in a circuit at resonance are substantially opposite in phase (the loss resistances affect the phasing slightly) and cancel each other. The voltage developed across the inductor, usually the lossiest component, compared with the voltage applied in series to the circuit, is a measure of the 'goodness' of the circuit. This ratio is often referred to as the magnification factor, Q, of the circuit. The Q factor is calculated from:

$$Q = \frac{\omega L}{R} \text{ or } \frac{1}{\omega CR}$$

where L and C are in henries and farads respectively.

5.3 Coupled (band-pass) resonant circuits

5.3.1 Methods of coupling

Radio signals carrying intelligence occupy a band of frequencies and circuits must be able to accept the whole of that band whilst rejecting all others. When tuned circuits are coupled together in the correct manner they form such a band-pass circuit. The coupling may be either by mutual inductance between the two inductances of the circuits as in Figure 5.3 or through discrete electrical components as in Figure 5.4.

Mutual inductance can be defined in terms of the number of flux linkages in the second coil produced by unit current in the first coil. The relationship is:

$$M = \frac{\text{flux linkages in 2nd coil produced by current in 1st coil}}{\text{current in 1st coil}} \times 10^{-8}$$

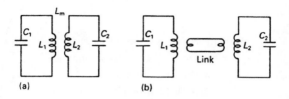

Fig. 5.3 *Mutual inductance (low impedance) coupling*

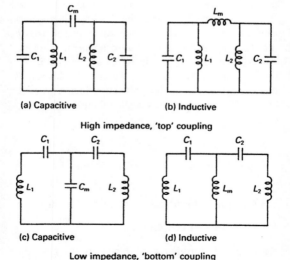

Fig. 5.4 *Electrical coupling*

where M = mutual inductance in henries. The e.m.f. induced in the secondary is $e_2 = -j\omega MI_1$ where I_1 is the primary current.

The maximum value of mutual inductance that can exist is $\sqrt{L_1 L_2}$ and the ratio of the actual mutual inductance to the maximum is the coefficient of coupling, k:

$$k = \frac{M}{\sqrt{L_1 L_2}}$$

The maximum value of k is 1 and circuits with a k of 0.5 or greater are said to be close coupled. Loose coupling refers to a k of less than 0.5.

An advantage of coupling using discrete components is that the coupling coefficient is more easily determined. Approximations for k for the coupling methods shown in Figure 5.4 are:

(a) $k = \dfrac{C_m}{\sqrt{C_1 C_2}}$ where C_m is much smaller than $(C_1 C_2)$

(b) $k = \dfrac{\sqrt{L_1 L_2}}{L_m}$ where L_m is much larger than $(L_1 L_2)$

(c) $k = \dfrac{\sqrt{C_1 C_2}}{C_m}$ where C_m is much larger than $(C_1 C_2)$

(d) $k = \dfrac{L_m}{\sqrt{L_1 L_2}}$ where L_m is much smaller than $(L_1 L_2)$

5.3.2 Response of coupled circuits

When two circuits tuned to the same frequency are coupled their mutual response curve takes the forms shown in Figure 5.5, the actual shape depending on the degree of coupling.

When coupling is very loose the frequency response and the current in the primary circuit are very similar to that of the primary circuit alone. Under these conditions the secondary current is small and the secondary response curve approximates to the product of the responses of both circuits considered separately.

As coupling is increased the frequency response curves for both circuits widen and the secondary current increases. The degree of coupling where the secondary current attains its maximum possible value is

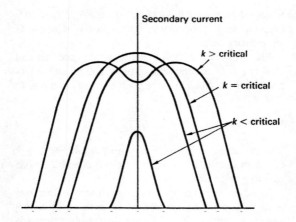

Fig. 5.5 *Effect of degree of coupling*

called the critical coupling. At this point the curve of the primary circuit shows two peaks, and at higher coupling factors the secondary response also shows two peaks. The peaks become more prominent and further apart as coupling is increased, and the current at the centre frequency decreases.

References

Langford-Smith, F. (1955). *Radio Designer's Handbook*, Iliffe and Sons, London.

Terman, F. E. (1943). *Radio Engineers' Handbook*. McGraw-Hill, London.

6

OSCILLATORS

6.1 Oscillator requirements

Oscillators generate the frequencies used in radio and electronic equipment. The performance of those which determine the operating frequencies of radio systems is tremendously important. Most oscillators must:

- Generate a precise frequency of high purity.
- Be highly stable, i.e. produce an output constant in frequency and level despite changes in temperature, supply voltage and load.
- Be tunable in frequency.
- Produce minimum noise and microphony (fluctuations in frequency with vibration).

These requirements conflict. A readily tunable oscillator cannot be precise and highly stable, and compromises must be made; either a less stringent specification must be accepted where permissible or the facility for tuning restricted.

Not all oscillators need to produce a pure output, devoid of spurious frequencies. The clock generators in digital circuitry, for instance, produce square waves, but a radio transmitter carrier generator and receiver local oscillator must produce a pure sine wave output if spurious radiations and receiver responses are to be avoided.

6.2 Tunable oscillators

An oscillator is an amplifier with a portion of the output fed back to the input. The feedback must be positive, i.e. in phase with the input, and the loop gain, input back to input via the feedback loop, must be sufficient to overcome the losses in the circuit.

Most radio frequency oscillators – and some audio ones – use inductance and capacitance (*LC*) tuned circuits as the frequency determining elements. Figure 6.1

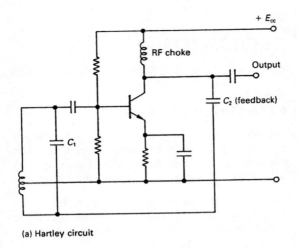

(a) Hartley circuit

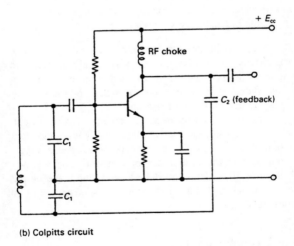

(b) Colpitts circuit

Fig. 6.1 *Hartley and Colpitts oscillators*

shows two commonly used basic circuits, the Hartley and the Colpitts.

The frequency is determined by the values of L and C_1 (the combined values of C_1 in the Colpitts circuit) and the amount of feedback by the collector choke and C_2. Such circuits produce a very pure output but, principally because the physical dimensions of the frequency determining components change with temperature, the accuracy of the set frequency is doubtful and is not very stable. Temperature compensation can be applied by selecting a capacitor for C_1 with the correct negative temperature coefficient (assuming that the inductance increases with temperature), inserting temperature compensation for the rise in collector current with temperature and stabilizing the supply voltage. In the design of equipment an oscillator should be built into an area of low temperature change.

6.3 Quartz crystal oscillators

The problems of frequency accuracy and stability are largely overcome by using a quartz crystal as the frequency determining element (see Chapter 7).

Figure 6.2(a) shows a circuit for an oscillator using the crystal's parallel resonant mode. In this circuit, the rising voltage developed across R_e on switch-on is applied via C_1 to the base accelerating the rise of current through the transistor. When saturation is reached the voltage across R_e becomes static and the voltage on the base falls, reducing the transistor current. The oscillations are only sustained at the parallel resonant frequency of the crystal where it presents a high impedance between base and collector. C_t enables the parallel resonance of the crystal to be adjusted to a precise frequency.

Figure 6.2(b) shows a series resonant crystal oscillator and here L_t is the tuning inductor of a Colpitts oscillator. The loop gain is adjusted so that the circuit will oscillate only at the series resonant frequency of the crystal where it presents a very low resistance. At other frequencies the crystal presents an increasing impedance in series with L_t, shunted by R which can be of a low value. When first setting the oscillator, L_t is adjusted, with the crystal short-circuited, for oscillation at a frequency close to that desired. The short-circuit is then removed and L_t used as a fine trimmer.

The same circuit will operate at the parallel resonant frequency of the crystal by making C_1 equal to the crystal load capacitance.

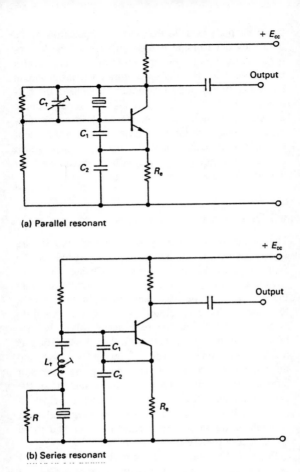

(a) Parallel resonant

(b) Series resonant

Fig. 6.2 *Quartz crystal oscillators*

The maximum frequency error permitted by the British Radiocommunications Agency specification MPT 1326 for mobile radio equipment designed for 12.5 kHz channel separation in the band 100–300 MHz is plus or minus 1.5 MHz. This is an overall accuracy of 0.0005% over the temperature range −10°C to +55°C. Well-designed standard crystal oscillators meet this specification, but higher stability can be obtained by operating the crystal in an oven at a constant higher temperature.

Until recently equipment which was required to change operating frequency quickly was fitted with several crystals, one for each operating frequency, and a change of frequency was made by selecting the appropriate crystal. Frequency synthesizer circuits are now normally used for such applications.

6.3.1 Overtone oscillators

Piezoelectric crystals can oscillate at more than one frequency. The oscillations of a crystal slab are in the form of *bulk acoustic waves* (BAWs), and can occur at any frequency that produces an odd half-wavelength of the crystal's physical dimensions (e.g. $1\lambda/2$, $3\lambda/2$, $5\lambda/2$, $7\lambda/2$, $9\lambda/2$, where the fundamental mode is $1\lambda/2$). Note that these frequencies are not exact harmonics of the fundamental mode, but are actually valid oscillation modes for the crystal slab. The frequencies fall close to, but not directly on, some of the harmonics of the fundamental (which probably accounts for the confusion). The overtone frequency will be marked on the crystal, rather than the fundamental (it is rare to find fundamental mode crystals above 20 MHz or so, because their thinness makes them more likely to fracture at low values of power dissipation).

The problem to solve in an overtone oscillator is encouraging oscillation on the correct overtone, while squelching oscillations at the fundamental and unde-sired overtones. Crystal manufacturers can help with correct methods, but there is still a responsibility on the part of the oscillator designer. It is generally the case that overtone oscillators will contain at least one $L-C$ tuned circuit in the crystal network to force oscillations at the right frequency.

6.4 Frequency synthesizers

Frequency synthesizers offer the stability of a quartz crystal oscillator combined with the facility to change operating frequency rapidly. They are essential for equipment operating on trunked or cellular networks where the frequency of the mobiles is changed very rapidly on instructions from the network.

6.4.1 Voltage controlled oscillators

The advent of the variable capacitance diode (varicap), where the capacitance across the diode varies according to the applied DC voltage, made the frequency synthesizer a practicality.

When a varicap diode replaces the tuning capacitor in an oscillator the circuit becomes a DC-voltage-controlled oscillator (VCO). The two varicaps in Figure 6.3 are used to minimize harmonic production and to obtain a greater capacitance change per volt.

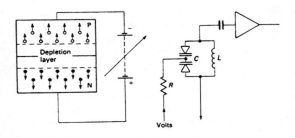

Fig. 6.3 *Voltage variable capacitance diode*

The VCO is the circuit that directly generates the output frequency of a frequency synthesizer but, by itself, inherits the problem of frequency stability. To overcome this, the frequency and phase of the VCO are compared with those of a crystal-controlled high stability oscillator. Any frequency or phase difference between the two oscillators creates a DC voltage of the correct sense to change the frequency of the VCO to agree with that of the crystal oscillator.

While the stability of the crystal oscillator is transferred to the synthesizer output, additional noise is produced close to the operational frequency and the elimination of microphony requires careful physical design.

6.4.2 Phase-locked loops

Figures 6.4(a) and (b) are diagrams of a simple phase-locked loop (PLL). The outputs of both the crystal oscillator and the voltage-controlled oscillator are fed

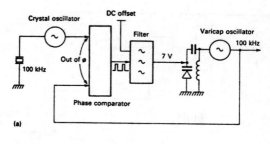

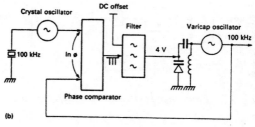

Fig. 6.4 *Simple phase-locked loop*

to the phase comparator which produces pulses whenever there is a frequency or phase difference between the two inputs. The pulses will be either positive or negative depending on the sense of the difference, and their width is dependent upon the magnitude of the differences. The pulses are then fed to a low pass loop filter which smooths them. If the time constant of the filter is sufficiently long it will completely remove the pulses and produce a DC output proportionate to the input pulse width which is applied to the VCO in the right sense to correct the frequency error. To enable the pulses to swing the VCO frequency in either direction, a small bias voltage of about 4 volts is applied to the varicap.

Identical frequencies have been selected for both oscillators in Figure 6.4 but in practice this is seldom the case. More frequently, the VCO runs at a higher frequency than the crystal oscillator and a divider is used to equate the frequencies applied to the phase comparator (Figure 6.5). Changing the division ratio provides a convenient means of tuning the oscillator.

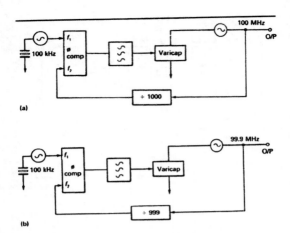

Fig. 6.5 *Frequency variable phase-locked loop*

While the division ratio is 1000, the VCO will run at 100 MHz but if the division ratio is changed to 999 the comparator will produce pulses which, when converted to a DC voltage by the loop filter, will change the frequency of the VCO to 99.9 MHz, and the loop will lock at this new frequency.

The design of the loop filter is critical. Too long a time constant lengthens the settling time when changing frequency, yet if it is too short any deliberate frequency modulation will be removed. In practice, a relatively long time constant is chosen which is shortened by a 'speed up' circuit introduced whenever a channel change is called for.

The above values would enable a radio operating on a system with a channel separation of 100 kHz to change channel, but mobile radio channel separations are 25 kHz, 12.5 kHz or even 6.25 kHz at frequencies from 50 MHz to at least 900 MHz. To change channel at these frequencies a synthesizer must use a high division ratio. With a reference frequency applied to the comparator of 6.25 kHz and an operating frequency of, say, 450 MHz, the frequency select divider must have a ratio of 72 000, and be programmable in steps of 1 with a minimum operating speed of at least 900 MHz. A problem is then that the technology capable of meeting these requirements, emitter coupled logic (ECL), is power hungry, and the preferred LSI low power technology, CMOS, has a maximum operating speed of about 30 MHz. A simple ECL pre-scaler to bring the VCO frequency to about 30 MHz needs a ratio of 20 (500 MHz to 25 MHz). However, every change of 1 in the CMOS divider ratio then changes the total division ratio by 20. The solution is to use a dual modulus pre-scaler.

6.4.3 Dual modulus pre-scaler

The division ratio of the dual modulus pre-scaler (Figure 6.6) is programmable between two consecutive numbers, e.g. 50 and 51 (P and $P + 1$) and, in conjunction with two CMOS dividers, $\div A$ and $\div N$, provides a fully programmable divider.

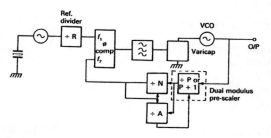

Fig. 6.6 *Programmable frequency synthesizer*

The A and N dividers are pre-loaded counters. These count down and when the count value reaches zero they produce an output which changes the division ratio of the pre-scaler. The total division ratio, N_t, is decided by the initial programmed contents of the A and N counters and the setting of the pre-scaler. The initial content of the A counter must be less than that of the N counter.

Consider the pre-scaler set to divide by $P + 1$. For every count of $P + 1$, the contents of the A and N counters are reduced by 1 until the contents of the A counter are zero. The difference between the original contents of the A and N counters, $N - A$, remains in the N counter, and the total count, N_t, up to now, is

$A(P + 1)$. At this point the division ratio of the pre-scaler is changed to P. Now, for every P count, the contents of the N counter are reduced by 1 until zero is reached. Under these conditions the total division ratio is given by:

$$N_t = A(P + 1) + (N - A)P$$
$$= AP + A + NP - AP$$
$$= NP + A$$

For example, let $P = 50$ so $P + 1 = 51$, let $N = 10$ and $A = 7$. Then:

$$N_t = 10 \times 50 + 7 = 507$$

Now, change A to 6:

$$N_t = 10 \times 50 + 6 = 506$$

a change of N_t by 1.

Programming a divider
Example:

> VCO frequency $= 455.6\,\text{MHz}$
> Reference frequency $= 12.5\,\text{kHz}$

Calculate N_t, and the numbers which must be programmed into the A and N counters, assuming $P = 50$:

1. Calculate $N_t = 455.6\,\text{MHz}/12.5\,\text{kHz} = 36\,448$.
2. Divide N_t by P: $36\,488/50 = 728.96$. Make $N = 728$.
3. For A, multiply fraction by P: $0.96 \times 50 = 48$.
4. Check $N_t = NP + A = 728 \times 50 + 48 = 36\,448$.
   Change A to 47: $NP + A = 728 \times 50 + 47 = 36\,447$.
   $36\,447 \times 12.5\,\text{kHz} = 455.5875\,\text{MHz}$, the adjacent 12.5 kHz channel.

6.4.4 Direct digital synthesis

A method of direct digital frequency synthesis replaces the voltage-controlled oscillator by a numerically controlled oscillator (NCO) where the function of the VCO is digitally synthesized.

The direct digital synthesizer generates an analogue sine wave from digital sine wave samples applied to a digital to analogue (d/a) converter. There are limitations to the method in terms of bandwidth and spectral purity.

6.5 Caesium and rubidium frequency standards

Where extra high stability is required for, say, laboratory standards or in quasi-synchronous wide area coverage systems, oscillators utilizing the atomic resonances of substances like caesium and rubidium, although expensive, may be employed.

Caesium oscillators are used to provide standard frequencies such as 1, 5 and 10 MHz with accuracies of $\pm 7 \times 10^{-12}$ over a temperature range of 0 to 50°C with a long-term stability of 2×10^{-12}.

Rubidium oscillators are used to provide secondary standards and in some quasi-synchronous radio systems. Their accuracy is less than that of caesium, the long-term drift being of the order of 1×10^{-11} per month.

References

Belcher, R. *et al.* (1989). *Newnes Mobile Radio Servicing Handbook*. Butterworth-Heinemann, Oxford.

Edis, E.A. and Varrall, J.E. (1992). *Newnes Telecommunications Pocket Book*. Butterworth-Heinemann, Oxford.

7

PIEZO-ELECTRIC DEVICES

7.1 Piezo-electric effect

When electrical stress is applied to one axis of a quartz crystal it exhibits the piezo-electric effect: a mechanical deflection occurs perpendicular to the electric field. Equally, a crystal will produce an e.m.f. across the electrical axis if mechanical stress is applied to the mechanical axis. If the stress is alternating – the movement of the diaphragm of a crystal microphone is an example – the e.m.f. produced will be alternating at the frequency of the movement. If the stress alternates at a frequency close to the mechanical resonance of the crystal as determined by its dimensions, then large amplitude vibrations result. Polycrystalline ceramics possess similar qualities.

Quartz crystals used for radio applications are slices cut from a large, artificially grown crystal. The slices are then ground to the appropriate size to vibrate at a desired frequency. The performance of an individual slice – the crystal as the end user knows it – depends upon the angle at which it was cut from the parent crystal.

Each crystal slice will resonate at several frequencies and if the frequency of the stimulus coincides with one of them the output, electrical or mechanical, will be very large.

The vibrations occur in both the longitudinal and shear modes, and at fundamental and harmonic frequencies determined by the crystal dimensions.

Figure 7.1A shows a typical natural quartz crystal. Actual crystals rarely have all of the planes and facets shown. There are three *optical axes* (X, Y and Z) in the crystal used to establish the geometry and locations of various cuts. The actual crystal segments used in RF circuits are sliced out of the main crystal. Some slices are taken along the optical axes, so are called Y-cut, X-cut and Z-cut slabs. Others are taken from various

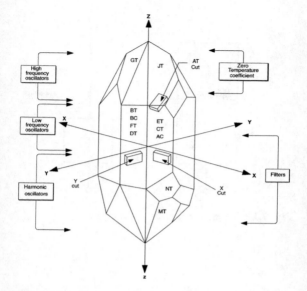

Fig. 7.1A *Natural quartz crystal*

sections, and are given letter designations such as BT, BC, FT, AT and so forth.

7.1.1 Equivalent circuit of a quartz crystal

A quartz crystal behaves similarly to a very high Q tuned circuit and the equivalent circuit of a crystal is shown in Figure 7.1B.

C_1 and L_1 are equivalent to the inductance and capacitance of a conventional tuned circuit and R_1 represents the losses in the quartz and the mounting arrangements. C_0, typically 3–15 pfd, represents the shunt capacitance of the electrodes in parallel with the can capacitance. If the oscillatory current is considered, the resonant frequency is decided by the values of C_0 in series with C_1, L_1 and R_1, and all crystals basically resonate in a series mode. Figure 7.2 illustrates the changes in impedance close to resonance. However when a high impedance, low capacitance, load is connected across the crystal terminals it behaves as a parallel tuned circuit exhibiting a high resistance at the resonant frequency. A crystal operating in the parallel

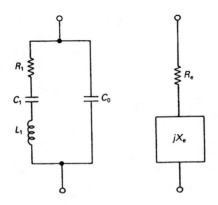

Fig. 7.1B *Equivalent circuit of a crystal*

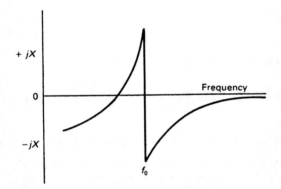

Fig. 7.2 *Crystal reactance close to resonance*

mode oscillates at a higher frequency than that of series resonance.

A crystal will resonate at its fundamental frequency or at one or more of its harmonics. As the desired resonant frequency is increased, a crystal slice operating at its fundamental frequency becomes extremely thin and fragile. Consequently, overtone crystals are composed of larger slices of quartz operating close to, but not necessarily at, an exact harmonic of the fundamental frequency. Crystals operating at the 3rd, 5th and 7th harmonics are often employed at frequencies above approximately 25 MHz.

7.2 Quartz crystal characteristics

7.2.1 Resonant frequency

The resonant frequency is determined by the mass of the finished crystal which can be adjusted by grinding and the deposition of gold or other metal onto the crystal faces during manufacture. The adjustment is made to suit the intended method of operation, series or parallel, and at a specific temperature, usually 25°C. When parallel mode is specified, allowance is made for the load or circuit capacitance, usually 20–30 pfd, in parallel with C_0.

7.2.2 Frequency stability

Temperature coefficient

A crystal's resonant frequency varies with temperature and this temperature coefficient is determined by the angle at which the slice was cut from the parent crystal. Commonly used cuts are AT and BT. Because of its better performance AT is the most common.

Typical examples of the temperature coefficients for these are shown in Figure 7.3.

The temperature coefficient is specified, usually in parts per million (ppm) per degree C, or as a percentage, over a defined temperature range. The standard European temperature range is −10°C to +60°C. A crystal designed for a restricted temperature range has a better stability over that range than one designed for operation over a wide temperature range will have when used over a restricted range.

For higher frequency stability crystals may be operated in a temperature-controlled oven operating at a more constant high temperature.

Common frequency tolerance specifications are ±0.005% or 0.0025% from −55°C to +105°C. These include the frequency errors from all sources, including the calibration tolerance; thus, the temperature coefficient is slightly better than these figures.

Ageing

The resonant frequency shifts with age from that set at production, following a curve similar to that in Figure 7.4. Initially the frequency shift for a given period of time is rapid but slows with

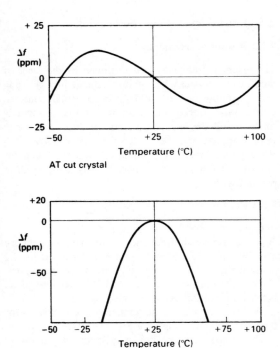

Fig. 7.3 *Frequency vs. temperature curves AT and BT cut crystals*

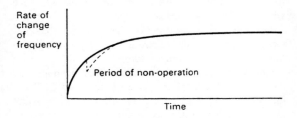

Fig. 7.4 *Effect of ageing*

age. The frequency may shift in either direction, and although it is possible to specify crystals ageing in one direction – high stability oscillators for quasi-synchronous transmission systems is an application – they are selected from a batch, not specifically manufactured. Once a crystal has been operated, a subsequent long period of inactivity can produce a glitch in the ageing curve followed by a higher rate of change for a short time.

7.2.3 Load capacitance and pullability

When a crystal is operated in the parallel mode across a low capacitance load the results are a higher frequency and larger output voltage to the load. Increasing the load capacitance causes a reduction in frequency approaching that of series resonance.

The change in frequency that can be achieved by varying the load capacitance – a small trimmer capacitor is often connected across the crystal for this purpose – is the crystal's pullability. A typical pullability is from -1 ppm/pfd to -20 ppm/pfd for a total shunt capacitance of 40 pfd ($C_0 + C_{load}$).

7.2.4 Activity, effective series resistance (ESR) and Q

All these characteristics are interrelated. A crystal's activity, its vibrational response, can be quoted in terms of the effective series resistance. A higher effective series resistance implies lower activity, lower output and lower Q. The usual range of ESRs is from $20\,\Omega$ to $100\,\Omega$ although higher values occur in some low frequency crystals. Some manufacturers may quote activity levels for crystals for use in a parallel mode as effective parallel resistance (EPR). The EPR is the value of the resistor which, if connected in lieu of the crystal in an oscillator, would give the same output level as the crystal. The higher the EPR, the greater the crystal activity and Q.

7.2.5 Spurious responses

Crystals will resonate at frequencies other than those of the fundamental and harmonic modes for which they were designed; Figure 7.5 shows the overtone (harmonic) and some typical spurious responses. The spurious responses of overtone crystals can occur with very little separation from the desired overtone frequency requiring very careful oscillator design if they are to be avoided.

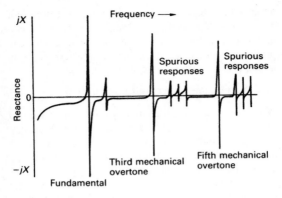

Fig. 7.5 *Overtone response of a quartz crystal*

7.2.6 Case styles

A wide range of mounting styles is available. The American military nomenclature is widely used to describe them and Figure 7.6 shows the outlines of some of the more popular styles.

7.3 Specifying quartz crystals

The details which must be specified when ordering crystals are:

1. *Frequency.* Normally specified in kHz up to 9999.999 kHz and in MHz from 10.0 MHz upwards except for integer values which are all specified in MHz. The frequency must be described to seven significant figures, otherwise any figure that might follow those given will be taken as zero.
2. *Mounting or holder style.*
3. *Frequency tolerance.* This is the cutting or calibration tolerance acceptable at 25°C. It should be borne in mind that cost rises with increased manufacturing accuracy and a slight adjustment (pullability) is possible in the circuit.
4. *Frequency stability.* Normally specified as a plus or minus value measured over a defined temperature range. A crystal designed for a restricted range has a better performance over that range than one

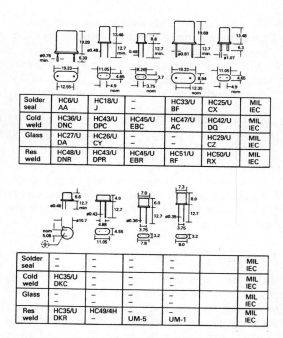

Fig. 7.6 *Crystal case styles*

designed for a wider range so it is important not to overspecify.

5. *Temperature range.* The range over which the crystal is required to operate and meet the performance specified in 4. Standard temperature ranges are:

    0 to 5°C
    −10 to 60°C
    −20 to 70°C
    −30 to 80°C
    −40 to 90°C
    −55 to 105°C
    −55 to 125°C

It is sufficient when ordering from some manufacturers to quote only the lower temperature limit.

For ovened operation the quoted figure, say 80°C, would denote the oven temperature.

6. *Circuit condition*. This specifies the shunt capacitance that the circuit will place across the crystal in parallel mode operation.
7. *Drive level*. The maximum power that the crystal can safely dissipate. 1 mW is a typical value for crystals used in radio transmitters and receivers.

A typical specification therefore reads:

16.66667 MHz	HC49	20	30	10	30
1	2	3	4	5	6

referring to the items listed above

When the crystal is for operation in series mode, it is usually sufficient to replace the last figure with 'S'. The drive level is not normally specified in the ordering details.

7.4 Filters

Both quartz and ceramic materials are used in the production of radio frequency filters. Ceramic filters do not have the same performance as quartz but have the advantages of a lower cost. They are used at lower frequencies and where the higher stability and lower spurious responses of quartz are not essential.

Crystal filters are obtainable at frequencies up to about 45 MHz. Most of these use either a number of discrete crystals arranged in the form of a lattice or a monolithic structure. A single crystal will behave as an extremely narrow band filter and it is possible to use a crystal bar in this way down to a very few kilohertz.

The characteristics of filters can be divided into groups affecting the performance (Figure 7.7).

7.4.1 Passband performance

- *Insertion loss*. The loss at centre frequency, in dB, resulting from the insertion of the filter in a transmission system.
- *Flat loss*. The insertion loss at the frequency of minimum loss within the passband.
- *Attenuation*. The loss of a filter at a given frequency measured in dB.
- *Passband (bandwidth, BW_1)*. The range of frequencies attenuated less than a specified value, typically 3 or 6 dB.

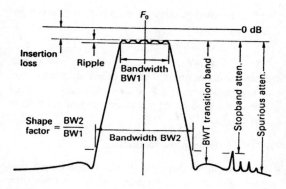

Fig. 7.7 *Filter characteristics*

- *Centre frequency* (f_0). The arithmetic mean of the passband limits.
- *Fractional bandwidth*. A specified frequency, typically the minimum loss point or F_0, from which all attenuation measurements are made.
- *Ripple*. The amplitude difference, in dB, between the maximum peak and minimum passband valley. Both the peak and the valley are defined by a surrounding change in slope, i.e. sign of the amplitude response. This is very important as a high ripple, particularly between a peak and the adjacent trough, produces rapid phase changes as the signal moves across the passband resulting in audio distortion and corruption in digital signals.

7.4.2 Stopband performance

- *Attenuation*. The output of a filter at a given frequency relative to the defined insertion loss reference.
- *Stopband*. The range of frequencies attenuated by a greater amount than some specified minimum level of attenuation.
- *Transition band (bandwidth, BW_2)*. The range of frequencies differently attenuated between the passband and stopband limits.
- *Shape factor*. The ratio of the bandwidth at some point within the transition region, typically 60 dB,

to the specified passband bandwidth. It is given by:

$$\text{Shape factor} \frac{BW_2}{BW_1}$$

- *Spurious attenuation.* The specified minimum level of attenuation received by all non-harmonic related resonances of each crystal resonator within the filter network.

7.4.3 Time domain performance

- *Insertion phase.* The phase shift at the output load (measured at the reference frequency) resulting from the insertion of the filter.
- *Differential phase.* The measurement of phase at a given frequency relative to the phase at the reference frequency.
- *Phase linearity.* The phase error in degrees between the phase points and a straight line drawn through the phase points.
- *Group delay.* The time by which a signal will be delayed before it appears at the filter output, i.e. the derivative of phase with respect to frequency.
- *Differential delay.* The measurement of delay at a given frequency relative to the reference frequency.

7.4.4 Source and load impedance

- *Source impedance.* The impedance of the circuit driving the filter, measured at the reference frequency.
- *Load impedance.* The impedance of the circuit terminating the filter at its output, measured at the reference frequency.

7.4.5 Non-linear effects

- *Maximum input level.* The driving point power, voltage or current level above which intolerable signal distortion or damage to the device will result.
- *Drive level stability.* The ability of the filter to return within a specified tolerance of its original insertion loss, at a specified drive level, after experiencing changing environmental and/or drive level conditions.

Table 7.1 Manufacturers' specifications for two stock 10.7 MHz filters

Centre freq.	Passband width (plus/minus) (6 dB)	Attenuation bandwidth	Ripple (max)	Ins. loss (max)	Term. impedance Ω S/pfd
10.7 MHz	3.75 kHz	8.75 kHz (45 dB)	2.0 dB	3.0 dB	1.5 k/1
10.7 MHz	7.5 kHz	12.5 kHz (60 dB)	2.0 dB	4.0 dB	3 k/1
		15.0 kHz (60 dB)			
		20.0 kHz (80 dB)			

- *Drive level linearity*. The maximum permissible variation in insertion loss, per dB change in drive level, measured over a specific dynamic range.
- *Inband intermodulation distortion*. The attenuation, in dB, of third and higher order signal products, inband, relative to the power level of two signals placed within the passband.
- *Out-of-band intermodulation distortion*. The attenuation, in dB, of third and higher order signal products, inband, relative to the power level of two signals placed in the stopband, or one signal in the transition region and the other in the stopband.

A manufacturer's specifications for two stock 10.7 MHz filters are given in Table 7.1.

References

Crystal Product Data Book (1993). International Quartz Devices Ltd, Crewkerne, Somerset.
Edis, E.A. and Varrall, J.E. (1992). *Newnes Telecommunications Pocket Book*. Butterworth-Heinemann, Oxford.

BANDWIDTH REQUIREMENTS AND MODULATION

8.1 Bandwidth of signals at base band

8.1.1 Analogue signals

The amount of information and the speed at which it is transmitted determines the base bandwidth occupied by a signal. For analogue signals, the base bandwidth is the range of frequencies contained in the signal; it is not the same as that occupied by a radio frequency carrier modulated by the signal. Examples of base bandwidths are given in Table 8.1.

Table 8.1 *Base bandwidths*

Application	Frequency range (Hz)
Speech	
High fidelity reproduction	15–15 000
Good fidelity	150–7000
Public address	200–5000
Restricted bass and treble	500–4000
Toll quality (good quality telephone line)	300–3400
Communications quality (radio communication)	300–3000
Mobile radio (12.5 kHz channel separation)	300–2700
Music (for FM broadcasting)	30–15 000
Video	60 Hz–4.2 MHz

8.1.2 Digital signals

Bit rate (b/s) and baud rate are terms used to specify the speed of transmitting digital information. Where the duration of all the signalling elements is identical the terms are synonymous, but not where the duration of the information bits differs.

As the term implies, the bit rate is the number of bits transmitted per second but the baud rate (after J.M.E. Baudot, the code's inventor) is the reciprocal of the length, in seconds, of the shortest duration signalling element. Figure 8.1(a) shows a binary code pattern where all the bits are of equal duration, in this case 1 millisecond; the bit rate is 1000 per second and the baud rate is $1/0.001 = 1000$ also.

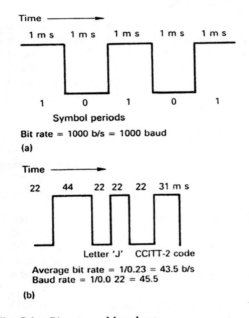

Fig. 8.1 *Bit rate and baud rate*

On telegraphy systems all the bits may not be of the same duration and Figure 8.1(b) shows the pattern for the letter 'J' in the CCITT-2 code as used for teletype. In this code a letter is composed of 5 elements, each of 22 ms duration, but each letter is preceded by a space of 22 ms and followed by a mark of 31 ms. The duration of each character is 163 ms – the time for 7.5 elements – but is comprised of only 7 bits.

The baud rate is $1/0.022 = 45.5$ baud.

The average bit duration is $163/7 = 23.29$ ms.

The average bit rate is $1/0.023 = 43.5$ bits per second.

If the bandwidth were under consideration the baud rate, being faster, would be the figure to use.

A stream of binary coded information is composed of pulses where, say, a pulse (mark) represents digit 1 and the absence of a pulse (space) represents digit 0. The highest frequency contained in the information is determined by the bit – or baud – rate. Because a series of 1s or 0s may be consecutive in the data stream, the pulse repetition rate will vary throughout the message although the bit rate will be constant, and over a long period of time as many spaces will be sent as pulses. Transmitting either a stream of spaces or of marks requires no bandwidth; it is only when a change of state occurs that frequencies are produced. In the duration of one cycle (Figure 8.1(a)) two bits may be carried and the maximum *fundamental* frequency contained in the wave is one-half the number of bits per second, i.e. the channel capacity, bits/second, equals twice the bandwidth in hertz.

8.1.3 Channel capacity – Hartley – Shannon theorem

Channel capacity as stated by Hartley's law is, in the absence of noise:

$$C = 2\delta f \log_2 N$$

where

C = channel capacity, bits per second
δf = channel bandwidth, Hz
N = number of coding levels (2 in binary system)

When noise is present, the channel capacity calculated according to the Hartley–Shannon theorem is:

$$C = \delta f \log_2(1 + S/N)$$

where $S/N =$ the ratio of total signal power to total noise power at the receiver input within the bandwidth, δf.

8.2 Modulation

For radio transmission, the low frequency information signal is carried on a radio frequency wave and it must change (modulate) that carrier. The modulation may change the amplitude, frequency or phase of the carrier. Modulation aims to achieve:

1. the transfer of information with the minimum distortion or corruption
2. the modulation of the carrier with the minimum loss of power
3. efficient use of the frequency spectrum.

8.3 Analogue modulation

8.3.1 Amplitude modulation (AM)

There are a number of methods of modulation where the amplitude of the carrier is varied by the information signal but the most commonly used is double sideband AM (DSB). Figure 8.2 shows a radio frequency carrier modulated by a low frequency signal.

The amount or depth of modulation is expressed as percentage ratio, $m\%$, of the maximum to minimum amplitude:

$$\text{Mod. depth} = m\%$$
$$= \frac{\text{max. amplitude} - \text{min. amplitude}}{\text{max. amplitude} + \text{min. amplitude}}$$
$$\times 100\%$$

When the modulation is increased to the point where the minimum amplitude falls to zero, 100% modulation occurs. Any further increase in modulation produces spurious, out-of-band frequencies (AM splash), a source of interference for other radio users. For this reason, the depth of amplitude modulation is usually limited to 70%.

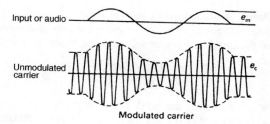

Fig. 8.2 *Amplitude modulation*

An alternative expression for modulation depth is in terms of a modulation index from 0 to 1. The peak carrier voltage in Figure 8.2 is E_c and the peak modulation voltage, E_m. The modulation index, m, is:

$$m = \frac{E_m}{E_c}$$

Amplitude modulation produces a band of frequencies above and below the carrier frequency – the upper and lower sidebands. The width of each sideband is equal to the highest modulating frequency so the bandwidth of an AM wave is $2 \times$ the highest modulating frequency. To conserve spectrum, the range of modulating frequencies is restricted. For example, radio communication quality speech is limited to 300 Hz to 3000 Hz. The bandwidth occupied by a double sideband, amplitude modulated carrier for this service is 6 KHz (Figure 8.3).

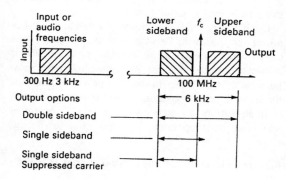

Fig. 8.3 *Amplitude modulation – sidebands*

Power relationships in an AM wave

The total power in an AM wave is the sum of powers of the carrier, the upper sideband and lower sideband:

$$P_t = \frac{E_{carr}^2}{R} + \frac{E_{lsb}^2}{R} + \frac{E_{usb}^2}{R}$$

where all values are RMS and R is the resistance in which the power is dissipated. From the peak voltages shown in Figure 8.2 the power in the unmodulated carrier is:

$$P_c = \frac{E_c/\sqrt{2}\,^2}{R} = \frac{E_c^2}{2R}$$

The power in the sidebands

$$P_{lsb} = P_{usb} = \left(\frac{mE_c/2}{\sqrt{2}}\right)^2 \div R = \frac{m^2 E_c^2}{8R}$$
$$= \frac{m^2}{4}\frac{E_c^2}{2R}$$

The ratio of the total power in the wave to the carrier power is:

$$\frac{P_t}{P_c} = 1 + \frac{m^2}{2}$$

As m cannot exceed 1, the maximum RMS power in the wave is $P_t = 1.5P_c$; but if m reaches 1, the peak sum of E_c and E_m is $2E_c$ and so the instantaneous peak power is $2P_c$. Circuitry must be capable of handling this power level without distortion.

Double sideband amplitude modulation wastes power and spectrum. Two-thirds of the power is in the carrier which conveys no information and one sideband is discarded in the receiver. Also, the modulation must be accomplished either in the final power amplifier of the transmitter necessitating a high power modulator, or in an earlier, low power stage when all subsequent amplifiers must operate in a linear, but inefficient, mode.

8.3.2 Double sideband suppressed carrier (DSBSC)

In an amplitude modulated wave the carrier conveys no information yet contains 2/3 of the transmitted power. It is possible to remove the carrier by using a balanced modulator (Figure 8.4), and improve the power efficiency by this amount.

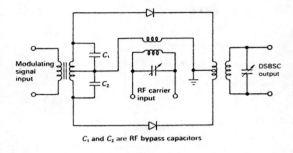

C_1 and C_2 are RF bypass capacitors

Fig. 8.4 *Balanced modulator*

In a balanced modulator, the modulating voltage is fed in push–pull to a pair of matched diodes or amplifiers while the carrier is fed to them in parallel. The carrier components in the output cancel leaving the two sidebands. The result is a double sideband suppressed-carrier (DSBSC) wave, which is not sinusoidal, formed by the sum of the two sidebands. The carrier must be re-introduced in the receiver and its accuracy in both frequency and phase is critical.

8.3.3 Single sideband suppressed carrier (SSB or SSBSC)

The advantages of single sideband suppressed carrier transmission over double sideband AM are:

- removal of the carrier saves 2/3 of the total power
- removal of one sideband saves 50% of the remaining power
- an SSBSC transmitter only produces power when modulation is present
- the occupied bandwidth is halved; a spectrum saving
- the received signal-to-noise ratio is improved by 9 dB for a 100% modulated carrier. Halving the bandwidth accounts for 3 dB, the remainder from the improved sideband power to total power ratio. The

S/N ratio improves further with lower modulation levels

- reduced susceptibility to selective fading and consequent distortion.

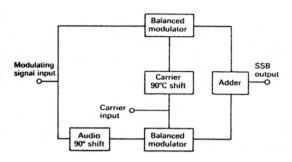

Fig. 8.5 *Phase shift production of single sideband*

Two methods of generating a single sideband wave are in general use. One filters out the unwanted sideband after removal of the carrier by a balanced modulator. The other is a phase shift method (Figure 8.5). Here, the modulating signal is fed to two balanced modulators with a 90° phase difference. The output from both modulators contains only the sidebands but, while both upper sidebands lead the input carrier voltage by 90°, one of the lower sidebands leads it by 90° and the other lags it by 90°. When applied to the adder, the lower sidebands cancel each other while the upper sidebands add.

A single sideband AM wave modulated by a sinusoid consists of a constant amplitude signal whose frequency varies with the frequency of the modulating wave. Note that this is not the same as FM: the frequency in SSB does not swing to either side of the carrier. It is higher than the carrier frequency if the upper sideband is transmitted, and lower if the lower sideband is selected. The single sideband waveform is sinusoidal and, although the frequency of the reintroduced carrier must be highly accurate (± 2 Hz), the phase is unimportant making a single sideband receiver less complex than one for DSBSC.

On some systems a pilot carrier is transmitted and the transmitter output power is then specified in terms of peak envelope power (pep), the power contained in a wave of amplitude equal to the pilot carrier and transmitted sideband power. Where no pilot carrier is transmitted, the power is specified as peak sideband power (psp).

8.3.4 Frequency modulation (FM)

Both frequency and phase modulation (both may be referred to as angle modulation) effectively vary the frequency of the carrier rather than its amplitude. Frequency modulation varies the carrier frequency directly but its amplitude remains constant regardless of the modulating voltage. Angle modulation is employed at VHF and above for both communications and broadcasting services.

When frequency modulated, a carrier frequency either increases or decreases when the modulation voltage is positive and varies in the opposite sense when the modulating voltage is negative (Figure 8.6). The amount of modulation, i.e. the 'deviation' of the carrier from its nominal frequency, is proportional to the amplitude, and not the frequency, of the modulating voltage. The modulation index, M, is defined as the deviation divided by the modulating frequency:

$$M = \frac{f_d}{f_m} \text{ or } \frac{\omega_d}{\omega_m}$$

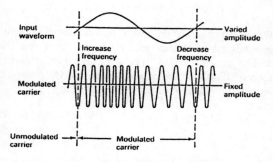

Fig. 8.6 *Frequency modulation*

129

where

f_d = deviation in hertz

f_m = modulating frequency in hertz

$\omega_d = 2\pi f_d$ = deviation in radians

$\omega_m = 2\pi f_m$ = modulating frequency in radians

The 'deviation ratio', D, is given by:

$$D = \frac{f_{d(max)}}{f_{m(max)}}$$

The peak deviation is the maximum amount of modulation occurring (equivalent to 100% amplitude modulation). With FM, this is limited only by the need to conserve spectrum; there is no technical limit where distortion occurs, as with 100% AM. The maximum permitted deviation for a service is determined by regulation.

The bandwidth of a frequency modulated signal is made up of the carrier and a series of sidebands, sometimes referred to as sidecurrents, spaced apart from each other at the modulating frequency. The number of sidebands is proportional to the modulation index and their amplitudes decrease with spacing from the carrier. It is generally considered satisfactory to transmit those sidebands $M + 1$ in number, with amplitudes greater than 10% of that of the carrier for that modulation index (Figure 8.7). The sidebands occur on both sides of the carrier and:

1st order sidebands = $f_c \pm f_m$

2nd order sidebands = $f_c \pm 2f_m$, etc.

When the modulation index approaches 6 a good approximation of the bandwidth (δf) required for an FM transmission is $2(f_d + f_m)$ Hz. For example, speech 300–3000 Hz, max deviation 15 kHz (relevant to a VHF, 50 kHz channel spacing system):

$$\delta f = 2(f_d + f_m) = 2(15 + 3)\,\text{kHz} = 36\,\text{kHz}$$
$$M = 15/3 = 5$$

The bandwidth is also given by $\delta f = f_m \times$ highest needed sideband $\times$ 2. From Bessel functions

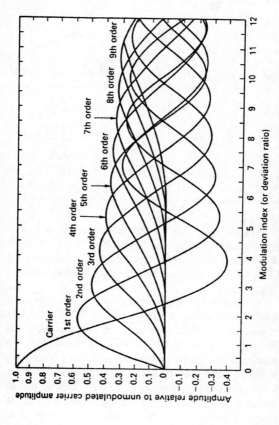

Fig. 8.7 *Amplitudes of components of FM wave with variation of modulation index*

(Figure 8.7), a modulation index of 5 requires the 6th order sideband to be transmitted ($M + 1$). Therefore:

$$\delta f = f_m \times 6 \times 2 = 3 \times 6 \times 2 = 36\,\text{kHz}$$

For specific values of M the carrier of an FM wave disappears. The successive disappearances and the modulation index are given in Table 8.2.

Table 8.2 *Successive disappearances and modulation index*

Order of disappearance	Modulation index M
1	2.40
2	5.52
3	8.65
4	11.79
5	14.93
6	18.07
$n(n > 6)$	$18.07 + \pi(n - 6)$

Systems where the modulation index exceeds $\pi/2$ are considered to be wide band FM (WBFM), those with a modulation index lower than $\pi/2$, narrow band FM (NBFM). The bandwidth of an NBFM signal is $2f_{m(max)}$.

Pressure on spectrum necessitates narrower channel spacings for communications and, at VHF, 12.5 kHz is normal with ±2.5 kHz as the maximum permitted deviation. Lower standards of performance with a restricted modulation index and a highest modulating frequency of 3 kHz have had to be accepted.

For transmitters used on 12.5 kHz channel spaced systems the highest modulating frequency is, in practical terms, 2700 Hz because the specification requires the frequency response to fall above 2.55 kHz.

The modulation index on such systems, assuming a highest modulating frequency of 3 kHz, is $2.5/3 = 0.8333$ ($<\pi/2$ and so system is NBFM) and the amplitude of the 2nd order sideband is <10% of carrier amplitude (Figure 8.7), so the bandwidth $= 2f_{m(max)} = 6\,\text{kHz}$.

8.3.5 Phase modulation

The end result of phase modulation is frequency modulation, but the method of achieving it and the definition of the modulation index is different. Phase modulation is used in VHF and UHF transmitters where the carrier frequency is generated directly by a crystal oscillator. The frequency of a crystal oscillator can be varied by only a few radians but if the oscillator frequency is multiplied to produce the final carrier frequency the phase variation is also multiplied to produce a frequency deviation.

In frequency modulation deviation is proportional to the modulating voltage, but in phase modulation the frequency deviation is proportional to both the modulating voltage and frequency.

The phase modulation, ϕ_d radians, equals the modulation index so, for phase modulation, the modulation index is $\phi_d = \omega_d/\omega_m$.

Phase modulation has, therefore, a frequency response for the deviation that rises at 6 dB per octave of the modulating frequency. The flat frequency response of FM can be produced on a phase modulated transmitter by installing a filter with a response falling at 6 dB per octave in the audio amplifier, and the rising response of PM can be produced in an FM transmitter with a rising response filter.

8.3.6 Pre- and de-emphasis

Phase modulation, and FM modified to give a rising frequency response (pre-emphasis), offer an improved signal-to-noise ratio in the receiver. The +6 dB per octave response produced in the transmitter is restored to a flat response in the receiver by a −6 dB per octave filter in the audio circuitry which reduces both the enhanced levels of the higher speech frequencies and the high frequency noise (de-emphasis).

8.3.7 Merits of amplitude and frequency modulation

The advantages and disadvantages of AM and FM are given in Table 8.3.

Table 8.3 *Advantages and disadvantages of AM and FM*

	Advantages	Disadvantages
AM	Simple modulators and de-modulators Narrower bandwidth than wide-band FM	Susceptible to man-made noise Audio strength falls with decreasing RF signal strength Inefficient power usage Limited dynamic range Transmitter output power not easily adjusted
FM	Less susceptible to noise Constant audio level to almost the end of radio range Capture effect in receiver	Wider bandwidth Capture effect may be undesirable, e.g. aviation communications
	More power-efficient Transmitter output power easily adjustable	

8.4 Digital modulation

8.4.1 Data processing

Filtering

A data pulse with a sharp rise- and fall-time produces harmonic frequencies and requires a wide bandwidth if it is to maintain its shape during transmission. Consequently, to transmit the data over a limited bandwidth the pulses must be shaped to reduce the harmonic content as much as possible without impairing the intelligibility of the signal. This is accomplished by the use of low-pass (Gaussian) filters of which the result is a string of smoother pulses, often referred to as 'tamed' (Figure 8.8(b)). Tamed FM permits high

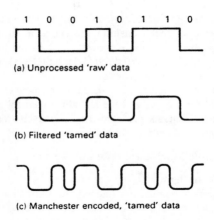

| 1 | 0 | 0 | 1 | 0 | 1 | 1 | 0 |

(a) Unprocessed 'raw' data

(b) Filtered 'tamed' data

(c) Manchester encoded, 'tamed' data

Fig. 8.8 *Data processing*

data rates within a limited channel bandwidth whilst maintaining acceptable adjacent channel interference levels.

Manchester encoding

A serious problem with the transmission of binary data is that unless the clocks in the transmitter and receiver are synchronous the digits become confused, particularly where a continuous string of 1s or 0s occurs. Manchester encoding makes 1s change state from 1 to 0, and 0s from 0 to 1 during each digit period (Figure 8.8(c)), facilitating the synchronization of the clocks and rendering the digits more easily recognizable.

Gray coding

For binary data the number of signalling levels is not restricted to two (mark and space). Multi-level signalling has the advantage that one signalling element carries the information for more than one information bit, thus reducing the bandwidth requirement.

If the number of levels is increased to m, where $m = 2^n$ (i.e. $n = \log_2 m$), the m-level data symbol is represented by n binary digits of 0 or 1 (see quaternary phase shift keying, Section 8.4.5). For example, in a quaternary data signal, $m = 4 = 2^2$ giving the binary sequences 00, 01, 10, 11 for each signalling level. This process is Gray coding where there is only one digit difference for each transition between adjacent levels. The number of levels is not restricted to four and 16 level quadrature amplitude modulation (QAM), where $n = 4$, is an efficient system.

In a Gray coded signal where each element contains n bits ($n = \log_2 m$) and the signalling element rate is B bauds (elements per second), the transmission rate is $B \log_2 m$ bits per second.

8.4.2 On/off and amplitude shift keying

On/off keying

The earliest modulation method. A continuous radio frequency wave (CW) is interrupted in a recognizable pattern (Morse code). To provide audibility the carrier is heterodyned with a beat frequency oscillator (BFO) in the receiver. The use of a modulated continuous wave (MCW) eliminates the need for a BFO but the bandwidth of the signal is increased. The problem with on/off keying is the lack of a reference level. If the signal strength temporarily falls below the sensitivity threshold of the receiver it appears to the operator as a series of spaces.

Binary amplitude shift keying (ASK or BASK)

This shifts the level of an audio frequency subcarrier which then modulates a radio frequency carrier.

Because the level of a subcarrier is changed, AM sidebands are produced. Also, because the keyed waveform is non-sinusoidal harmonics occur. The occupied sub-carrier bandwidth for ASK is:

Bandwidth $= 2B$

where B = bit repetition rate (bits/second).

When the RF carrier is modulated its bandwidth is $2(f_c + B)$ where f_c = subcarrier frequency.

8.4.3 Frequency shift keying (FSK)

Although used for conveying digital information, frequency shift keying in reality employs frequency modulation. In its original form, developed for HF transmission, FSK changes the carrier frequency to indicate a 1 or a 0 but retains the nominal carrier frequency as a reference and to represent a mark. A downwards shift of carrier frequency by 170 Hz represents a space in the HF radio system.

Modern FSK uses two different modulation frequencies to represent 1s and 0s. If intersymbol interference (ISI) is to be avoided the separation of the tones must be more than half the bit rate, and a factor of 0.7 is often used. Standard CCITT frequencies for various bit rates are given in Table 8.4.

Table 8.4 *Standard CCITT frequencies for various bit rates*

Bit rate (bps)	Frequencies		Subcarrier frequency
	0	1	
600	1700	1300	1500
1200	2100	1300	1700
up to 300	1180	980	1080
	1850	1650	1750

The base bandwidth requirement is:

$$\text{Bandwidth} = f_2 - f_1 + 2B$$

The bandwidth of a modulated carrier is:

$$\text{RF bandwidth} = 2(f_2 + B)\text{(narrow band FM)}$$

Minimum shift keying (MSK) is a form of FSK where the frequency deviation is equal to half the bit rate.

Gaussian minimum phase shift keying (GMSK)
Similar to MSK, the Gaussian filters improve the adjacent channel performance against a small cost (approximately 1%) in ISI while achieving high data transmission rates.

8.4.4 Fast frequency shift keying (FFSK)

Fast frequency shift keying may either amplitude or frequency modulate the carrier. In binary FFSK, the data is changed in a modem to tones of 1800 Hz to represent binary 0 and 1200 Hz to represent binary 1. During transmission a binary 1 consists of 1 **cycle** of 1200 Hz, f_1, and a 0, $1\frac{1}{2}$ cycles of 1800 Hz, f_2, i.e. a bit rate of 1200 bps. For acceptable intersymbol interference the distance between the tones cannot be less than half the bit rate and the 600 Hz separation in FFSK represents the fastest signalling speed – hence the description – and minimum bandwidth. For this reason it is sometimes called minimum frequency shift keying (MFSK). The base bandwidth is the same as for FSK, but the RF bandwidth depends upon the system deviation. For example, on a 12.5 kHz channel spaced system carrying FFSK and complying with Radiocommunications Agency Code of Practice MPT 1317, deviation = 60% of system deviation = 1.5 kHz:

$$f_1 = 1200\,\text{Hz}$$
$$f_2 = 1800\,\text{Hz}$$
$$B = 1200\,\text{bps}$$

$f_{max} = f_2 + B = 3\,\text{kHz}$ and the mod. index $m = $ max. dev./$f_{max} < \pi/2$ so the system is NBFM and the bandwidth $= 2(f_2 + B) = 6\,\text{kHz}$.

Where more data states than binary are to be transmitted, multistate FSK (M-ary FSK) is also possible where M may be up to 32 states.

Both FFSK and M-ary FSK are well suited to radio transmission as the change of state occurs while the signals are passing through zero, avoiding sudden phase changes (Figure 8.9(a)).

The minimum distance between the tones used in FFSK of 0.5 times the bit rate is not ideal for immunity to intersymbol interference (ISI). A separation of 0.7 times the bit rate would improve the ISI but increase the bandwidth.

8.4.5 Phase shift keying

There are several variants of phase shift keying. Binary phase shift keying (BPSK or PSK) changes the phase of the carrier by 180° at the zero crossing point (Figure 8.9(b)). No carrier frequency is present with PSK as half the time the carrier is multiplied by +1 and the other half by −1 and cancels out, but the reference phase of the carrier must be re-inserted at

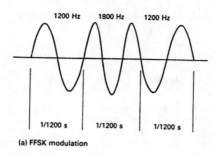

Logic 0 three half cycles 1800 Hz
Logic 1 two half cycles 1200 Hz

1200 Hz   1800 Hz   1200 Hz

1/1200 s   1/1200 s   1/1200 s

(a) FFSK modulation

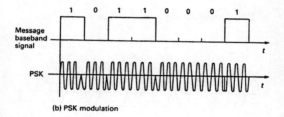

1 0 1 1 0 0 0 1

Message
baseband
signal

PSK

(b) PSK modulation

Fig. 8.9 *FFSK and PSK*

the receiver. The bandwidths occupied are the same as for ASK, i.e.:

Baseband $= 2B$

RF bandwidth (AM or NBFM) $= 2(f_c + B)$

Differential phase shift keying (DPSK) advances the phase 90° or 270° at each change of logic state (Figure 8.10). Changing phase only at a change of logic state saves bandwidth which, for DPSK, is equal to the bit rate.

An important advantage of both FFSK and PSK over FSK is that because the moment of change is predefined it is possible to recover data more accurately. However, the transition between signalling states is not smooth requiring large and rapid phase shifts. Multilevel systems with less phase shift between elements are preferable.

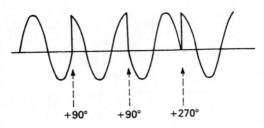

Data	Phase change
0	+90°
1	+270°

(a) DPSK modulation

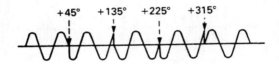

Dibit	Phase change
00	+45°
01	+135°
11	+225°
10	+315°

(b) QPSK modulation

Fig. 8.10 *DPSK and QPSK*

Quaternary, or quadrature, phase shift keying (QPSK) is a four-level Gray-coded signalling method with 90° phase shift between adjacent signalling elements (Figure 8.10(b)). If the signal is considered as a vector the points at $\pi/4$ (45°), $3\pi/4$ (135°), $5\pi/4$ (225°) and $7\pi/4$ (315°) represent the transition points between states and the binary data.

8.5 Spread spectrum transmission

The spread spectrum technique spreads the carrier containing the information over a very wide bandwidth, typically 1.25 MHz, using pseudo-noise generation techniques as described in Chapter 12. The transmitter uses what is in effect a digital key to spread the bandwidth and the receiver is equipped with an identical key for despreading. A number of users with different keys can occupy the same band at the same time. The system operates well in poor signal-to-noise or high interference environments.

A continuous wave (CW) transmission concentrates all the radiated energy on a single frequency (Figure 8.11). Amplitude modulation and narrow band FM widen the radiated bandwidth, reducing the energy at the carrier frequency and per kHz. Wide band FM carries the process a stage further until with spread spectrum the band width is increased to the extent that the signal almost disappears into the noise floor.

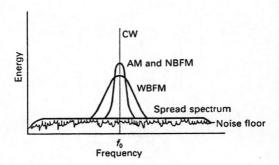

Fig. 8.11 *Comparative energy dispersal*

Spreading of the bandwidth is achieved by multi-plying the digitally modulated signal by a spreading code at a much higher bit rate (100–1000 times the signal bit rate). This is done by combining the signal with the output of a random code generator running at 2 or 3 orders of magnitude faster than the binary signal rate. Figure 8.12 is a block diagram of a spread spectrum system. A clock running at the spreading rate R_c is used to drive both the spreading generator and, after frequency division, the data encoder. The carrier is first BPSK modulated by the encoded data and then

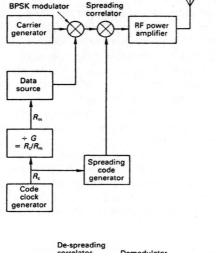

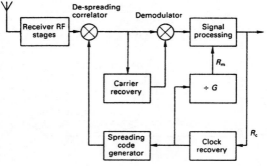

Fig. 8.12 *CDMA (spread spectrum) system*

in a balanced modulator (the spreading correlator) by the high rate code from the spreading generator. The resultant transmitted bits are referred to as chips to distinguish them from data bits. In the receiver, the clock pulses R_c are recovered and used to drive both the despreading generator and decoder.

Many users can be accommodated by allocating each a unique spreading code. It is common to use a pseudo-noise (PN) generator to multiply the bit rate and then to modulate the carrier with either FSK or PSK.

Although spread spectrum is a digital system, in quality of signal there are similarities with analogue:

Analogue	Digital
Signal/noise ratio	Energy per bit, E_b/N_o
Intelligibility, signal/noise + distortion	Bit error rate

References

Belcher, R. *et al.* (1989). *Newnes Mobile Radio Servicing Handbook.* Butterworth-Heinemann, Oxford.

Edis, E.A. and Varrall, J.E. (1992). *Newnes Telecommunications Pocket Book.* Butterworth-Heinemann, Oxford.

Kennedy, G. (1977). *Electronic Communications Systems*, McGraw-Hill Kogashuka, Tokyo.

9

FREQUENCY PLANNING

9.1 International and regional planning

The International Telecommunications Union (ITU) administers the planning and regulation of the radio frequency spectrum on a worldwide basis through the World Administrative Radio Conferences.

For planning purposes the world is divided into three regions as shown in Figure 9.1. The boundaries are formed by geographical features suited to the purpose such as seas, high mountain ranges or uninhabited remote areas.

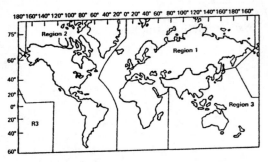

Fig. 9.1 *ITU defined regions. For purposes of international allocations of frequencies the world has been divided into three regions*

The World Administrative Radio Conferences decide the use of blocks of the spectrum, e.g. sound broadcasting, television, marine communications, and the countries permitted to use the blocks for those allotted purposes.

9.2 National planning

Having been allocated blocks of frequencies for a particular type of use, the administration of each country determines the method of allocating frequency bands from each block to the user categories within their country. Every country has a radio regulatory department within its administration. In the UK this is the Radiocommunications Agency of the Department of Trade and Industry.

9.2.1 The role of the Radiocommunications Agency

The aims of the Radiocommunications Agency in the UK, and, in general, the radio regulatory bodies of other countries are:

1. To ensure that the radio frequency spectrum is used in ways which maximize its contribution to national social and economic welfare, having regard to safety of life factors.
2. To make the maximum amount of spectrum available for commercial use.
3. To provide an expert service to government as a whole in the field of radio regulation.

The first stage of national planning is the assignment of the radio frequencies (channels) within a geographic area. Both the allocation of the blocks to users and the geographic assignment of channels if not wisely carried out can result in spectrum pollution – intermodulation products are one source (see Chapter 19) – and unacceptable interference between services and users. The incorrect allocation of blocks may not only cause interference in the home country but, even at VHF and UHF, between adjacent countries. Incorrect assignment of channels causes a more local problem but, depending on the frequencies involved, the undesirable effects can spread over a wide area.

The second stage of national planning is the assignment of discrete channels for use on multi-user sites where the selection of incompatible frequencies causes interference, receiver de-sensitization and, possibly, blocking, and intermodulation products.

Not all channels are allocated directly by the regulatory body. Blocks of channels, usually comprised of two frequencies, may be issued to responsible user organizations: the Home Office for the police and fire services is an example. These organizations then become responsible for the frequency planning and allocation within their user group.

Additional to the allocation of frequencies, the Radiocommunications Agency through the licensing procedure regulates the use of base station sites, the maximum transmitter power, and antenna directivity for each service. It also prepares and publishes technical specifications with which all equipment must comply. A list of the current specifications, of which single copies can be obtained, is available from The Information and Library Service, Room 605, Radiocommunications Agency, Waterloo Bridge House, Waterloo Road, London SE1 8UA.

The Radio Investigation Service (RIS) is the branch of the Agency which, in addition to investigating interference, inspects all radio installations prior to commissioning and, if in order, issues an inspection certificate. No station is permitted to operate without a certificate and may not be modified subsequent to the issue of the certificate.

Current policy throughout Europe is leading towards the de-regulation of radio communications while safe-guarding the protection from interference. Allocation of frequencies by pricing is also under consideration on the basis that a scarce resource, the spectrum, will be allocated to the users having the greatest need.

9.3 Designations of radio emissions

Radio emissions should be expressed in a three-symbol code form, which defines the exact nature of carrier, signal and transmitted information. The first symbol defines the carrier, the second symbol defines the signal, and the third symbol defines the information.

First symbol

A   Double-sideband amplitude-modulated
B   Independent sideband amplitude-modulated
C   Vestigial sideband amplitude-modulated
D   Amplitude- and angle-modulated simultaneously, or in a predefined sequence
F   Frequency modulated
G   Phase modulated
H   Single-sideband, full carrier
J   Single-sideband, suppressed carrier

K	Amplitude-modulated pulse sequence
L	Width-modulated pulse sequence
M	Position phase modulated pulse sequence
N	Unmodulated carrier
P	Unmodulated pulse sequence
Q	Pulse sequence in which carrier is angle-modulated during the pulse period
R	Single-sideband, reduced or variable level carrier
V	Pulse sequence with a combination of carrier modulations, or produced by other means
W	Carrier is modulated by two or more of angle, amplitude, and pulse modes, simultaneously or in a defined sequence
X	Other cases

Second symbol

0	No modulating signal
1	Digital signal without modulating sub-carrier
2	Digital signal with modulating sub-carrier
3	Analogue signal
7	Two or more channels with digital signals
8	Two or more channels with analogue signals
9	Composite system with one or more channels of digital signals and one or more channels of analogue signals
X	Other cases

Third symbol

A	Aural telegraph
B	Automatic telegraph
C	Facsimile
D	Data
E	Telephony (and sound broadcasting)
F	Television
N	No information transmitted
W	Combination of any of the above
X	Other cases

9.4 Bandwidth and frequency designations

A four symbol code should be used to express bandwidth and frequency to three significant figures. A letter to denote the unit of frequency is placed in the position of the decimal point, where the letters and bandwidths are:

Letter Bandwidth

H	Below 1000 Hz

K Between 1 and 999 kHz
M Between 1 and 999 MHz
G Between 1 and 999 GHz

So, a frequency of 120 Hz is 120H, while a frequency of 12 Hz is 12H0 etc.

9.5 General frequency allocations

VLF, LF, MF (frequency in kHz)

10.0	140.5	Fixed; maritime; navigation
140.5	283.5	Broadcast
255.0	526.5	Radio navigation; fixed
526.0	1606.5	Broadcast
1606.5	1800.0	Maritime and land mobile; fixed
1810.0	1850.0	Amateur (shared in UK)
1850	2000	Amateur
1850	2045	Fixed; mobile
2045	2173.5	Maritime mobile; fixed
2160	2170	Radiolocation
2173.5	2190.5	Mobile
2190.5	2194	Maritime
2194	2625	Fixed; mobile
2300	2498	Broadcast
2625	2650	Maritime mobile
2650	2850	Fixed; mobile
2850	3155	Aero mobile

HF (frequency in kHz)

3155	3400	Fixed; mobile
3200	3400	Broadcast
3400	3500	Aero mobile
3500	3800	Amateur, fixed; mobile
3800	4000	Amateur (region 2 only)
3800	3900	Fixed; mobile
3800	3950	Aero mobile
3950	4000	Fixed; broadcast
4000	4063	Fixed; maritime mobile
4063	4438	Maritime mobile
4438	4650	Fixed; mobile
4650	4750	Aero mobile
4750	5060	Fixed; mobile; broadcast
5060	5480	Fixed; mobile
5450	5730	Aero mobile
5730	5950	Fixed; mobile
5950	6200	Broadcast
6200	6525	Maritime mobile
6525	6765	Aero mobile
6765	7000	Fixed; mobile

7000	7100	Amateur
7100	7300	Amateur (region 2 only)
7100	7300	Broadcast (regions 1 and 3)
7300	8195	Fixed
8100	8815	Maritime mobile
8815	9040	Aero mobile
9040	9500	Fixed
9500	10 000	Broadcast
10 000	10 100	Aero mobile
10 100	11 175	Fixed
10 100	10 150	Amateur
11 175	11 400	Aero mobile
11 400	11 650	Fixed
11 650	12 050	Broadcast
12 050	12 230	Fixed
12 230	13 200	Maritime mobile
13 200	13 360	Aero mobile
13 360	13 600	Fixed
13 600	13 800	Broadcast
13 800	14 000	Fixed
14 000	14 350	Amateur
14 350	15 000	Fixed
15 000	15 100	Aero mobile
15 100	15 600	Broadcast
15 600	16 360	Fixed
16 360	17 410	Maritime mobile
17 410	17 550	Fixed
17 550	17 900	Broadcast
17 900	18 030	Aero mobile
18 030	18 068	Fixed
18 068	18 168	Amateur
18 168	18 780	Fixed
18 780	18 900	Maritime mobile
18 900	19 680	Fixed
19 680	19 800	Maritime mobile
19 800	21 000	Fixed
21 000	21 450	Amateur
21 450	21 850	Broadcast
21 850	21 870	Fixed
21 870	22 000	Aero mobile
22 000	22 855	Maritime mobile
22 855	23 200	Fixed; mobile
23 200	23 350	Aero mobile
23 350	24 890	Fixed; mobile
24 890	24 990	Amateur
25 010	25 070	Fixed; mobile
25 070	25 210	Maritime mobile
21 210	25 550	Fixed; mobile

25 550	25 670	Radio astronomy
25 670	26 100	Broadcast
26 100	26 175	Maritime mobile
26 175	28 000	Fixed; mobile
28 000	29 700	Amateur
29 700	30 000	Fixed; mobile

VHF, UHF (frequencies in MHz)

30.0	50.0	Fixed; mobile
47.0	68.0	Broadcast (TV)
50.0	52.0	Amateur (UK)
50.0	54.0	Amateur (regions 2 and 3)
68.0	74.8	Fixed; mobile
70.0	70.5	Amateur (UK)
74.8	75.2	Aero navigation
75.2	87.5	Fixed; mobile
87.5	108	Broadcast (FM)
108	118	Aero navigation
118	137	Aero mobile
137	138	Spacecraft; satellites
138	144	Aero mobile; space research
144	146	Amateur
146	148	Amateur (regions 2 and 3 only)
146	174	Fixed; mobile
156	174	Maritime mobile
174	230	Broadcast (TV)
220	225	Amateur (USA)
230	328.6	Fixed; mobile
328.6	335.4	Aero navigation
335.4	400	Fixed; mobile
400	410	Space research; meteorology
410	430	Fixed; mobile
430	440	Amateur; radiolocation
440	470	Fixed; mobile
470	855	Broadcast (TV)
855	1300	Fixed; mobile
902	928	Amateur (USA)
934	935	Citizens band (UK)
1240	1325	Amateur
1300	1350	Aero navigation
1350	1400	Fixed; mobile
1400	1429	Space (uplink); fixed
1429	1525	Fixed; mobile
1525	1600	Space (downlink)
1600	1670	Space (uplink)
1670	1710	Space (downlink)
1710	2290	Fixed; mobile
2290	2300	Space (downlink); fixed

2300	2450	Amateur; fixed
2310	2450	Amateur (UK)
2300	2500	Fixed; mobile
2500	2700	Fixed; space (downlink)
2700	3300	Radar
3300	3400	Radiolocation; amateur
3400	3600	Fixed; space (uplink)
3600	4200	Fixed; space (downlink)
4200	4400	Aero navigation
4400	4500	Fixed; mobile
4500	4800	Fixed; space (downlink)
4800	5000	Fixed; mobile
5000	5850	Radio navigation; radar
5650	5850	Amateur
5850	7250	Fixed; space (uplink)
7250	7900	Fixed; space (downlink)
7900	8500	Fixed; mobile; space
8500	10 500	Radar; navigation
10 000	10 500	Amateur
10 700	12 700	Space (downlink); fixed
12 700	15 400	Space (uplink); fixed
17 700	20 000	Space (up/down); fixed
24 000	24 250	Amateur

9.6 Classes of radio stations

AL	Aeronautical radionavigation land station
AM	Aeronautical radionavigation mobile station
AT	Amateur station
AX	Aeronautical fixed station
BC	Broadcasting station, sound
BT	Broadcasting station, television
CA	Cargo ship
CO	Station open to official correspondence exclusively
CP	Station open to public correspondence
CR	Station open to limited public correspondence
CV	Station open exclusively to correspondence of a private agency
DR	Directive antenna provided with a reflector
EA	Space station in the amateur-satellite service
EB	Space station in the broadcasting-satellite service (sound broadcasting)
EC	Space station in the fixed-satellite service
ED	Space telecommand space station
EE	Space station in the standard frequency-satellite service

EF	Space station in the radiodetermination-satellite service
EG	Space station in the maritime mobile-satellite service
EH	Space research space station
EJ	Space station in the aeronautical mobile-satellite service
EK	Space tracking space station
EM	Meteorological-satellite space station
EN	Radionavigation-satellite space station
EO	Space station in the aeronautical radionavigational-satellite service
EQ	Space station in the maritime radionavigation-satellite service
ER	Space telemetering space station
ES	Station in the intersatellite service
EU	Space station in the land mobile-satellite service
EV	Space station in the broadcasting-satellite service (television)
EW	Space station in the earth exploration-satellite service
EX	Experimental station
EY	Space station in the time signal-satellite service
FA	Aeronautical station
FB	Base station
FC	Coast station
FL	Land station
FP	Port station
FR	Receiving station only, connected with the general network of telecommunication channels
FS	Land station established solely for the safety of life
FX	Fixed station
GS	Station on board a warship or a military or naval aircraft
LR	Radiolocation land station
MA	Aircraft station
ME	Space station
ML	Land mobile station
MO	Mobile station
MR	Radiolocation mobile station
MS	Ship station
ND	Non-directional antenna
NL	Maritime radionavigation land station
OD	Oceanographic data station
OE	Oceanographic data interrogating station
OT	Station open exclusively to operational traffic of the service concerned

PA	Passenger ship
RA	Radio astronomy station
RC	Non-directional radio beacon
RD	Directional radio beacon
RG	Radio direction-finding station
RM	Maritime radionavigation mobile station
RT	Revolving radio beacon
SM	Meteorological aids station
SS	Standard frequency and time signal station
TA	Space operation earth station in the amateur-satellite service
TB	Fixed earth station in the aeronautical mobile-satellite service
TC	Earth station in the fixed-satellite service
TD	Space telecommand earth station
TE	Transmitting earth station
TF	Fixed earth station in the radiodetermination-satellite service
TG	Mobile earth station in the maritime mobile-satellite service
TH	Earth station in the space research service
TI	Earth station in the maritime mobile-satellite service at a specified fixed point
TJ	Mobile earth station in the aeronautical mobile-satellite service
TK	Space tracking earth station
TL	Mobile earth station in the radiodetermination-satellite service
TM	Earth station in the meterological-satellite service
TN	Earth station in the radionavigation-satellite service
TO	Mobile earth station in the aeronautical radionavigation-satellite service
TP	Receiving earth station
TQ	Mobile earth station in the maritime radionavigation-satellite service
TR	Space telemetering earth station
TS	Television, sound channel
TT	Earth station in the space operation service
TU	Mobile earth station in the land mobile-satellite service
TV	Television, vision channel
TW	Earth station in the earth exploration-satellite service
TX	Fixed earth station in the maritime radionavigation-satellite service

TY	Fixed earth station in the land mobile-satellite service
TZ	Fixed earth station in the aeronautical radionavigation-satellite service

9.7 Radio wavebands

Frequency band	Frequency	Wavelength	Waveband definition
VLF	3 to 30 kHz	100000 to 10000 m	myriametric
LF	30 to 300 kHz	10000 to 1000 m	kilometric
MF	300 to 3000 kHz	1000 to 100 m	hectometric
HF	3 to 30 MHz	100 to 10 m	decametric
VHF	30 to 300 MHz	10 to 1 m	metric
UHF	300 to 3000 MHz	1 to 0.1 m	decimetric
SHF	3 to 30 GHz	10 to 1 cm	centimetric
EHF	30 to 300 GHz	1 to 0.1 cm	millimetric
EHF	300 to 3000 GHz	0.1 to 0.01 cm	decimillimetric

References

Pannell, W.M. (1979). *Frequency Engineering in Mobile Radio Bands.* Granta Technical Editions, Cambridge.

10

RADIO EQUIPMENT

10.1 Transmitters

10.1.1 Transmitter functions

The functions of all transmitters and the terminology used to describe them, irrespective of the modulation method, are:

1. To generate the radio frequency carrier and amplify it to an appropriate power level; the RF power output.
2. To modulate the carrier with the intelligence to the pre-determined level; the modulation depth for AM, the deviation for FM or PM. The process must introduce the minimum noise and distortion, and prevent the modulation from exceeding the permitted level.
3. Radiate the minimum signals at frequencies outside the permitted bandwidth. Out-of-band or spurious radiation is strictly controlled by the Radiocommunications Agency MPT specifications.

10.1.2 Amplitude-modulated transmitters

Figure 10.1 is a block diagram of an amplitude modulated transmitter; in this case a quartz crystal oscillator generates the carrier frequency, although a frequency synthesizer could equally well be used. The carrier frequency in an AM transmitter is usually generated either at the transmitted frequency or one of its subharmonics; 2nd, 3rd or 6th subharmonic frequencies are typical choices depending on the final frequency.

The output of the oscillator is amplified to the level of the specified power output and if the oscillator runs at a subharmonic a frequency multiplier stage will be included, as in Figure 10.1, in the amplifier chain before the final stage, the power amplifier (PA). A tuned filter in the aerial circuit removes from the output unwanted frequencies which might cause interference

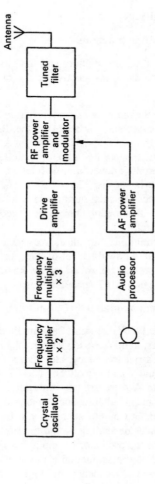

Fig. 10.1 *Crystal controlled AM transmitter*

with other users. A matching circuit correctly matches the impedance of the filter circuit to that of the aerial to ensure maximum power transfer.

In the audio circuits, the speech input from the microphone is processed by controlling the range of frequencies it contains and limiting its amplitude. This eliminates the risk of over-modulation and the production of out-of-band frequencies. Over-modulation produces out-of-band frequencies in any transmitter, but with AM once 100% modulation is exceeded frequencies are produced across a wide range of the spectrum: a disastrous situation for other users. After processing, the audio is amplified in an AM transmitter to a high level and applied to the RF power amplifier to vary its output in the form illustrated in Figure 8.2.

AM transmitters are power inefficient. The RF PA cannot operate in class C. It must be linear so as not to distort the speech, and the audio, when it is the power amplifier which is modulated, must be amplified to a high power. If the modulation is applied at a lower power level (in an early stage of the amplifier chain) all the subsequent amplifiers must operate in a linear, but inefficient, manner.

The power output of an AM transmitter is normally specified in terms of the RMS value of the carrier power but the average and peak powers will depend on the depth of modulation, 100% modulation producing a peak power of twice the carrier peak power (see Chapter 8).

10.1.3 Angle modulated transmitters

Figure 10.2 is a block diagram of a frequency modulated transmitter using a frequency synthesizer for carrier generation. The frequency is generated at the final frequency and RF amplifiers raise the power level to that specified for the transmitter output. When a crystal oscillator is used for carrier generation, it must operate at a very low frequency because a quartz crystal oscillator can be frequency modulated by a few radians only. Several stages of the RF amplifier chain then operate as frequency multipliers. Similar filter and aerial matching circuits to those of an AM transmitter are necessary in the output arrangements of the transmitter.

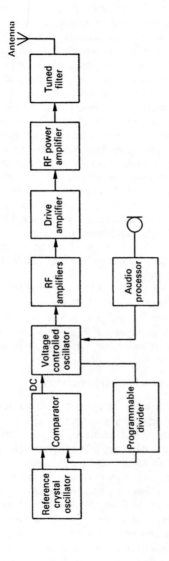

Fig. 10.2 *Synthesizer controlled FM transmitter*

The audio processing circuitry is similar to that for AM transmitters but the modulation (deviation) is applied to a stage operating at a very low RF power level (directly to the VCO in a synthesizer-equipped transmitter and immediately following the crystal oscillator in a direct crystal controlled one); high power audio is not necessary. An additional simple circuit in an FM transmitter may be included to enhance the higher audio frequencies at a rate of 6 dB/ octave. This is pre-emphasis and has the merit of improving the level of speech to noise at the receiver (signal-to-noise ratio). The audio level must still be limited because, while the effect of over-deviation is not as disastrous as over-modulation in an AM transmitter, increasing deviation produces a steadily increasing range of frequencies, known as side-currents, outside the permitted bandwidth.

Angle modulated transmitters are more power efficient than amplitude modulated because the modulation is applied at a low power level and also, as no audio frequencies are directly present in the RF amplifier and PA stages, these can operate efficiently in class C.

Phase modulated (PM) transmitters, because the phase shift with modulation is very small, generate the carrier at a very low frequency and use direct crystal control. The frequency is then multiplied many times, thirty-two to thirty-six is common, up to the final frequency. After multiplication, the phase modulation which was originally a few radians has effectively become frequency modulation. The phase modulation process produces pre-emphasis inherently. As far as the user is concerned there is no practical difference between phase modulation and frequency modulation with added pre-emphasis.

10.1.4 Transmitter specifications

The Radiocommunications Agency issues specifications with which all equipment must comply. These specifications are concerned principally with the prevention of interference and obtaining the maximum use of the frequency spectrum. The characteristics defined in the MPT specifications and other features which affect the user, apart from the physical dimensions, are:

- *Supply voltage.*
- *Operational frequency band.*
- *Modulation method.*
- *Channel separation.*
- *RF power output and impedance.* Output ranges from about 0.5 W to 5 W for hand-portables and 5 W to 25 W for mobiles. The maximum power permitted on a system will be specified in the Licence. Output impedance is commonly 50 Ω.
- *Spurious emissions.* The level of these is critical for the prevention of interference with other users on different frequencies. The limit for VHF and UHF is a maximum of 0.25 μW.
- *Residual noise.* Not always quoted by manufacturers, it is the noise level existing on an unmodulated carrier. A typical figure is better than 40 dB referred to full deviation.
- *Audio frequency distortion.* Typically <3% and usually measured with a modulating frequency of 1 kHz at 60% modulation.
- *Audio frequency response.* This is the variation of modulation level over the audio frequency spectrum. Typically within +1 dB to 3 dB over a frequency range of 300 to 3000 Hz (2.55 kHz for 12.5 kHz channel spacing equipment). It may be quoted with reference to a pre-emphasis curve.
- *Switching bandwidth.* This is the frequency range over which the transmitter will operate without retuning and without degradation of performance. Much equipment is now specified to cover a complete frequency band, e.g. 146–174 MHz, without retuning.

10.2 Receivers

10.2.1 Receiver functions

A receiver's functions are:

- Detect a weak signal; the minimum level, which may be as low as 0.25 microvolts, defines the receiver sensitivity.

- Amplify a received signal and maintain the information contained in a minimum strength signal at a minimum of 12 dB above the electrical noise level (signal-to-noise ratio). If the audio distortion produced in the receiver is also taken into account the above figure becomes the signal-to-noise + distortion (Sinad) ratio. As the signal is increased, the ultimate Sinad should attain 50 to 55 dB.
- Separate the wanted signal from any unwanted ones which may be very close in frequency (the adjacent channel may be 12.5 kHz away at UHF); the selectivity.
- Recover the information from the carrier; demodulation.
- Amplify the audio information to a level suitable for operating a loudspeaker; the audio power output. The audio amplification must introduce the minimum distortion.
- Disenable the audio amplifiers in the absence of signal to cut out the electrical noise. This is done by the mute or squelch circuit.

10.2.2 Types of receiver

It is possible to amplify directly the incoming RF signal to a level suitable for demodulation. This is done in a tuned radio frequency (TRF) receiver, but these are seldom used today because of the problems of obtaining sufficient selectivity and gain at one radio frequency, and the difficulty of retuning a number of RF stages to change frequency. Almost all receivers designed for analogue communications now operate on the superheterodyne principle where the incoming radio frequency is converted to a lower, more manageable, intermediate frequency (IF). The fixed IF means that only the oscillator and, possibly, one RF amplifier stage need retuning for a change of channel. The lower frequency of the IF facilitates the acquisition of adequate gain with stability and selectivity.

There is little difference in the layout of receivers for AM and FM except that the circuits perform their functions differently. Figure 10.3 is a block diagram of a typical FM receiver with a crystal controlled local oscillator.

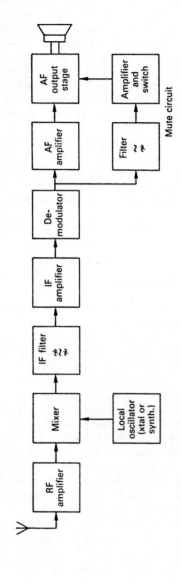

Fig. 10.3 *FM single superheterodyne receiver*

When two frequencies are applied to a non-linear circuit such as the mixer, they combine to produce other frequencies, their sum and difference being the strongest. The superheterodyne mixes a locally generated frequency with the received signal to produce a, usually lower, frequency retaining the modulation of the received signal. Commonly used values for this intermediate frequency (IF) are 465 kHz for MF and HF receivers and 10.7 MHz for VHF and UHF. At these fixed frequencies the necessary high amplification with low noise and stability and the required selectivity are easier to obtain. The local oscillator (injection) frequency = signal frequency ± IF frequency.

Superheterodyne receivers are susceptible to a particular form of interference. Assume a local oscillator frequency of 149.3 MHz is mixed with a wanted signal of 160 MHz to produce the IF frequency of 10.7 MHz. A signal of 138.6 MHz would also combine with the local oscillator to produce 10.7 MHz. The frequency of this spurious response is the image, or second channel frequency. The IF amplifier cannot discriminate against it so some degree of selectivity must also be provided in the RF amplifier and input circuitry. Intermediate frequencies are chosen which are a compromise between ease of obtaining adjacent channel discrimination and image frequency rejection.

The IF amplifier contains a block filter, either crystal or ceramic (see Figure 7.7), necessary to discriminate between channels adjacent in frequency. The design of the filter is crucial. It must be wide enough to accommodate the band of frequencies present in the modulation plus an allowance for frequency drift and its response over this band must be uniform with minimal ripple, particularly if data is to be received, yet its response must be of the order of 100 dB at the frequency of the adjacent channel.

Apart from the demodulator the main difference between AM and FM receivers lies in the operation of the IF amplifier. The IF stages in an AM receiver are linear, although their gain is variable. Part of the IF amplifier output is rectified and used to control the gain to provide automatic gain control (AGC). An increase of signal above a predetermined level, with delayed

AGC, causes a reduction in IF gain maintaining a sensibly constant audio output level.

The IF amplifier in a FM receiver possesses a very high gain, some 100 dB, and is non-linear. On receipt of a signal, or even with only the receiver noise, it runs into limitation, cutting off both positive and negative peaks of the signal or noise. This gives FM its constant level audio output over a wide range of signal levels. It also produces the capture effect where a strong signal, fully limiting, completely removes a weaker signal. A signal difference of some 6 dB is required to provide effective capture.

Most receivers employ only one change of frequency, single superheterodynes, but double superheterodynes are occasionally used at VHF and above. A double superheterodyne changes the frequency twice, perhaps to 10.7 MHz for the first IF and then, using a fixed frequency crystal second local oscillator, to a lower, often 1.2 MHz or thereabouts, second IF. The result is greater gain and selectivity but the incorporation of a second oscillator and mixer increases the number of possible spurious responses.

A variation of a very old receiver circuit, the Autodyne, forgotten in about 1914, is now finding favour in receivers for digitally modulated signals. Under its new names of Homodyne or zero-IF receiver it lends itself to the purpose. In this type of receiver the local oscillator runs at the same frequency as the incoming signal, hence the zero-IF. Frequency or phase shifts of the incoming signal representing the data emerge from the mixer at base band and are applied directly to the processing circuits.

In a communications receiver the noise generated in the aerial and RF stages in the absence of a signal is amplified to what may be, when demodulated, an unacceptable level. To eliminate the annoyance the loudspeaker is switched off during no-signal periods by a squelch or mute circuit. The mute circuit rectifies the high frequency noise at the demodulator and uses it to switch off the audio amplifier. When a signal is received the limiting action of the IF amplifier of an FM receiver depresses the noise in favour of the signal. Some AM mobile receivers use additional FM circuitry to provide improved mute action.

10.2.3 Noise figure

An ideal receiver would generate no noise and the signal-to-noise ratio, in a receiver of given bandwidth, would be determined by the level of the signal at the base of the antenna compared with the noise produced in the antenna. The noise factor of the ideal receiver is the number of times the signal power must exceed the antenna noise power to produce a 1:1 ratio at the receiver. It is given by (see Section 1.5.1):

$$\frac{e^2_{antenna}(e.m.f.)}{4kTBR}$$

When the receiver input impedance is matched to the antenna impedance, half the power is dissipated in the antenna and the noise factor is 2 (3 dB). In a practical receiver, the noise generated in the RF amplifier is the most significant, and the noise figure is the sum of the RF amplifier noise plus all the preceding losses. Figures between 4 and 6 dB are common and the higher the noise figure, the worse the receiver sensitivity.

Signal-to-noise ratio (SNR or S_n)

Receivers are evaluated for quality on the basis of *signal-to-noise ratio* (S/N or 'SNR'), sometimes denoted S_n. The goal of the designer is to enhance the SNR as much as possible. Ultimately, the minimum signal level detectable at the output of an amplifier or radio receiver is that level which appears just above the noise floor level. Therefore, the lower the system noise floor, the smaller the *minimum allowable signal*.

Noise factor, noise figure and noise temperature

The noise performance of a receiver or amplifier can be defined in three different, but related, ways: *noise factor* (F_n), *noise figure* (NF) and *equivalent noise temperature* (T_e); these properties are definable as a simple ratio, decibel ratio or Kelvin temperature, respectively.

Noise factor (F_n). For components such as resistors, the noise factor is the ratio of the noise produced by a real resistor to the simple thermal noise of an ideal resistor.

The noise factor of a radio receiver (or any system) is the ratio of output noise power (P_{no}) to input noise power (P_{ni}):

$$F_N = \left[\frac{P_{NO}}{P_{NI}} \right]_{T=290/K}$$

In order to make comparisons easier the noise factor is usually measured at the standard temperature (T_o) of 290 K (standardized room temperature); although in some countries 299 K or 300 K are commonly used (the differences are negligible). It is also possible to define noise factor F_N in terms of the output and input signal-to-noise ratios:

$$F_N = \frac{S_{NI}}{S_{NO}}$$

where

> S_{NI} is the input signal-to-noise ratio
> S_{NO} is the output signal-to-noise ratio

Noise figure (NF). The *noise figure* is the frequency used to measure the receiver's 'goodness', i.e. its departure from 'idealness'. Thus, it is a *figure of merit*. The noise figure is the noise factor converted to decibel notation:

$$NF = 10 \log(F_N)$$

where

> NF is the noise figure in decibels (dB)
> F_N is the noise factor
> log refers to the system of base-10 logarithms

Noise temperature (T_e). The noise 'temperature' is a means for specifying noise in terms of an equivalent temperature. That is, the noise level that would be produced by a resistor at that temperature (expressed in degrees Kelvin). Evaluating the noise equations shows that the noise power is directly proportional to temperature in degrees Kelvin, and also that noise power collapses to zero at the temperature of Absolute Zero (0 K).

Note that the equivalent noise temperature T_e is *not* the physical temperature of the amplifier, but rather a theoretical construct that is an *equivalent* temperature that produces that amount of noise power in a resistor. The noise temperature is related to the noise factor by:

$$T_e = (F_N - 1)T_0$$

and to noise figure by

$$T_e = KT_0 \log^{-1}\left[\frac{NF}{10}\right] - 1$$

Noise temperature is often specified for receivers and amplifiers in combination with, or in lieu of, the noise figure.

Noise in cascade amplifiers

A noise signal is seen by any amplifier following the noise source as a valid input signal. Each stage in the cascade chain amplifies both signals and noise from previous stages, and also contributes some additional noise of its own. Thus, in a cascade amplifier the final stage sees an input signal that consists of the original signal and noise amplified by each successive stage plus the noise contributed by earlier stages. The overall noise factor for a cascade amplifier can be calculated from *Friis' noise equation*:

$$F_N = F_1 + \frac{F_2 - 1}{G1} + \frac{F_3 - 1}{G1G2} + \cdots$$
$$+ \frac{F_N - 1}{G1G2 \cdots G_{N-1}}$$

where

F_N is the overall noise factor of N stages in cascade
F_1 is the noise factor of stage-1
F_2 is the noise factor of stage-2
F_N is the noise factor of the nth stage
$G1$ is the gain of stage-1
$G2$ is the gain of stage-2
G_{n-1} is the gain of stage $(n-1)$.

As you can see from Friis' equation, the noise factor of the entire cascade chain is dominated by the noise contribution of the first stage or two. High gain, low noise radio astronomy RF amplifiers typically use low noise amplifier (LNA) circuits for the first stage or

167

two in the cascade chain. Thus, you will find an LNA at the feedpoint of a satellite receiver's dish antenna, and possibly another one at the input of the receiver module itself, but other amplifiers in the chain might be more modest (although their noise contribution cannot be ignored at radio astronomy signal levels).

The matter of signal-to-noise ratio (S/N) is sometimes treated in different ways that each attempts to crank some reality into the process. The signal-plus-noise-to-noise ratio (S + N/N) is found quite often. As the ratios get higher, the S/N and S + N/N converge (only about 0.5 dB difference at ratios as little as 10 dB). Still another variant is the SINAD (signal-plus-noise-plus-distortion-to-noise) ratio. The SINAD measurement takes into account most of the factors that can deteriorate reception.

10.2.4 Signal-to-noise ratio and bandwidth

To obtain the maximum signal-to-noise ratio the bandwidth of every circuit must be designed to admit its operational band of frequencies only. The wider its bandwidth the more noise a circuit admits, and the more the bandwidth exceeds that needed, the worse becomes the signal-to-noise ratio. There is a linear ratio between bandwidth and noise power admitted: doubling the bandwidth doubles the noise power.

Improvements in local oscillator crystal frequency stability has resulted in improved signal-to-noise ratios by reducing the necessary width of the IF filter.

Demodulation, the recovery of the audio from the IF bandwidth affects the signal-to-noise ratio. The relationship is complex, but consider two examples. First, for 12.5 kHz channel spacing FM:

> The audio frequency range is 300 Hz to 3000 Hz
> Bandwidth, $b = 2700$ Hz
> The modulation index $M = f_{d(max)}/f_{m(max)}$
> Signal/noise out $= 3M^2/2b \times$ signal/noise in
> For a 12.5 kHz PMR channel:

> $M = 2500/3000 = 0.83$
> $b = 2.7$ kHz
> $3M^2/2b = 0.38$

A 0.38 times reduction in power is 4.8 dB. Demodulation in this case worsens the signal-to-noise ratio by some 5 dB, and if a signal-to-noise ratio of 12 dB is required at the loudspeaker, 17 dB is needed at the input to the demodulator.

When channel separations were 25 kHz, and deviation 5 kHz, the situation was:

Audio frequency range, 300 Hz to 3000 Hz
Bandwidth, $b = 2700$ Hz
For a 50 kHz channel:

$M = 5000/3000 = 1.66$
$b = 2.7$ kHz
$3M^2/2b = 1.54$

A 1.54 times gain in power is $+1.8$ dB. In this case the demodulation improved the signal-to-noise ratio slightly. Reducing the bandwidth would have the same effect.

For AM, the signal-to-noise ratio is dependent on the modulation depth. The demodulation process reduces the signal-to-noise ratio by 6 dB but the recovered audio is less than the IF bandwidth by a factor of 3:1 which compensates for the demodulation loss. Reducing the modulation depth degrades the signal-to-noise ratio by 6 dB for every halving of the modulation depth.

10.2.5 Receiver specifications

The important features of receiver specifications are:

- *Sensitivity*. The minimum signal to which a receiver will respond. For an AM receiver, the generally accepted standard is the signal (30% modulated with sinusoidal tone, either 400 Hz or 1 kHz) required to provide an audio output of 50 mW. For an FM receiver, the standard is the unmodulated signal required to produce a 20 dB reduction in noise. A typical figure is 0.25 μV (p.d.) for 20 dB quieting. However, sensitivity is often quoted in terms of either the signal-to-noise ratio or Sinad so, in modern parlance, sensitivity and signal-to-noise ratio are sometimes considered to be synonymous.

- *Signal-to-noise ratio (may be quoted as Sinad, signal-to-noise and distortion).* Typically 0.3 µV (p.d.) for 12 dB Sinad.
- *Spurious response attenuation.* Typically better than 80 dB.
- *Adjacent channel selectivity.* Better than 65 dB at 12.5 kHz channel spacing.
- *Cross modulation.* The modulation, in the receiver, of a wanted signal by a stronger, unwanted signal. It is usually caused by non-linearity in the receiver RF stages.
- *Blocking and de-sensitization.* The reduction in sensitivity of a receiver due to overloading of the RF stages when a strong signal is applied. A blocked receiver may take an appreciable time to recover.
- *Audio frequency response.* Typically within +1 dB to 3 dB of a 6 dB/octave de-emphasis curve from 300 Hz to 3000 Hz (2.55 kHz for 12.5 kHz channel spacing, above which the response falls more rapidly).
- *Audio output.* Typically 3–4 W. For hand-portables, 100–500 mW. (Distortion may also be quoted, typically better than 5%.)
- *Switching bandwidth.* As for transmitter.
- *Duplex separation.* For a receiver/transmitter combination, the minimum separation between the receiving and transmitting frequencies which will permit duplex operation with minimal degradation of the receiver performance.

10.3 Programmable equipment

The current trend is for many of the functions of radio equipment – even down to the control buttons on the front panel – to be software controlled. In one manufacturer's equipment the only screwdriver-adjustable control is that for setting maximum deviation; this is insisted upon by the Radiocommunications Agency.

Software control enables radio sets to be cloned so that once a single piece of equipment is programmed with its frequencies, power output, selective calling, etc., the whole of a fleet of mobiles can be identically programmed in a short time. Some of the functions can be programmed by the user, but not all, and different manufacturers permit different degrees of programming.

The functions are allocated degrees of priority which determine which functions can be changed by each class of person. Again, some of the priorities may be altered by the user but the highest priority functions are installed in a programmable read only memory (PROM) by the manufacturer and can only be altered by replacing the PROM, in most instances a job for the manufacturer.

The essential equipment for major reprogramming is an IBM or equivalent computer and the equipment manufacturer's software which is supplied as a package complete with hardware interface and instructions.

Cloning may be carried out using an interconnecting cable between the equipment, or restricted reprogramming via a unit supplied by the equipment manufacturer for the purpose.

The functions which may be programmed from a computer are:

- Channel frequencies.
- Transmitter power and deviation.
- Receiver squelch setting (referred to Sinad ratio). Selective calling details including encode or decode only, type of signalling, extension of first tone and call sign.
- Transmission time out timer.
- Channel busy light. Transmitter inhibit on busy channel.
- Alert tones. Tone decoder indicator.
- Low battery indicator. Battery saver.

Restricted reprogramming may include:

- Channel frequencies, including any offsets.
- Lock out of any channels.
- Channel spacing.
- Channel search priority.
- Frequency stability.
- Signalling type.
- Control functions.
- Synthesizer reference crystal (the frequency can be trimmed to compensate for ageing).
- Transmitter power.

- CTCSS (encode and decode).
- Timers.
- Alerts.

Pagers are also programmable. The functions which can be changed include:

- Code number.
- Alert tone/vibrator/repeat.
- Urgency.
- Out of range warning.
- Printer on/off.
- Language of display.

References

Belcher, R. *etal.* (1989). *Newnes Mobile Radio Servicing Handbook*. Butterworth-Heinemann, Oxford.

Langford-Smith, F. (1955). *Radio Designer's Handbook*, Iliffe, London.

11

MICROWAVE COMMUNICATION

11.1 Microwave usage

Microwaves are loosely considered to be those at frequencies between 1 GHz (30 cm) and 100 GHz (0.3 cm). Principal ground communications usage is for point-to-point links carrying information and control signals on systems such as multiplexed communication – including data networks, telemetry and mobile radio. The propagation of the higher frequencies, 30–100 GHz, is being studied for possible future use in micro-cellular radio-telephone systems. These higher frequencies are also used for satellite communications and radar applications. Microwave frequencies up to 3 GHz have now been reserved for mobile use.

11.2 Propagation

The path loss is higher at microwave frequencies than at VHF and UHF: Figure 1.5 charts the free space loss between isotropic radiators for microwaves at frequencies of 1, 2, 3.5 and 7 GHz over distances of 10 to 1000 km. The free-space loss between isotropic radiators is given by

Free space loss, $dB = 32.4 + 20 \log_{10} d + 20 \log_{10} f$

where d is in km and f in MHz.

The free-space loss between practical antennas is given by:

$$\text{Loss, dB} = 10 \log_{10} \left(\frac{(4\pi d)^2}{\lambda^2} \frac{1}{G_t G_r} \right)$$

where

d = path length, metres
λ = wavelength, metres

G_t = power gain of transmitting antenna
G_r = power gain of receiving antenna

The antenna gains are expressed relative to an isotropic radiator (not in dB).

Absorption varies with atmospheric humidity, and as energy is also absorbed by the ground the path height therefore has an effect on the losses. Absorption by rain is a factor at the higher frequencies but, although considered to be insignificant below about 3 GHz, it has an indirect effect. Wet foliage, for example, produces considerable absorption at frequencies as low as 450 MHz.

Waves of millimetric lengths are special cases and narrow bands of very high absorption due to resonance effects exist at 22 and 183 GHz for water vapour, and 60 and 119 GHz for oxygen. Non-resonant attenuation occurs due to scatter from rain, hail and snow. The attenuation increases with frequency as the wavelength approaches the dimensions of a raindrop. Bands of very low absorption, 'atmospheric windows', where the water vapour and oxygen attenuations are very low occur at 37, 97, 137 and 210 GHz.

Objects close to a path may severely affect the received signal and a proposed path must be examined for the likely effects of these at the planning stage. The effects may be due either to diffraction bending the wave away from the line of sight between the antennas, or reflection causing multi-path signals.

The additional losses in a microwave path caused by objects either intruding into the first Fresnel zone or close to it, and which exhibit 'knife-edge' or 'smooth-sphere' characteristics, are shown in Figure 11.1. A negative F/F_1 indicates an intrusion. When the path clearance exceeds 0.6 times the first Fresnel zone radius, the free space loss is achieved.

Refraction also affects microwave propagation and temperature inversions may cause ducting resulting in a loss of signal and, possibly, interference. When ducting occurs, a layer of air of low refractive index is formed between two highly refractive layers and a wave may be trapped between them. Under these conditions a signal can be carried an abnormal distance before it can return to Earth resulting in loss of signal

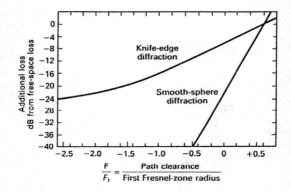

Fig. 11.1 *Knife-edge and smooth-sphere diffraction*

at the intended receiver and possibly interference at a distant one on the same frequency.

11.3 *K* factor

The degree and sense of refraction of wave is related to a factor known as K. Refraction normally bends the wave downwards, extending the range, but the refractive index of the atmosphere varies from place to place and with time and height. At times, during an atmospheric inversion for instance, and in some places on the earth the effect is reversed:

> When no bending occurs, $K = 1$.
> When $K > 1$, the bending is downwards, effectively increasing the Earth's radius.
> When $K < 1$, the bending is upwards, effectively reducing the Earth's radius.

For most of the time $K > 1$, 1.33 being accepted as the normal factor, but for small periods K may be less than one.

For a point-to-point link, the path between the transmitting and receiving sites should be a clear line of sight, although by making allowances for refraction, the Earth's radius has, up to now, generally been considered to be effectively increased by the factor of 1.33. However, in the interests of reliability a decrease to

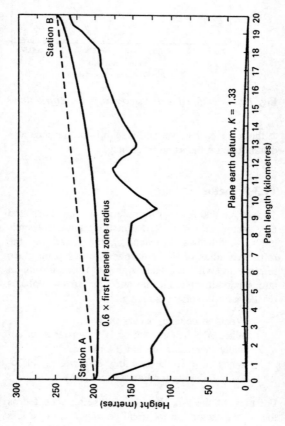

Fig. 11.2 Example of path profile

0.7 times the radius (the minimum K factor considered likely to occur) is now often used in link planning. When a link has been planned using a higher K factor, a temporary reduction of K not only reduces the radio horizon but effectively raises objects close to the path, possibly to the point where they become significantly close to the first Fresnel zone.

It is usual when planning a link to plot the profile of a path on paper with curved horizontal graduations to represent the amended Earth's radius. The radio beam can then be drawn as a straight line between the antenna locations. Figure 11.2 is an example.

11.4 Fresnel zones, reflections and multi-path fading

Signals which arrive at the receiver by more than one path as the results of reflection or diffraction may arrive in any phase relationship to the direct wave. When they arrive anti-phase to the direct wave, cancellations result. The intensity and phase of the spurious signal may not be constant, thus providing random multi-path fading.

Where a carefully drawn profile of a link path shows there to be a clear line of sight, the effect of waves reflected or diffracted from objects close to the line of the direct wave must then be considered. The effect of these indirect waves can be predicted by calculating where the reflection occurs in relation to a series of ellipsoids which can be drawn around the line-of-sight path between the transmitting and receiving antennas. These ellipsoids, known as the Fresnel zones, contain the points where reflected waves will follow a path of constant length, as shown in Figure 11.3.

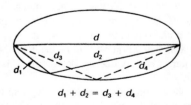

$$d_1 + d_2 = d_3 + d_4$$

Fig. 11.3 *Fresnel zone: reflected path lengths*

Waves reflected at the odd-numbered Fresnel zones will travel an odd number of half-wavelengths further than the direct wave but, because a 180° phase change usually occurs in the reflection process, will arrive at the receiver in phase with the direct wave. Waves reflected at even-numbered zones will arrive anti-phase to the direct wave with a cancelling effect. The effect of reflected waves diminishes with reflections from the higher order zones. The radius of a Fresnel zone in metres at the point of intrusion is given by:

First zone, F_1:

$$F_1 = 31.6\sqrt{\frac{\lambda d_1 d_2}{d}}$$

or

$$F_1 = 548\sqrt{\frac{d_1 d_2}{fd}}$$

where

$\quad F_1$ $\quad$ = radius in metres at point of intrusion
$\quad d_1 + d_2 = d$ (path length in km)
$\quad \lambda$ $\qquad$ = wavelength in metres
$\quad f$ $\qquad$ = frequency in MHz

Second zone:

$$F_2 = \sqrt{2} \times F_1$$

Third zone:

$$F_3 = \sqrt{3} \times F_1$$

and so on.

The degree of reflection from an object depends on its nature, the greatest reflection occurring from smooth flat ground or water. Where a path lies over the sea, variations in the path length of a reflected wave due to tides may render a path unusable. When the height of the antenna closest to the sea is varied, the *effect* of the reflected wave passes through a series of minima and maxima and adjustment of the height of that antenna can reduce or, occasionally, overcome the effect.

Atmospheric conditions change giving rise to fading and variations of the multi-path effects. The reliability of a link may be crucial to the success of a complete system and, where a critical path in terms of performance exists, long-term tests are advisable to ensure that variations of propagation do not reduce the reliability to an unacceptable level. Paths which contain obstacles in the line of sight which will cause additional losses are obviously suspect. So are those where objects or large stretches of water or flat ground which might produce diffraction or reflections of the wave lie close to the line of sight.

11.5 Performance criteria for analogue and digital links

The transmission quality for analogue modulated systems is based on the signal-to-noise ratio. The noise is specified relative to a standard test tone level and is commonly expressed as either picowatts psophometrically weighted (pWp) or decibels (dB) of C-message weighted noise above a reference noise level of 90 dBm (defined as 0 dBrnc0). Typical objectives range from 28 dBrnc0 for long-haul routes to 34 dBrnc0 for short-haul routes. When the signal fades the noise increases until the threshold noise level is reached. When the threshold (typically 55 to 58 dBrnc0) is exceeded, the transmission quality is considered unacceptable.

With digitally modulated systems the bit error rate (BER) is the measure of transmission quality. The bit error rate is the number of bit errors per total received averaged over a period of time. If the transmission rate is 10 Mbits per second and 100 bit errors occur over a 100 second period, the BER is 10^7, an average of 1 error in 10^6 bits. The acceptable level of transmission is determined by the type of traffic.

For PCM voice traffic, bit errors manifest themselves as clicks and a threshold of 10^6 (1 click approximately every 15 seconds) is usually considered acceptable. At this threshold the speech is intelligible, but beyond it the clicks become annoying and intelligibility falls rapidly.

For data with error correction a higher BER of 10^8 is normally acceptable (Communications International, 1989).

11.6 Terminology

A number of different units are used worldwide to define the performance and transmission levels of a radio relay system (for decibel definitions see Chapter 2). Important international definitions and units are:

- *Zero transmission reference point.* This is a point arbitrarily established in a transmission circuit, with all other levels in the circuit being stated with reference to this point. Its relative level is 0 dBr.
- *Standard test tone.* The standard test tone for use at audio circuit points is defined as a power of 1 milliwatt (0 dBm) at a frequency of 1000 Hz applied at the zero transmission reference level point.

11.7 Link planning

Planning a link involves producing a profile of the path and calculating the net loss in the system to arrive at a transmitter output power which will produce the designed signal-to-noise ratio at the receiver. A simple example using the profile of Figure 11.2 is shown in Section 11.8.

Transmission lines and waveguides are discussed in Chapter 3 and microwave antennas in Chapter 4.

11.8 Example of microwave link plan

Frequency: 2000 MHz

Antenna type, station A: P6F-17C	height agl.		20 m
Antenna type, station B: P6F-17C	height agl.		20 m
Feeder type, station A: LDF5P50A	loss, dB/100 m		6.46
Total length, antenna to equipment			30 m
Feeder type, station B: LDF5P50A	loss, dB/100 m		6.46
Total length, antenna to equipment			30 m

Performance

Path length: 20 km, therefore clear path loss	124.0 dBi
Obstruction loss	0.0 dB
Feeder loss, station A:	1.9 dB
Feeder loss, station B:	1.9 dB
Feeder tail loss, total for link	1.5 dB
Connector loss, total for link	1.5 dB
Total loss	**130.8 dB**
Gain, antenna A	28.6 dB
Gain, antenna B	28.6 dB
Total gain	**57.2 dB**

Nett loss (total loss − total gain)	**73.6 dB**
Receiver threshold for max. signal/noise	125.0 dBW
Design fade margin	+30.0 dB
Design receiver input level	**95.0 dB**
(threshold − fade margin)	
Transmitter output power	**21.4 dBW**
(receiver input − nett loss)	

Reference

Krzyczkowski, M. (1989). *Communications International,* August.

12

INFORMATION PRIVACY AND ENCRYPTION

12.1 Encryption principles

Radio communication was never secret, but since the advent of fast frequency scanning receivers the ability to overhear, even on cellular radio telephones, is within easy reach of anyone. Privacy systems are, however, available which will deter the casual listener and gain time against the determined eavesdropper.

Messages, which may be either speech or data, are encrypted to prevent both eavesdropping and the injection of spurious information. The aim is to make the encryption and decryption as easy and inexpensive as possible for authorized users and time-consuming, difficult and costly for the eavesdropper (cryptanalyst). Figure 12.1 shows a standard cryptographic channel.

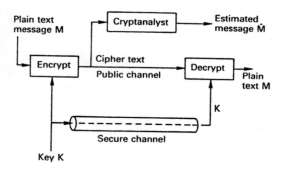

Fig. 12.1 *Cryptographic channel*

A plain text message M (speech, written or digital) is encrypted by mixing with a key K to produce a cipher text. The cipher text may be transmitted over a channel which is accessible to the public and hence to the cryptanalyst. The key is issued via a secure channel to the authorized recipient who uses it to decipher the message. The cryptanalyst without access to the key attempts to derive the maximum information from the cipher text to enable him or her to estimate the content of the message.

One key may be used continuously or for long periods or, to increase the cryptanalyst's confusion, the key may be changed frequently, perhaps even for each character of the message. A sequence of key changes which repeats after a fixed number of characters produces what is known as periodic encryption.

Encryption may be either symmetrical or asymmetrical. Symmetrical encryption uses the same key for both encryption and decryption. Asymmetrical encryption uses a different key for each process, thus providing for different levels of authorization. Encryption keys may be supplied to many persons who are authorized to transmit encrypted messages but decryption keys may be issued to only a few authorized recipients.

12.2 Speech encryption

The encryption of speech offers fewer possibilities than does the encryption of written or digital data messages. The simplest method of encrypting speech is scrambling by inverting the speech frequencies; Figure 12.2 shows this process.

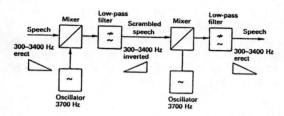

Fig. 12.2 *Speech inversion*

The speech, contained in the band 300–3400 Hz, is mixed with a key frequency of 3700 Hz producing an erect, upper side band from 4000–7100 Hz, and an inverted, lower side band where the 300 Hz components of speech have become 3400 Hz and the 3400 Hz have been inverted to 300 Hz. The upper side band is rejected by a lowpass filter and the inverted lower side band is transmitted. In the receiver the scrambled speech is mixed again with 3700 Hz to produce an erect side band – the original non-inverted speech message.

Simple inverted speech is easily unscrambled. There is little choice of key frequency and if the eavesdropper uses a slightly different frequency the pitch is changed but the speech is readable. Also, if the centre frequencies only of the inverted band are selected by means of a band-pass filter, inverted speech becomes intelligible.

A sophistication which renders the speech more secure divides the speech band into sections and transmits them separately using a different key frequency for each band (audio frequency hopping). The divisions of the speech band may also be treated as blocks and transposed in time according to a user-programmable pattern to create further confusion in the mind of the eavesdropper.

The most up-to-date methods of speech scrambling convert the speech into digital form by either pulse code modulation (PCM) or some other method. The digits corresponding to the speech may then be either transmitted as frequency modulation, e.g. FFSK, on an analogue radio system or, possibly after further encryption, transmitted directly on a digital system. Digitized speech creates improved security not only by the digitization itself but by offering the higher encryption capabilities of data.

The price to be paid for security with analogue encryption is a degradation of the received signal-to-noise ratio by 9 dB, effectively reducing the range of a transmitter by approximately 40%.

12.3 Data encryption

Digital data may be encrypted by changing the digits, perhaps adding superfluous digits, and transmitting the resultant cipher message either in blocks of a fixed size or as a stream.

Block encryption treats the blocks in the same way as in the encryption of speech, with different keys being used for each block – or each character – and the blocks re-distributed in time.

Stream encryption has no fixed block size and each plain text bit, M_i, is encrypted with the ith element, K_i, of a pseudo-random, sometimes called pseudo-noise (PN), key.

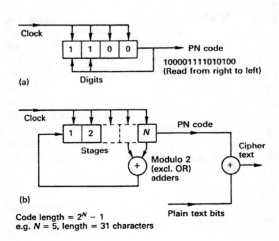

Code length = $2^N - 1$
e.g. $N = 5$, length = 31 characters

Fig. 12.3 *Pseudo-noise (PN) generation*

Figure 12.3 shows two methods of generating pseudo-random keys or pseudo-noise. The first, Figure 12.3(a), operates as follows. At each clock pulse the contents of the pre-loaded four-stage shift register are stepped forward from left to right. Immediately after the shift, the output bit is fed back into the 1st and 2nd stages. It introduces a new bit into stage 1, and is added by modulo 2 addition to the new content of stage 2, producing a new set of contents. The initial loading of 1100 emerges as a pseudo-random 15 bit sequence which then repeats. The periodicity of the sequence is given by:

$$\text{Sequence length, characters} = 2^N - 1$$

where N = the number of stages in the shift register.

For a four-stage register, therefore, the sequence repeats after 15 bits and is shown in Figure 12.3(a).

A more commonly used method combines the outputs of two or more of the earlier stages in a modulo 2 adder and feeds the result back to the input of the register as in Figure 12.3(b).

To form the cipher text the resultant pseudo-random key is mixed with the original data message in a second modulo 2 adder. If the clock rate for the shift register is the same as the bit rate of the plain text message, the plain text bits are exchanged for those of the modulo 2 sum, but if the shift register runs faster than the plain text bit rate, additional bits are added into the cipher text. This is more common and extends the time taken by a cryptanalyst to estimate the message. The price to be paid for the improved security is either a slower effective bit rate for the message or a higher overall bit rate and hence an increased bandwidth requirement.

Mixing the cipher text with the output from an identical PN generator in the receiver recovers the original text.

Modulo 2 addition

A modulo 2 adder is an exclusive-OR gate which produces a logic 1 output whenever either of the inputs is at logic 1 and the other is at logic 0. When both inputs are identical, the exclusive-OR gate produces a logic 0 output.

The truth table for an exclusive-OR gate is:

Input		Output
A	B	Y
0	0	0
0	1	1
1	0	1
1	1	0

Modulo addition is not limited to two inputs. Any quantity of binary numbers may be added: if there is an odd number of logic 1s in a column, the adder produces a logic 1 output, if an even number, i.e. no remainder in the binary addition, the output is logic 0.

Synchronous encryption

The key is generated independently of the message from a previously loaded register. If a character is lost during transmission of a synchronous text, resynchronization of transmitter and receiver key generators is necessary before transmission can continue.

Self-synchronous encryption
The key for each character is derived from a fixed number of previous characters of the plain text message. For example, the shift register is pre-loaded with the plain text characters so that in a four-stage register the key used for encrypting the 4th character will be the 4th previous message character. If a self-synchronized transmission loses a character, the system automatically re-synchronizes the same number of characters (in this case four) later.

Written messages may be encrypted using one of the classical mechanical methods of rearrangement of the letters before digital encryption.

When sufficient RF channels exist frequency hopping is a further possibility, and the spread spectrum technique, where the signal energy is spread over a very wide band of frequencies, not only offers very high security but also makes detection of the signal difficult. The shift register techniques described above are also used for generation of the frequency hopping sequence and the spreading of the base band frequencies.

12.4 Code division multiple access (CDMA) or spread spectrum

The extension of pseudo-random key or noise generation is code division multiple access or spread spectrum transmission, described in Chapter 8. The spread spectrum technique provides an extremely high level of security by reducing the radiated energy at any one frequency to very little above the ambient noise level by spreading the transmission over a very wide band. The transmitter uses what is in effect an extended digital key to spread the bandwidth and the receiver is equipped with an identical key for de-spreading. The transmission almost disappears into the noise and, without the appropriate key, the existence of a spread spectrum signal is very difficult to detect.

12.5 Classification of security

Unconditionally secure
Those systems where the cryptanalyst has insufficient information to estimate the content of the cipher regardless of the amount of time and computation facilities available. This is only realistic when using what is known as a one-time pad where the key is used once and once only.

Computationally secure
Encryption systems are specified in terms of the amount of time taken by a cryptanalyst to estimate the cipher's contents using the state of the art techniques. Unless an extremely long periodicity is used for a progressive key – months and even years in some instances – requiring many stages in the shift register, it is possible for a cryptanalyst who knows, or can estimate, a small part of the message to calculate all the parameters necessary to decipher the message. However, stream encryption with a pseudo-random key approaches perfect secrecy for a finite number of messages.

References

Chambers, W.G. (1985). *Basics of Communications and Coding*. Oxford University Press, Oxford.
Sklar, B. (1988). *Digital Communications*. Prentice Hall, Englewood Cliffs, NJ.

13

MULTIPLEXING

Multiplexing enables several information (speech or data) channels to be carried simultaneously over one bearer, a wide band, single frequency microwave radio link for example. Both frequency and time division multiplex are common methods. On trunked radio systems where channels are allocated to users on demand the multiplexing is referred to as frequency- or time-division multiple access (FDMA, TDMA).

13.1 Frequency division multiplex (FDM)

Frequency division multiplex divides a broad band of frequencies into slots, each wide enough to accommodate an information channel. This is achieved by amplitude modulating a higher frequency subcarrier with each speech signal to form groups of channels.

Each speech channel contains frequencies between 300 and 3400 Hz plus, in some systems, an out-of-band signalling tone of 3825 Hz and a guard band. Each channel modulates a base band subcarrier spaced at 4 kHz intervals upwards from 64 kHz. This produces an upper and lower side band from each channel (Figure 13.1). The carrier and upper side band are removed by filters and the lower (inverted) side band is transmitted. At the receiver, the base band frequencies are again mixed with the same subcarrier frequency to restore the original speech. The subcarrier frequencies are maintained to an accuracy of ± 1 Hz which creates the ± 2 Hz frequency translation error quoted in some telephone line specifications.

Twelve such channels form a CCITT basic group B occupying the band between 60 and 108 kHz. This basic group B may now be mixed with 120 kHz to produce a lower side band of 12 to 60 kHz, now basic group A. Filters leave 60 to 108 kHz free for a new basic group B.

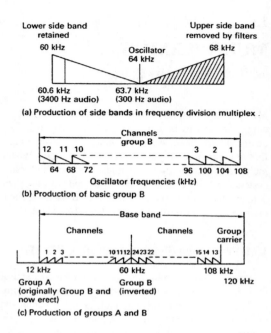

(a) Production of side bands in frequency division multiplex

(b) Production of basic group B

(c) Production of groups A and B

Fig. 13.1 *Frequency division multiplex*

The process may be repeated by using five basic groups to modulate still higher-frequency carriers, to produce super- and hyper-groups.

For FDM data communications, the bearer circuit bandwidth of 3000 Hz is divided into 12 channels each of 240 Hz bandwidth. Data is transmitted at 110 bits per second allowing a send-and-receive channel in each block of 240 Hz.

13.2 Time division multiplex (TDM)

A time division multiplex system conveys digital data, and speech must first be converted to data. TDM allocates short-duration time slots within a wider time frame to each information channel. For example, a continuous stream of data sent over a link at a rate of 2400 bit/s could convey the information contained in four 600 bit/s channels in short sequential bursts.

If the duration of one input bit is 1/600 s or 1.666 ms, a seven-bit character occupies 11.66 ms and 85 such characters can be sent per second. If the transmitted rate can be speeded up each bit sent at 2400 bit/s has a duration of 416 μs and 343 seven-bit characters can be sent per second. Such a system is shown in Figure 13.2. The data is stored in the buffers at the transmitter and the clock pulses are applied to each store/gate sequentially allowing one character from each data channel to be transmitted at a rate of 2400 bit/s. Perfect synchronization must be maintained between all channels and the transmitter and receiver to avoid data errors.

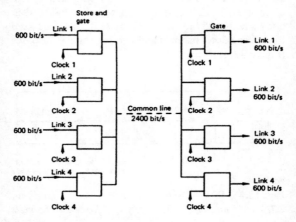

Fig. 13.2 *Time division multiplex*

The principle is illustrated in Figure 13.3 where a 1 s time frame at 2400 bit/s contains 343 time slots, each of 2.92 ms and containing a character from a specific information channel. Every second, therefore:

channel A would occupy slots 1, 5, 9, ..., 337
channel B would occupy slots 2, 6, 10, ..., 338
channel C would occupy slots 3, 7, 11, ..., 339
channel D would occupy slots 4, 8, 12, ..., 340

This leaves three blank slots; in practice slots are also allocated for preamble, address and synchronization purposes.

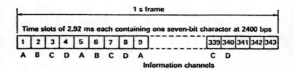

Fig. 13.3 *TDM time frame*

The digital base band signal adopted in Europe operates at 64 kbit/s and the multiplexed signal at 2048 kbit/s. An eight-bit word or sample of a PCM voice channel (see Chapter 14) occupies 3.9 µs, and the interval between successive samples of a channel is 125 µs, the time frame duration. Therefore, the number of channels (time slots) that can be accommodated in one frame is $125/3.9 = 32$. Thirty of the slots are used for information channels and two for control purposes (Figure 13.4).

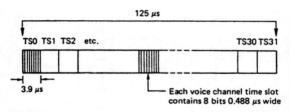

Fig. 13.4 *TS0, TS16 are used for signalling; TS1 – TS15 and TS17 – TS31 are used for voice channels*

13.3 Code division multiple access (CDMA)

Spread spectrum transmission (described in Chapter 8) is a form of multiplexing. In addition to high security it permits multiple occupation of the wide – typically 1.25 MHz – frequency band. A number of users possessing keys of low correlation can occupy the same band at the same time. The system operates well in poor signal-to-noise or high interference environments.

Reference

Edis, E.A. and Varrall, J.E. (1992) *Newnes Telecommunications Pocket Book.* Butterworth-Heinemann, Oxford.

SPEECH DIGITIZATION AND SYNTHESIS

14.1 Pulse amplitude modulation (PAM)

The digitization of analogue waveforms by pulse code modulation is accomplished in two stages. First the waveform is sampled to produce pulse amplitude modulation (PAM). Short-duration samples are taken at regular intervals which are long compared with the sampling time but short in relation to the highest signal frequency. The result is a train of pulses whose amplitude envelope is the same as the analogue waveform. The envelope (Figure 14.1) will contain:

- clock frequency f_c, the sampling rate
- all the signal frequencies contained in the waveform from the lowest, f_1, to the highest, f_2
- upper and lower side bands, $(f_c - f_2)$ to $(f_c - f_1)$ and $(f_c + f_1)$ to $(f_c + f_2)$
- harmonics of f_c and the upper and lower side band frequencies
- a DC component equal to the mean value of the PAM waveform.

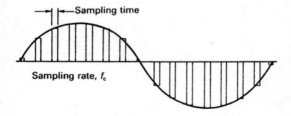

Fig. 14.1 *Pulse amplitude modulation (PAM)*

The envelope contains the original signal frequencies and can be demodulated by a low-pass filter which will pass f_2 but not the clock frequency. The clock frequency must therefore be higher than $2f_2$. For line communications in the UK a clock frequency of 8 kHz is used with a maximum modulating frequency of 3400 Hz.

14.2 Pulse code modulation (PCM)

To overcome the susceptibility of PAM signals to corruption of the amplitude waveform by noise and interference, the waveform is processed further to produce pulse code modulation (PCM) before transmission.

In this process the magnitude of the PAM samples with respect to a fixed reference is quantized and converted to a digital (data) signal. Quantizing rounds off the instantaneous sample pulse amplitude to the nearest one of a number of adjacent voltage levels. Figure 14.2 illustrates the process for an eight-level system. In the figure the amplitude at t_0 is 2, between 4 and 5 at t_1, between 5 and 6 at t_2, etc. After quantization the values would be 2 at t_0, 5 at t_1, 5 at t_2 and so on. The difference between the amplitude levels and the rounded-off values is the quantization noise or distortion.

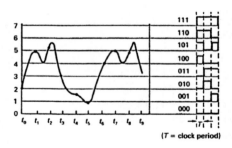

Fig. 14.2 *Eight-level pulse code modulation*

The binary pulse train – leaving a one-bit synchronizing space between each number – for this example would be as in Figure 14.3.

The number of quantizing levels is 2^n and the highest denary number represented is $(2^n - 1)$ where n is

s = synchronization space

Fig. 14.3 *Binary pulse train (from Figure 14.2)*

the number of bits used to represent each sample. If
each train of pulses is accompanied by one synchronizing bit, the number of bits per sample is $(n + 1)$.
If the sampling rate is f_c, the transmitted bit rate is
$(n + 1)f_c$. For example, an eight-bit word – including
the synch. bit – is commonly used to represent a sample, so, with a clock frequency of 8 kHz:

$$\text{number of quantizing levels} = 2^7 = 128$$

$$\text{sampling rate, } f_c = 8 \text{ kHz,}$$

$$\text{therefore transmitted bit rate} = (n + 1)f_c$$

$$= 64 \text{ kb/s.}$$

The maximum frequency of the pulses will be when
transmitting alternate 1s and 0s, and the occupied
bandwidth, $\frac{1}{2} \times$ bit rate, 32 kb/s.

Quantization as described above is linear, i.e. the
spacing of the quantization levels is the same over the
range of pulse amplitudes. This produces a poor signal-
to-noise ratio for low-level signals which is improved
by the use of more closely-spaced levels at small signal
amplitudes than at large amplitudes. Where non-linear
quantization is used the most significant bit of the
sample character identifies the sense of the signal.

14.3 Delta modulation

Delta modulation is basically differential PCM and is
a method of reducing the bandwidth by transmitting
information only when there is a change between
adjacent samples. If the value of a sample is greater
than the previous sample a binary 1 is transmitted; if
it is less then a binary 0 is sent. A typical sampling
rate is 32 kHz, but each eight-bit word is replaced by
one binary digit.

14.4 Speech synthesis

The synthesis of speech from noise waveforms uses less bandwidth for transmission than PCM. The characteristic sounds of speech, vowels, consonants, nasals, fricatives (f and th) and plosives (p) can be extracted from a noise waveform by filters. They can be transmitted by sending only the coefficients of the filters used to select them. The system of filters is a speech codec (code/decode). Dudley's vocoder, circa 1936, illustrates the principle. Dudley divided the speech band into ten 300 Hz wide channels and extracted the low frequency components from each band by narrow band low-pass filters. Within each band the system detected a hiss or a buzz. In the receiver the transmitted code switched a noise generator to produce a hiss or a buzz in the appropriate frequency band. The system created intelligent speech with a 10:1 reduction in bandwidth, i.e. a 3000 Hz speech band was compressed to 300 Hz. The effectiveness of speech codecs varies between languages and needs optimizing for some languages.

An example of the bandwidth saving with synthesis is the Pan-European (GSM) digital radio-telephone network where the filter coefficients are transmitted at 3.6 kbps plus an excitation sequence at 9 kbps, a data rate of 13 kbps compared with 32 kbps for PCM.

An alternative method of reducing the bandwidth requires a library of sample voice sounds and the transmission of the coordinates of the location of each sound within the library. The system saves bandwidth but requires a large store in both the transmitter and receiver.

References

Edis, E.A. and Varrall, J.E. (1992). *Newnes Telecommunications Pocket Book*, Butterworth-Heinemann, Oxford.

Kennedy, G. (1977). *Electronic Communications Systems*, McGraw-Hill Kogashuka, Tokyo.

15

VHF AND UHF MOBILE COMMUNICATION

Mobile communication operating throughout the VHF and UHF bands is expanding rapidly and, although the fastest growth has been in the radio-telephone field, interest in private mobile radio (PMR) is undiminished. Most of the procedures described for PMR are applicable to many other branches of radio communication, so traditional PMR is considered here. Other systems are discussed in later chapters.

A private mobile radio system, comprising a base station and mobiles, is one that is effectively owned by the user and, under the conditions of the licence, may only be operated by the user's own staff for his or her own business. Airtime cannot be leased to other persons.

15.1 Operating procedures

Frequencies are normally allocated in pairs, one for the up-link to the base station and one for the down-link to the mobile (not to be confused with a radio link used for control purposes). Such a pair of frequencies, spaced sufficiently apart to permit simultaneous transmission and reception by a station, comprises the radio channel. Occasionally, for special purposes and for small, low-power, and possibly temporary, operations a single frequency only may be allocated. The mountain rescue teams are an example.

The methods of operating are as follows.

Single-frequency simplex

This method uses a common frequency for transmission and reception by all stations operating on the system (Figure 15.1). Transmission and reception cannot take place simultaneously at a station, and a receiver is switched off whenever the transmit switch is operated. This prevents blocking of the receiver by

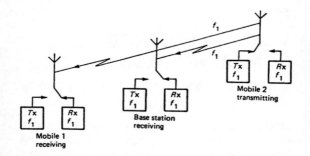

Fig. 15.1 *Single-frequency simplex*

the transmitter and acoustic feedback occurring. The method allows all stations within range to hear both sides of a conversation and to relay messages to more distant stations; an obvious advantage for mountain rescue.

At the end of a transmission an operator must say 'over' and switch off the transmitter to hear the reply. A conversation must end with the word 'out' so that other stations are aware that the system is unoccupied.

Two-frequency simplex

Separate frequencies are used for transmission and reception but whilst a station is transmitting its receiver is still switched off (Figure 15.2). Mobiles hear only the base station and, therefore, the relaying of mes-

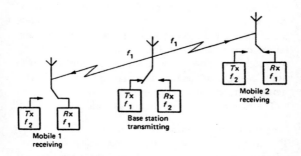

Fig. 15.2 *Two-frequency simplex*

sages is not possible. A further disadvantage is that because mobiles hear only the base station, they may be unaware of the system occupancy and transmit, interfering with an existing conversation. The advantage of two-frequency simplex is the avoidance of receiver blocking or de-sensitization, not only from the associated transmitter but also, at base stations where several channels within the same band are located, from nearby transmitters.

Duplex and semi-duplex

Separate frequencies are used for transmission and reception and, in full duplex, all stations can transmit and receive simultaneously as in a two-way telephone conversation (Figure 15.3). While a station is transmitting, its receiver audio output is switched from the loudspeaker to an earpiece to prevent acoustic feedback. A mobile cannot receive other mobiles directly but full duplex enables all stations to break in on a conversation in an emergency or to query part of a message; it also facilitates the use of talk-through where mobiles can speak to each other via the base station. To maintain awareness of system occupancy the base station may transmit a series of pips as an engaged signal during pauses in the despatcher's speech.

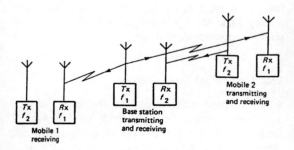

Fig. 15.3 *Duplex*

Many systems operate semi-duplex where only the base station operates a duplex procedure and the mobiles use a simplex procedure. This avoids the higher cost of duplex mobiles and offers most of the facilities of duplex, except that a despatcher cannot break in on a transmitting mobile.

Open channel/'All informed'

All mobiles hear all the calls from control, i.e. no selective calling is in operation.

Selective calling

Mobile receivers remain quiescent until specifically addressed; the opposite of open channel working. Individual mobiles, groups and a whole fleet may be addressed.

Auto-acknowledgement

When selectively called a mobile automatically transponds, sending the code for its address and, possibly, its status information. Mobiles can only acknowledge when individually addressed; auto-acknowledgement on group and fleet calls is prevented to avoid mobiles transponding simultaneously.

Status updating

The transmission, automatically or manually, of the data denoting the mobile's current status.

Call stacking

The storage of calls from mobiles and their presentation in call order to the despatcher. Arrangements are usually made to raise urgent calls to the top of the stack with an enhanced display.

15.2 Control of base stations

Where adequate radio signals over the desired service area of the system can be provided by one base station sited at the control point, it can be easily controlled, either directly from the front panel or over a multi-way cable.

In many instances, to obtain adequate coverage, the base station must be sited on high ground remote from the control point. Then, either a land line or a radio link, both of which will today probably be digital and the link microwave, must be used for control. A land line will most likely be rented but radio links are favoured by those users who insist on the complete system being under their direct control.

15.3 Common base station (CBS) operation

An economic method of providing mobile communication for users with a small quantity of mobiles and light traffic is a common base station or community repeater system. Although still referred to as private mobile radio, the base station is shared by several users who pay a fixed subscription for the service irrespective of airtime used. The station is controlled by radio using tone-controlled talk-through or what is sometimes called reverse frequency trigger (not normally permitted on a true PMR system). Figure 15.4 shows the layout of a single-station CBS system. Each participant's office contains a fixed transmitter/receiver operating on the same frequencies as a mobile and equipped with CTCSS. Different CTCSS tones are assigned to each user to ensure privacy. All office stations must use a directional antenna so that they access only the base station to which they subscribe.

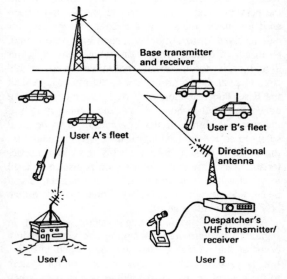

Fig. 15.4 *Common base station system*

When a user transmits, from either the office station or a mobile, the base station, on receipt of a signal containing a valid CTCSS tone, enters a talk-through mode. The caller's CTCSS tone is retransmitted and, in turn, opens the mute on all that user's mobiles allowing communication via talk-through. Mobiles are equipped with 'busy' lamps to indicate whenever the system is engaged, and transmission time-out timers to prevent excessively long calls excluding other users.

15.4 Wide area coverage

Where the desired service area for one radio channel is larger than can be covered by a single base station the way in which the base stations are to be controlled requires special consideration. Crucial aspects are the presentation of the best received signal from a mobile to the control operator or despatcher and the selection of the transmitter most likely to provide the best signal at the location of the mobile.

Receiver voting

Well-proven receiver voting circuits which present the best received signal to the despatcher have been used for many years. These circuits sample the signals received from a mobile at each base station and, by means of coded information – which may be either digital or in the form of continuous tones – enable equipment at the control centre to automatically select the best. The selection may be made by comparing either the signal-to-noise ratio or the signal strength of the received signal. If the information is to be used solely to select the best signal in terms of readability, the signal-to-noise ratio is probably the better characteristic to use, but if the information is also to be used to select a base transmitter, the signal strength could be considered more satisfactory. Some systems utilize both types of information.

A typical 3-station voting system is shown in Figure 15.5 where signal sampling and vote encoding occur at the receiving sites and the coded information is passed over the base station control system to the control centre. This method is necessary when the selection is made on a signal strength basis, but where the signal-to-noise ratio is used for the selection the sampling and encoding can be done at the control centre taking into account the noise occurring

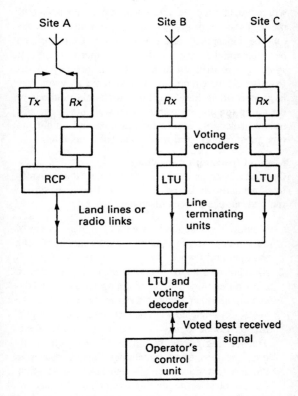

Fig. 15.5 *Radio scheme with one transmit/receive plus two receiver-only stations with voting*

in the control network. Receiver voting systems operate very quickly, and changes of the selected receiving site may occur several times during a message without the despatcher's awareness.

The broadcast transmit and receive paths are not always reciprocal, for instance when low power hand-portables are integrated with higher power vehicular mobiles. In these circumstances the use of additional receiver-only fill-in stations is an economic and satisfactory proposition.

Base transmitter control poses a problem much more difficult to resolve. The selection of a base transmitter to communicate with a mobile whose precise whereabouts may be unknown, the broadcasting of messages to all, or groups of, mobiles whose locations may be widespread, and the provision of talk-through between mobiles are all facilities required on major systems, and difficult to provide satisfactorily. Apart from trunking, which can be economic only on very large networks, there are three traditional methods of operating the base transmitters on wide area schemes: manual selection, automatic selection and simultaneous transmission from more than one transmitter.

Manual transmitter selection

On many systems the despatchers select the transmitters manually. It is the simplest and least expensive method but has serious disadvantages:

1. Making the selection may entail trying a number of transmitters before sending the message, increasing the operator's work load and wasting air time.
2. Mobiles outside the service area of the selected transmitter may call, interrupting an existing conversation, either because they are unaware that the system is engaged or have received a poor signal that they believe may have been intended for them. Transmitting bursts of 'engaged' pips sequentially over the unselected transmitters during pauses in the despatcher's speech alleviates the first situation; the use of selective calling overcomes the second.
3. Broadcast messages must be transmitted on each transmitter in turn and talk-through between mobiles which are not in the service area of the same transmitter is not practicable.

Automatic transmitter selection

Selecting the transmitters automatically, or semi-automatically, is an improvement over manual selection. On a system where the mobiles are not equipped with selective calling and automatic acknowledgement of a call, automatic selection of transmitters can only occur on receipt of a call from a mobile. The transmitter through which to reply is then selected by the receiver voting system at the same time as it selects the signal to present to the despatcher. The selection is made at the start of the call and, because of the switching times involved, the transmitter selected is retained for the duration of the call. On these open channel systems, a calling despatcher must initially manually select a transmitter.

On systems where selective calling and auto-acknowledgement of a call are provided, the system can be made fully automatic by transmitting the data corresponding to a mobile's call sign from each transmitter in turn until a satisfactory acknowledgement is received. The successful transmitter is then retained for the duration of the conversation and, at its conclusion, is usually the one used to commence another call.

Such a system overcomes the disadvantage of the need to manually select the transmitter but the problems with mobiles outside the service area of the selected transmitter, and of broadcasting and talk-through, remain.

Simultaneous transmission

Operationally, simultaneous transmission from all sites is ideal and, under various names such as Spaced Carrier, Simulcast and Quasisynch., has been around since the mid-1940s. Its operational value is proven but systems require special care in their planning, adjustment and subsequent maintenance.

An early form of simultaneous transmission was the amplitude modulated spaced carrier system. Used very successfully at VHF on systems using 25 kHz channel separation, the transmitter carrier frequencies were separated by 7 kHz – above the mobile receivers' audio pass-band. With the reduction of channel spacings to 12.5 kHz, spaced carrier operation on this basis was no longer possible and alternatives are to either synchronize, or very nearly synchronize, the carrier frequencies. There are, however, undesirable effects of synchronous and quasi-synchronous transmission but, with care, these can be reduced to an acceptable level. They are:

1. The beat note between transmitters being audible in the mobile receiver.
2. Variations in signal level due to interference patterns between signals from more than one transmitter.
3. Distortion due to audio phase differences and differing modulation levels when signals of

comparable strength are received from more than one transmitter.

The beat note is easily dealt with. It is rendered unobjectionable either by placing it outside the mobile receiver audio pass-band – which is the usual method – or by synchronizing the transmitter carrier frequencies so that no beat note is produced. Synchronization, however, raises other problems, which are particularly severe at VHF but less so at UHF.

The spaced carrier system placed the beat note above the receiver pass-band, but modern systems place it below. This means that the beat note between the lowest and highest carrier frequencies must be below about 150 Hz if it is to be unobtrusive. Tests have shown the optimum carrier separation to be from 0.5 Hz to 4 Hz between any two transmitters using amplitude modulation, and from 5 Hz to 40 Hz between adjacent frequency transmitters using angle modulation.

Two transmitters on the same or closely-spaced frequencies produce deep nulls in the received signal in the areas where they provide almost equal strength signals. This is a natural phenomenon and the effect can only be reduced by the correct siting of stations and antenna configurations. Because of the longer wavelength the effect is more detrimental at VHF than UHF. Where the frequencies are quasi-synchronous the interference pattern is continually moving, but with synchronized carriers the pattern is virtually stationary and at low band VHF wavelengths it is possible to stop a vehicle in a place of semi-permanent zero signal. While moving slowly in an area of equal signal strengths from two transmitters, the cancellations become very objectionable. At 450 MHz the distance between the nulls is so short that it is almost impossible to remain in one, and while moving they are unnoticed. Strong signals in the overlap areas minimize the time that the signals fall below the receiver noise threshold at each cancellation. They are the key to reducing the annoyance from the cancellations. Amplitude modulated systems have an advantage in that the receivers produce less severe bursts of noise during the signal nulls.

The audio distortion, provided the equipment does not introduce significant additional harmonic distortion, is attributable to audio phase differences in the signals received from more than one transmitter. For the distortion to be severe, the received signals must differ by less than 6 dB on an angle modulated system; capture effect in the receiver removes the audible effects at greater differences.

The phase differences arise from differing audio characteristics in all the circuits in the path including land lines, radio links and control equipment. Common to all systems are the phase differences due to the different path lengths from the control centre to a receiving mobile, but where land lines or multiplexed circuits are used for the control of the base stations, variable bulk and group delays and frequency translation errors can present serious difficulties. Tests have shown AM to be slightly more tolerant than angle modulation in respect of this type of distortion, phase delays of 100 µs being acceptable with AM compared with 70 µs with FM.

Differing path lengths between the control centre to the transmitting sites can be equalized by installing audio delay circuits at the transmitters. The path lengths from each transmitting site to all the places where equal signals occur must then be less than about 21 km (equivalent to 70 µs) for acceptable quality. Modern techniques enable delays to be dynamically equalized to compensate for variations in the path.

Figure 15.6 shows the layout of a multi-station scheme including the audio delays. However, signals do not confine themselves to neat circles and the worst situation is where the signals from two transmitters arrive at a mobile more or less equal in strength and phase. In this situation, apart from the ripple caused by the carrier frequency offset of the transmitters, a satisfactory signal would still be received but, occasionally, the presence of a weaker signal from a third, distant station (C in Figure 15.6) intrudes during the cancellation periods and, because of its long path length, introduces severe distortion. The area where this situation occurs is often small and the areas of overlap can usually be moved slightly by adjustment of either the transmitter power or antenna directivity.

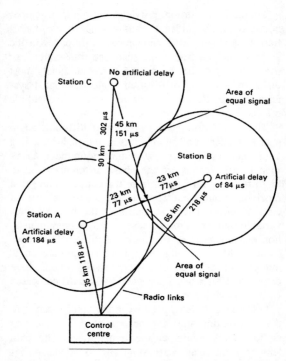

Fig. 15.6 *Quasi-synchronous transmitter system*

SIGNALLING

Many radio receivers are required to respond to instructions sent to them over the radio channel to which they are tuned. Several methods of signalling these instructions are in use, from the continuous tones used for controlling the receiver mute operation to the rapid data to initiate channel changes and transmitter power control on cellular radio-telephones.

16.1 Sub-audio signalling

The slowest signalling system uses continuous sub-audio frequency tones. Known as continuous tone controlled signalling (previously squelch) system (CTCSS) its performance, in the UK, is specified in Radio-communications Agency specification MPT 1306. The most common use for the system is to control receiver mute opening. Permitting a mute to open only on receipt of an authorized signal, its use enables privacy between users to be maintained on shared systems, common base station systems for example, and reduces the annoyance factor from interference in the absence of a signal. Thirty-two tones are permitted and assigned by the Radiocommunications Agency (Table 16.1).

Table 16.1 *CTCSS tones and modulation levels*

67.0	110.9	146.2	192.8
71.9	114.8	151.4	203.5
77.0	118.8	156.7	210.7
82.5	123.0	162.2	218.1
88.5	127.3	167.9	225.7
94.8	131.8	173.8	233.6
103.5	136.5	179.9	241.8
107.2	141.3	186.2	250.3

The tones are transmitted at very low modulation levels (Table 16.2).

Table 16.2 *Modulation levels*

System Channel spacing	Amplitude modulation Modulation depth (%)	Angle modulation Peak deviation ± Hz
25 kHz	10 to 20	400 to 800
12.5	10 to 20	200 to 400

16.2 In-band tone and digital signalling

16.2.1 Dual tone multi-frequency (DTMF)

Developed for telephone work, dual tone multi-frequency signalling has crept into some radio systems but is slow to operate compared with other in-band systems. Two tones are transmitted simultaneously to represent each digit. In the UK they conform to the MF4 code (Table 16.3).

Table 16.3 *MF4 code of DTMF tones*

Digit	Frequencies (Hz)	
1	697	1209
2	697	1336
3	697	1477
4	770	1209
5	770	1336
6	770	1477
7	852	1209
8	852	1336
9	852	1477
0	941	1336
*	941	1209
#	941	1477

Some push button telephones signal to the local exchange by generating a combination of two frequencies. Each row and each column of the push button keypad is connected to an oscillator of set frequency. When any push button is pressed, tones corresponding to its row and column frequencies are therefore generated. The row and column oscillator frequencies are as shown below:

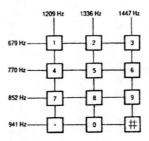

16.2.2 5-tone signalling

So-called 5-tone signalling uses a series of tones – usually five but other quantities are used in some systems. The system is popular for selective calling and addressing purposes (Table 16.4). There are a range of tones established by various organizations but, although the tones in each series may be assigned to each digit, equipment manufacturers sometimes amend the code and the usage. The relevant Radiocommunications Agency document is MPT 1316, *Code of Practice, Selective Signalling for use in the Private Mobile Radio Services*.

16.3 Digital signalling

FFSK

Many analogue modulated systems use digital signalling in the form of 1200 bit/s fast frequency shift keying which is faster than 5-tone but less robust.

DCS

Digitally coded squelch (DCS) is a digital, but slower, alternative to CTCSS. It is not in current use in the UK. DCS continuously repeats the constant binary bit pattern of a 23 bit word. Golay error coding reduces the number of data bits to 12 of which 3 initiate the Golay sequence leaving 9 information bits, i.e. a 9 bit word. With a 9 bit word, $2^9 = 512$ codes are possible but, because of problems with the frequencies produced, the codes with 7 or more 1s or 0s in succession are

Table 16.4 *5-tone codes*

Standard	ZVEI	DZVEI	EIA	CCIR/ EEA	CCITT	EURO
Tone duration (ms)	70	70	33	100/ 40	100	100
Pause duration (ms)	0	0	0	0	0	0
Code/digit	Tone frequency (Hz)					
0	2400	970	600	1981	400	980
1	1060	1060	741	1124	697	903
2	1160	1160	882	1197	770	833
3	1270	1270	1023	1275	852	767
4	1400	1400	1164	1358	941	707
5	1530	1530	1305	1446	1209	652
6	1670	1670	1446	1540	1335	601
7	1830	1830	1587	1640	1477	554
8	2000	2000	1728	1747	1633	511
9	2200	2200	1869	1860	1800	471
Repeat	2600	2400	459	2110	2300	1063
Alarm	2800	2600		2400		
Free-tone						1153
Group tone	970	825	2151	1055		

ZVEI Zuverein der Electronisches Industrie. Designed to operate on 20 kHz channel spaced systems. On 12.5 kHz channel spaced systems transmission of 2800 Hz creates difficulty and depressed ZVEI (DZVEI) was adopted and recommended in MPT 1316.

EIA Electrical Industries Association.

CCIR Committee Consultatif International Radio Communication. The longer tone duration offers robustness against corruption but is slow to operate. Originally designed for marine use, each digit was transmitted twice to ensure reliability.

EEA Electronic Engineering Association. Recommended in MPT 1316.

CCITT International Telegraph and Telephone Consultative Committee.

discarded leaving 104 'clean codes'. The transmitted rate is 134 bit/s and the highest fundamental frequency is therefore 67 Hz. The lowest fundamental frequency is produced when 6 zeros follow 6 ones = 1/6th of the fundamental frequency = 11.7 Hz. The important harmonics are:

$$\text{third} = 201 \text{ Hz}$$
$$\text{fifth} = 335 \text{ Hz}$$
$$\text{seventh} = 469 \text{ Hz}$$

The seventh harmonic is well within the audio pass band of most receivers but at 3 octaves from the fundamental is relatively weak.

16.4 Standard PSTN tones

Tone	UK		USA	
	Frequency	Sequence	Frequency	Sequence
Number unobtainable	400 Hz	Continuous	500 + 600 Hz	Interrupted every second
Number busy	400 Hz	0.75 s on, 0.75 s off,	480 + 620 Hz	Interrupted once per second
System busy	400 Hz	0.4 s on, 0.35 s off, 0.225 s on, 0.525 s off	480 + 620 Hz	Interrupted twice per second
Dial tone	50 Hz 33.3 Hz		350 + 440 Hz	
Ringing tone	400 Hz, or 400 + 450 Hz, modulated by 17, 25, or 50 Hz	0.4 s on, 0.2 s off, 0.4 s on, 2 s off	440 + 480 Hz	Interrupted once per second
Pay tone	400 Hz	0.125 s on, 0.125 s off		
Time tones	900 Hz	0.15 s on, 0.85 s off, Three times every three minutes		
Recorder warning			1400 Hz	One burst, every 15 s

References

Belcher, R. *et al.* (1989). *Newnes Mobile Radio Servicing Handbook*. Butterworth-Heinemann, Oxford.

Edis, E. A. and Varrall, J. E. (1992). *Newnes Telecommunications Pocket Book*. Butterworth-Heinemann, Oxford.

17

CHANNEL OCCUPANCY, AVAILABILITY AND TRUNKING

17.1 Channel occupancy and availability

The usefulness of any communications network depends on successful calls. Assuming that no calls fail for technical reasons, the success rate depends on the amount of traffic on the channel and how well it is managed. Channel occupancy is a measure of the amount of traffic carried and is affected by:

1. the number of subscribers (mobiles on a radio network)
2. the number of calls per subscriber in a given time
3. the average duration of each message.

The occupancy of any system varies hourly, daily and possibly seasonally. To be successful the network must be able to cope with the traffic during the periods of highest occupancy and, when considering occupancy it is usual to examine the conditions over one hour, the 'busy hour'. The occupancy of a channel can be expressed as a percentage:

$$\text{Channel occupancy} = \frac{nct}{36}\%$$

where

$n =$ number of subscribers (on a mobile radio system, mobiles on watch)

$c =$ call rate (number of calls per subscriber in 1 hour)

$t =$ average message duration in seconds

Occupancy may also be expressed in erlangs where continuous traffic, i.e. 100% occupancy over a specified period, is 1 erlang. Less than 100% occupancy is expressed as a decimal fraction, e.g. 30% occupancy is 0.3 erlang. Demand may also be expressed in traffic units, A:

$$A = \frac{Ct}{T} \text{erlangs}$$

where C is the number of calls in time T and t is the mean call duration.

The amount of congestion occurring on a system can be defined by the grade of service it offers. Grade of service is expressed as a percentage in terms of the probability of meeting either congestion or waiting more than a specified delay before a call can be established. Erlang's formula is:

Grade of service
$$= \frac{A^N/N!}{1 + A + A^2/2! + A^3/3! + \cdots + A/N!}$$

where A is the number of traffic units (erlangs) and N is the number of outlets (trunked channels).

The probability that delayed calls will have to wait more than a specified time $(W > t)$ is:

$$P(W > t) = \exp\left(-\frac{(C - A)t}{H}\right)$$

where

A = traffic units (erlangs)
C = number of trunked channels
H = mean call duration in busy hour

The above assumes that:

1. the number of users is infinite
2. the intervals between calls are random
3. call durations are random
4. call set-up times are negligible
5. delayed calls are queued so that they are dealt with on a first come, first served basis.

The number of potential radio users is continuously growing and any technique that makes more effective use of the finite radio spectrum is welcome. Whenever technology has permitted, higher frequencies and reduced channel spacings have been employed in the attempt to meet the demand. A method of making more effective use of the available spectrum is trunking.

17.2 Trunking

The generally accepted term for the automatic allocation of communication channels to subscribers on demand is trunking, and a 'trunked radio system' is one in which many users share the use of a common pool of radio channels. Channels from the pool are allocated to users on demand and as they become unoccupied; no channels are allocated to specific users or groups of users.

When making a telephone call, provided the speech quality is good it is immaterial which route the call takes. The same applies to radio; providing the message reaches the right person quickly, the radio frequency which carries it does not matter.

Most trunked radio systems use a calling channel on which calls are requested. When a call is established and accepted a working channel from the pool is allocated for the duration of the call and at its termination the channel is returned to the pool for re-allocation to another user. The important principle is that any user has access to any free channel. This has the advantage that more mobiles per channel can be accommodated, or a better grade of service, i.e. less waiting time for access, can be provided. Basic trunked systems are unsuitable where everyone using the system needs to be a party to every call but the latest trunked mobile radio system arrange, on receipt of a call, to assign a working channel to a particular group of mobiles where they operate within the service area of one base station. They then operate on an 'all informed' basis.

Trunking provides the most benefit when the number of channels required is greater than the number assigned to the system. The effective gain in the number of mobiles that can be accommodated on a trunked system is illustrated by the following example:

A single radio channel with an efficient signalling system offering a 30% grade of service can accommodate 90 mobiles with a mean waiting time of 20 seconds if each mobile makes 1 call. Twenty non-trunked channels cater for 1800 mobiles. A 20 channel trunked system under the same conditions can accommodate 3430 mobiles with a mean waiting time of 16.4 seconds, a trunking gain of over 170 mobiles per channel.

The mobile transmitter/receivers used on trunked systems are more complex than those used on systems where channels are permanently allocated because the mobiles must either scan the channels searching for an available one or one containing a call directed to them, or listen continuously on a channel designated for calling. In the latter case when a call is initiated, a complex 'handshake' procedure occurs and the mobile is automatically switched to the channel allocated for that conversation. The band III public access mobile radio (PAMR) networks use a calling channel and many common base station (CBS) systems use channel scanning methods with no calling channel.

Working channels may be assigned for the duration of conversation, which is message trunking, or for the duration of a transmission, which is transmission trunking. With either type of trunking the channel is not returned to the pool immediately the signal from the mobile is lost; some waiting time is essential to cater for short-term signal dropouts but with message trunking the channel is retained for a longer period of time to allow the complete conversation to take place. On a very busy system this is wasteful and, on a system with very rapid signalling, transmission trunking may be used. Then the use of the working channel is lost very soon after the cessation of transmission and a new working channel assigned on its resumption. Some networks automatically migrate from message trunking to transmission as traffic increases.

17.3 In-band interrupted scan (IBIS) trunking

IBIS trunking is a method which is finding favour for common base station applications. All channels are used for communication with no dedicated calling channel, thus making the most effective use of the channels available.

On a CBS system all users access the base station, which operates talk-through, by radio. Each user is allocated his own continuously coded squelch system (CTCSS) tone to access the base station and to prevent overhearing by other users. A trunked common base station contains the transmitters and receivers for a number of radio channels and when the system is quiescent transmits bursts of an in-band (speech band) free-tone – 1953 Hz is one that is used – on one channel only. The mobiles, which are fitted with decoders to recognize both the free-tone and their CTCSS tone, scan all channels looking for one containing free-tone onto which they lock. Should the mobile operator press the transmit button the transmitter will only be enabled when the mobile receiver is on a channel containing either the free-tone or its CTCSS code. As soon as the channel becomes occupied the base station commences to radiate free-tone on another channel and mobiles of other user groups re-lock onto the new free channel. When either a mobile or the control station, which is a fixed mobile, makes a call it is on the channel radiating free-tone and the transmission contains the CTCSS code allocated to that user group. The repeater re-transmits the CTCSS code which activates the calling mobile and all other mobiles within that group. Mobiles of other user groups sensing the alien CTCSS tone leave the channel and continue scanning for a free channel. If all channels are engaged, the caller hears a 'busy' tone and must try again later.

17.4 Trunking to MPT 1327 specification

Specification MPT 1327 produced by the Radiocommunications Agency lays down a signalling standard for trunked private land mobile radio systems. The system uses fast frequency shift keying (FFSK) at 1200 bit/s with time division random access. The mobiles operate two-frequency semi-duplex and the base stations or trunking system controller (TSC) operate full duplex. A calling channel is used for call establishment but may be either dedicated or nondedicated. After a call is established messages are transmitted over a traffic (working) channel. The basic components of the signalling format, in sequence, are:

1. The link establishment time (LET), a period of transmission of undefined modulation.
2. A preamble comprising a sequence of bit reversals 1010...10, minimum 16 bits and ending with a binary zero for the receiver data demodulator to achieve synchronization.
3. A message. A contiguous transmission of a synchronization sequence, address codeword and, where appropriate, data codewords.
4. A hangover bit, H, of either binary zero or binary one, added to the last transmitted message identifying the end of signalling transmission.

The message synchronization sequence for both control and message channels consists of 16 bits, the first bit being a binary one on a control channel and binary zero on a traffic channel.

Messages, whether address or data, are transmitted in 64 bit codewords. The first bit, bit 1, indicates the status of the codeword. Binary one denotes an address word, binary zero, a message word. Bits 2–48 are the information field and bits 49–64 are check bits.

References

Edis, E.A. and Varrall, J.G. (1992). *Newnes Telecommunications Pocket Book*. Butterworth-Heinemann, Oxford.

Macario, R.C.V. (1991) *Personal and Mobile Radio Systems*. Peter Peregrinus, London.

Parsons, J. D. and Gardiner, J. G. (1989). *Mobile Communication Systems*. Blackie, London and Glasgow.

Radiocommunications Agency (1991). MPT 1327, *A Signalling Standard for Trunked Private Mobile Radio Systems*.

Radiocommunications Agency (1993). MPT 1318, *Engineering Memorandum, Trunked Systems in the Land Mobile Radio Service*.

18

MOBILE RADIO SYSTEMS

18.1 Paging

Where an immediate reply to a message is not essential, paging systems, which are basically a one-way path from controller to mobile, are an economic option. Paging is also spectrally economic, only one frequency being required and messages being of very short duration. The recipient of a call may be merely alerted by means of a tone emitted by the paging receiver or, if silent alert is required, a vibration. Alternatively, pagers displaying, and storing, alpha-numeric messages are readily available. Message handling services are also offered on some paging systems.

On-site paging systems are usually owned or rented by the user and consist of either a radio base station or, in the case of a non-radio system, an inductive loop around the perimeter of the service area. The details of on-site paging systems and the Radiocommunications Agency specifications governing them are:

RF frequencies:

  on-site, 26 MHz and 31 MHz bands, MPT 1365
  49.0–49.5 MHz, MPT 1335
  160–161 MHz
  458–459 MHz, MPT 1305

Channel spacings:

  12.5 kHz in the VHF bands and 25 kHz in UHF band

Modulation:

  FM, FFSK data

Wide area, national or international systems are generally owned by a network provider to whom the user subscribes for the service. There are frequencies allocated in the high VHF and UHF bands. The appropriate specifications are MPT 1308 for receivers and MPT 1325 for transmitters.

The Post Office Code Standardization Group (POC-SAG) standard which has been adopted by the CCITT allows the transmission of messages at a rate of 10 calls per second and a capacity of 1 million pagers with up to four alternative addresses per pager. The recommended bit speed is 512 bit/s with direct FSK deviating the transmitter frequency by ±4.5 kHz. Positive deviation indicates a binary zero and negative deviation a binary one.

In America the Golay sequential code is widely used for paging addressing. The format for tone-only paging consists of 12 address bits followed by 11 parity bits for each 23 bit word. Two bit rates are used, 300 bit/s for address codewords and 600 bit/s for message codewords.

There are European paging and message systems in operation including Euromessage and the frequency agile European Radio Message Service (ERMES).

18.2 Cordless telephones

CT1
The first generation of cordless telephone, still very popular in the UK, is designed to serve the domestic environment with a range of 100 m. The base station transmits on one of eight frequencies between 1.6 MHz and 1.8 MHz, and the handset on one of eight frequencies in the 47 MHz band. Frequency modulation is used with a deviation of ±4 kHz at the base station and 2.5 kHz at the handset. MPT 1322 applies and permits an erp of 10 mW at both the base station and the handset.

CT2
The second generation of cordless telephone. The services, under various names, telepoint, phonezone, etc., were planned to provide for the general public a lower cost alternative to the cellular radio-telephone networks which were seen at the time as a businessman's preserve. However, the low market uptake caused their demise in the UK although services are thriving in some European and Far Eastern countries.

One hand-portable transmitter/receiver can operate to a local base station installed in the home or office, or through one of a number of multi-channel base stations with a range of approximately 200 m installed in public places. While away from the local base station the subscriber must initiate a call: calls cannot be made from the PSTN to a CT2 subscriber's handset.

The operational frequencies are in the band 864.1 to 868.1 MHz and employ time division multiple access. Speech is digitized at 32 kbit/s, stored, and then transmitted at 64 kbit/s in 1 ms slots. This leaves the alternate 1 ms slots available for the digitized and stored speech of a reply. Duplex operation is achieved in this way on a single radio frequency.

Inter-operability between networks was to be ensured by specification MPT 1375 *Common Air Interface*.

Digital European Cordless Telephone (DECT)

A pan-European system complying with an ETSI standard, DECT operates in the 1880 to 1900 MHz band. It offers data handling facilities and the ability for a subscriber to receive calls while away from the local base station. The techniques are similar to those used for GSM although, because the mobile is virtually stationary, the constraints on data transmission are less severe and no hand-off is required. The 20 MHz RF bandwidth is divided into 13 carriers spaced at 1.7 MHz intervals, each carrier containing 12 TDMA channels with GMSK modulation.

CT3 is developed from DECT and operates in the band 800 to 900 MHz. Each 8 MHz section of that band is divided into 1 MHz blocks, each containing sixteen 1 ms time slots.

18.3 Trunked radio

Bank III trunked radio

So-called because it occupies part of the now redundant television band III, 174 to 225 MHz, and often referred to as private mobile radio, this is a subscriber access system. The network has almost nationwide coverage in the UK, arranged on both local and regional bases. The network permits radio communication between a fixed office station and mobiles, mobile to mobile, and limited access from mobiles to the PSTN.

The relevant Radicommunications Agency specifications are:

MPT 1323   *Angle-modulated radio equipment for use at Base and Mobile Stations Private Mobile Radio Service operating in the frequency band 174–225 MHz.*

MPT 1327   *A signalling standard for trunked Private Land Mobile Radio Systems.*

MPT 1343   *System interface specification for radio units to be used with commercial trunked networks operating in Band III sub-band 2.*

MPT 1347   *Radio interference specification for commercial trunked networks operating in Band III sub-band 2.*

MPT 1352   *Test Schedule. For approval of radio units to be used with commercial trunked networks operating in Band III sub-band 2.*

Frequency modulation is used with FFSK signalling, a channel spacing of 12.5 kHz and TDMA techniques (Section 17.4).

Trans-European Trunked Radio (TETRA)

A new European standard for digital trunking. The system is aimed at PMR users and will employ TDMA with four users sharing a 25 kHz radio channel with the option of two users on a 12.5 kHz channel. Speech will be digitally encoded at 4.8 kbit/s and transmitted at 36 kbit/s (gross bit rate including error correction overhead) using $\pi/4$ differential quaternary phase shift keying (DQPSK).

Digital Short Range Radio (DSRR)

This utilizes trunking techniques but with no infrastructure of control or switching centres; mobiles work to mobiles via a locally installed repeater. It offers great potential for short-term hire for local events. The band 933 to 935 MHz has been allocated to the service and accommodates 76 traffic and 2 control channels separated by 25 kHz. The operating procedure may be either single-frequency simplex or two-frequency semi-duplex. Mobiles remain on standby waiting to receive their selective signalling code (SSC) on one of the two control channels; mobiles with even serial numbers normally listen on control channel 26 and those with odd serial numbers on control channel 27. A mobile wishing to initiate a call first scans for a free traffic channel and, having obtained one, selectively addresses the intended recipients of the call over the relevant control channel. The recipient mobiles are then instructed automatically by the system to go to the appropriate traffic channel.

The modulation is GMSK at 4 kbit/s for addresses and 16 kbit/s for speech and data messages. The maximum transmitter erp is 4 W.

18.4 Analogue cellular radio-telephone networks

In the UK there are two, full duplex, analogue cellular networks operating to the total access communications system (TACS) standard, Cellnet and Vodaphone. Since their introduction the traffic has grown to the extent where, to reduce congestion, frequencies have needed to be borrowed from other planned services to meet the demand. These are known as extended TACS (ETACS) channels. Both networks operate in the frequency band 890–905 MHz (872–905, ETACS) base to mobile (down-link) and 935–950 MHz (917–950, ETACS) mobile to base (up-link) with a down-link to up-link separation of 45 MHz. They provide national coverage with roaming facility and hand-off from cell to cell as a mobile travels across the network, calling for more complex signalling than local trunking.

The control system must know at all times the location of all operational mobiles. It does this by continuously monitoring signal strength and instructing the mobile to change channel as it crosses cell boundaries.

To permit the maximum re-use of frequencies mobiles are instructed by the system controller to reduce transmitted power to the minimum necessary for acceptable communications in either eight or five steps, depending on the mobile classification, to 0.01 W. Mobiles are classified by their maximum effective radiated power output:

Class 1	10 W erp, vehicle mounted
Class 2	4 W erp, transportable
Class 3	1.6 W erp, transportable
Class 4	0.6 W erp, hand-portable

The channel separation is 25 kHz with a peak speech deviation of ±9.5 kHz. Signalling is by Manchester encoded PSK at 8 kbit/s with 6.4 kHz deviation.

To discriminate between stations transmitting on the same RF channel an FM supervisory audio tone (SAT) is transmitted concurrently with speech at ±1.7 kHz deviation. There are three such tones: 5970, 6000 and 6030 Hz.

An FM signalling tone (ST) of 8 kHz transmitted for 1.8 seconds (deviation $= \pm6.4$ kHz) indicates the clear-down of a conventional telephone call.

The advanced mobile phone service (AMPS) is the American cellular system designed for 30 kHz channel spacing with a 10 kbit/s signalling rate. TACS is based on the AMPS system but modified for the reduced UK channel spacing. Development of narrower channel spacing systems (NAMPS and NTACS) is taking place.

18.5 Global system mobile (GSM)

Global system mobile (GSM), formerly known as Groupe Speciale Mobile, is the pan-European digital cellular radio-telephone service. The operational requirements for GSM are severe, e.g. it must operate satisfactorily to a person walking, or in a slowly moving vehicle, in a street where much of the furnishings will introduce multi-path fading, and operate to a train travelling at 250 km/hr where Doppler frequency shift becomes significant. To reduce the corruption, a high degree of error detection and correction must be applied which increases the occupied bandwidth. To compensate, to some extent, for this, the system takes full advantage of the redundancy in speech to reduce the bandwidth during synthesis.

GSM operates full duplex in the band 890–915 MHz, up-link and 935–960 MHz, down-link.

A combination of FDMA and TDMA is employed. Each allocated band of 25 MHz is divided into 125 carriers spaced 200 kHz apart. The subdivision of each transmitted bit stream into 8 TDM time slots of 540 µs gives 8 channels per carrier and 1000 channels overall. Modulation is Gaussian minimum shift keying (GMSK).

Speech synthesis is by speech codec using linear predictive coding. It produces toll quality speech at 13 kbit/s.

18.6 Personal communication network (PCN)

A digital alternative to the TACS network which operates in the 1800–1900 MHz band. Currently entering service in Britain and mainland Europe.

18.7 Private mobile radio (PMR)

These systems are privately owned or rented and may be used only by the owner's own staff for the purpose of the owner's business. The leasing of airtime is forbidden. Common base station or community repeater systems where several users subscribe to, and use, the station is often also referred to as PMR.

18.8 UK CB radio

27 MHz band: 27.60125 to 27.99125 MHz

40 channels at 10 kHz spacing.
Max. erp, 2 W; max. transmitter output power, 4 W.
Antenna: single rod or wire, 1.5 m overall length, base loaded.
If mounted higher than 7 m transmitter output to be reduced at least 10 dB.
Modulation: FM only, deviation ±2.5 kHz max.

934 MHz band: 934.025 to 934.975 MHz

20 channels at 50 kHz spacing (may be reduced later to 25 kHz).

If synthesizer used spacing may be 25 kHz on precise channel frequencies specified.

Max. erp, 25 W; max. transmitter output power 8 W. If antenna integral, max. erp, 3 W.

Antenna: may have up to four elements, none exceeding 17 cm.

If mounted higher than 10 m, transmitter output to be reduced at least 10 dB.

Modulation: FM only, deviation ±5.0 kHz max.

Spurious emissions: for both bands, not exceeding 0.25 µW except for specified frequency bands where the limit is 50 n W.

For latest full specifications see publications MPT 1320 (27 MHz band) and MPT 1321 (934 MHz band) obtainable from the Information and Library Service of the Radiocommunications Agency.

References

Belcher, R. *et al.* (1989). *Newnes Mobile Radio Servicing Handbook*. Butterworth-Heinemann, Oxford.

Edis, E.A. and Varrall, J.E. (1992). *Newnes Telecommunications Pocket Book*. Butterworth-Heinemann, Oxford.

Macario, R.C.V. (1991). *Personal and Mobile Radio Systems*. Peter Peregrinus, London.

BASE STATION SITE MANAGEMENT

The information in this chapter applies mainly to sites for stations operating at VHF and above. Some conditions will be different for stations working at lower frequencies due to the generally physically larger transmitters and antennas.

19.1 Base station objectives

The objectives of a radio base station are:

1. To provide an adequate service over the whole of the desired service area in a well-defined manner. Special consideration must be given to those areas where, because of their location, weak signals or a high ambient electrical noise level might be expected.
2. To create the minimum interference with other radio users and be designed to receive the minimum from other users.
3. To be electromagnetically compatible with neighbouring installations, possibly on the same site.
4. To create the minimum impact on the environment.

19.2 Site ownership or accommodation rental?

The question of whether to privately develop a base station site or rent space on someone else's site may be resolved by the planning authorities rather than the developer. Before planning consent will be given the developer must normally prove that no existing site will meet the requirements. However, assuming that land is available and planning consent might be forthcoming, other factors to consider are costs, both initial and recurring, and access for services,

Table 19.1 *Factors affecting the decision to own or rent*

	Ownership	*Rental*
Initial costs	Legal charges	Legal charges
	Land clearance	Connection fee
	Provision of services	
	Erection of buildings and mast or tower	
	Installation of equipment	Installation of equipment
	Provision of standby power	Provision of standby power
Recurring costs	Maintenance of site	Rental
	Power	Normally included in rental
	Insurance (other than radio equipment)	Normally included in rental
Other considerations	Very long commissioning time	Very short commissioning time
	Site under user's control	Site under landlord's control
	Space may be leased to other users creating revenue	Accommodation will be shared with other users (install equipment in locked cabinet to preserve security)

materials during building and tower erection and for maintenance during bad weather (Table 19.1).

19.3 Choice of site

The first stage in selecting a site is to look for an existing site which might appear to cover the service area. Information may be available from the operators of such sites regarding their coverage, and coverage tests may be made or computer predictions obtained. For a receiving site the ambient noise level at the base station must also be considered. It should be realized that a site, or an antenna, can be too high, particularly when coverage is required at very short range or if interference from other users is likely to occur. Antennas radiate the minimum signal along the line of their elements so that vertically polarized antennas, i.e. with the E plane vertical, provide very little, if any, signal immediately below them. This creates a hole in the coverage, 'the Polo mint effect', close to the antenna.

Where there is capacity on an existing site the possibilities are:

1. Sharing an existing lightly used radio channel, e.g. common base station.
2. Sharing equipment space in an existing building or providing own building or extension on a communal site.
3. Sharing an antenna by means of multi-coupling or erecting own antenna on existing structure.

19.4 Masts and towers

Apart from short, unguyed poles mounted on the side of buildings, antenna supports fall into two types: guyed masts and self-supporting towers.

A guyed mast may be a single pole, or a square or triangular section lattice structure. The range of single-pole masts includes those of about 4 m maximum height mounted on chimney stacks and secured by wire lashings, and poles supported on the ground or the roof of a building. The latter types may be up to 30 m high. They are usually constructed from steel or aluminium hollow tubular sections – the bottom section is invariably steel – with a set of guys at about every 5 m. Mast erection is specialist work but to estimate the approximate height that a site can accommodate, a four-guy, single-pole mast requires a minimum spacing for the anchorage points in the form of a square with sides equal to half the height of the mast. The guys may be of steel, possibly plastic-coated, or more likely of synthetic fibre. Steel guys may affect the directivity of antennas in close proximity, and if the guys corrode at their fixing points, intermodulation interference may result.

Self-supporting towers come in all heights up to many hundreds of feet and may be made of a variety of materials from steel sections to concrete.

The factors determining the type of structure to be erected are the number and types of antenna, and the site conditions.

Antenna considerations:

- Physical dimensions and space requirements.
- Weight.

- Wind loading.
- Directivity. For example, the beam width of a 6.5 GHz, 3.0 m dia. microwave dish is 0.9 degrees. A tower supporting such an antenna must twist, therefore, less than about 0.3 degrees in the highest wind likely to be experienced at the site.
- Access for riggers to antenna mounting positions.

Site conditions:

- Availability of space for tower footings, and guys and anchorages if a mast is contemplated.
- Stability and type of ground.
- Weather conditions. High winds. Accumulation of ice and snow. Build-up of ice increases wind loading of antenna – and of mast or tower – considerably. There is no point in using grid microwave antennas where icing is likely to occur.
- Aesthetic and planning consent considerations.

19.5 Installation of electronic equipment

For radio equipment, safety is one of the first priorities and in addition to the electrical considerations, physical hazards must be avoided. Adequate working space must be provided around the equipment racks. Much equipment is now wall-mounted but tall, floor-standing cabinets must be firmly bolted down to prevent toppling when units are partially withdrawn for servicing and cables need to be routed safely. Overhead cable trays are generally considered the most satisfactory method but, to avoid interaction, it is recommended that cables for the various functions be segregated to reduce cross-talk and interference. Where cabling enters a cabinet at floor level, mounting the cabinet on a hollow plinth and running the cables through the plinth avoids damage to the cables and persons tripping over them.

For low voltage installations (see classifications below) the IEE wiring regulations apply. For high voltage installations, recommendations are given in BS 6701: part 1: 1990 which also covers accommodation, lighting levels and access arrangements.

Classification of installations:

- *Extra low voltage.* Normally not exceeding 50 V AC or 120 V DC.
- *Low voltage.* Exceeding 50 V AC or 120 V DC but not exceeding 1000 V AC or 1500 V DC between conductors, or 600 V AC or 900 V DC between conductors and earth.
- *High voltage.* Exceeding 1000 V AC or 1500 V DC between conductors, or 600 V AC or 900 V DC between conductors and earth.

19.6 Earthing and protection against lightning

Recommendations for earthing are given in BS 7430: 1991. This document covers the earthing of equipment and the principles of earthing for protection against lightning. All equipment metalwork must be bonded together and connected to the electricity supply earth point, the main earth terminal. In addition a connection must be made to the earthing system provided for protection against the effects of lightning.

Antenna systems by their nature are vulnerable to lightning strikes. Nearby taller structures may reduce the risk but precautions must still be taken. The zone of protection – a cone with its apex at the tip of the protecting structure and its base on the ground forming an angle of 45° to the perpendicular – does not necessarily protect structures above 20 m high. BS 6651: 1985 is the *Code of Practice for the Protection of Structures against Lightning.* Lightning protection begins at the top of the mast or tower where, ideally, the highest item should be a finial bonded to the tower. However, a finial mounted alongside an omni-directional antenna will affect is radiation pattern. Grounded antennas are preferred and the outer conductor of each feeder cable must be bonded to the tower at the top and bottom of its vertical run. Grounding kits for the purpose are obtainable. The feeder cables must also be bonded to an earthing bar at their point of entry to the equipment building. The codes of practice should be consulted regarding the routes to be taken by earthing tape and the methods of jointing. An important point is that neither very sharp bends nor 'U' bends are created.

The legs of the tower must be bonded together with earthing tape and each leg connected to an earthing system which may consist of a buried plate or rods. Several rods may be necessary for each subsystem to attain the specified earth resistance of not more than 10 ohms. The rods should, except in rock, be driven into ground which is not made up or backfilled or likely to dry out. Where several rods are necessary to achieve the specified earth resistance these should be spaced well apart, the reduction in earth resistance being small with parallel rods closely spaced. Joints to facilitate testing of each subsystem separately must be provided either above ground or in a purpose-built pit or chamber. Figure 19.1 shows a typical earthing system.

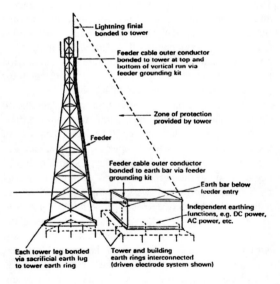

Fig. 19.1 *Typical example of good earthing practice (© Crown Copyright 1991, Radiocommunications Agency)*

The system should be tested regularly, at least annually, and the results recorded in a lightning protection system log book. The test comprises:

1. Measurement of the resistance to earth of each termination network and each earth electrode.
2. Visual inspection of all conductors, bonds and joints or their measured electrical continuity.

BS 7430 recommends the method of testing.

Other vulnerable systems are overhead cables for either the electricity supply or telephone and control purposes.

19.7 Erection of antennas

The installation of the antenna probably influences the performance of a radio system more than any other part. The heights involved render supervision of the installation difficult and, although antennas are relatively inexpensive, labour charges raise the cost of their replacement, and bad weather may delay rigging work. It is therefore important that good quality antennas are specified and that the initial installation is correct. In addition to the physical mounting arrangements, the dispositions of antennas on the structure are important. Some aspects are detailed below and MPT 1331 contains more information.

19.7.1 Directivity

The radiation pattern diagrams provided in the antenna manufacturers' catalogues are produced by an isolated antenna. They can be greatly affected by the proximity of the support structure, e.g. if a dipole is mounted close to the leg of a tower, the leg behaves as a reflector and the radiation pattern of the antenna approaches that of two-element Yagi. The less directional the antenna, the worse the effect. Omni-directional antennas should not be mounted on the side of a mast or tower if it can be avoided. If an omni-directional antenna must be so mounted it should be spaced two wavelengths from it. This is difficult to achieve at the lower frequencies considering the antenna dimensions, weight and windloading. The rear element of a Yagi antenna must be one wavelength clear of the structure. To obtain a nominally omni-directional pattern a better, albeit expensive, solution is to mount a number of antennas, phased correctly, around the tower.

All antennas should point directly away from the structure, i.e. at right angles to the side of the tower.

The proximity of metalwork also adversely affects the VSWR of an antenna, causing standing waves with their additional voltage stresses on the feeder and possibly producing interference.

19.7.2 Practical aspects of antenna installation

All installation components must be made from compatible materials, electrically dissimilar metals, e.g. steel and brass, brass and aluminium, being avoided at all costs; they will corrode and cause intermodulation. Preferred materials are aluminium, and galvanized or stainless steel. Mounting brackets must be secure – remembering that antennas and feeders vibrate – but not overtightened so as to distort and weaken antenna support booms.

Connectors of the correct impedance, and preferably with a captive centre pin, are vital. Type N connectors are available at either 50 or 70 ohms impedance, are robust and can be used at frequencies up to 10 GHz. UHF 83 series connectors, although very robust, are of imprecise impedance and, from first-hand experience of interference aggravated by a UHF 83 used to connect an antenna feeder to a receiver tail, should not be used on base station installations. Assembly instructions for a range of connectors are shown in Chapter 23. All joints must be waterproofed, preferably by first wrapping with a self-amalgamating tape for two inches either side of the joint, then a layer of PVC tape covered by a further layer of Sylglass or Denso tape. When installing feeder cables the required length should be loosely coiled and taken up the tower and paid out as it is secured, working from top to bottom. It must not be dragged from a drum on the ground. Feeders must be cleated at the intervals specified by the manufacturer. Route cables where they are easily accessible but least likely to suffer physical damage.

Health and safety

Apart from the physical risks to riggers, there are radiation hazards. Large amounts of RF power may be radiated from antenna systems and research is continuing into its effects on health. The most recent recommendations of the National Radiological Protection Board and the Department of Trade and Industry should be sought and followed. Riggers should be aware of the power levels present on a structure before climbing, and equipment switched off where levels are considered unsafe.

19.7.3 Antenna checking and fault finding

The majority of VHF and UHF antennas present a short-circuit across the antenna and to ground as a means of reducing their vulnerability to static charges. An ohmmeter applied to the lower end of the feeder will indicate a circuit through the feeder and antenna but may also indicate a short-circuit or low resistance fault, water in the feeder for instance. Most antenna system faults increase the VSWR present on the feeder cable. A good antenna, operating within its design bandwidth, exhibits a VSWR of 1:1. When connected to a length of co-axial cable this may be raised and a VSWR of 1.5:1 for the system is generally considered acceptable. However, faults can be masked by conditions occurring on the feeder, and a measurement taken with a standing wave, or forward and reflected power meter at the bottom of the feeder, may not indicate the true condition of the antenna. A measurement taken with an accurate dummy resistive load connected to the top of the feeder will prove the cable, and a measurement of the VSWR should also be taken directly at the antenna. A record of the VSWR at installation and any subsequent measurements is helpful.

19.8 Interference

There are two sources of radio interference induced by antenna systems. One occurs from strong signals radiated by either the system's own transmitter antenna or from a co-sited system operating on a close neighbouring frequency. The other source is intermodulation, 'rusty bolt effect', the mixing of two or more signals to produce the interfering frequency.

19.8.1 Antenna isolation

One solution to the problem of direct radiation is to space the antennas so that there is sufficient isolation between them. Because minimum radiation is present immediately below and above the ends of the elements of vertically polarized antennas, maximum isolation occurs when such antennas are mounted in a vertical line. Maximum radiation, and hence minimum isolation, occurs when antennas are mounted side-by-side. The degree of isolation depends on the spacing but a figure of 40–45 dB between a transmitter and a receiver antenna, and 20–25 dB between transmitting antennas, should be the target. To achieve 25 dB isolation requires a vertical separation of 0.9 wavelengths between the centres of vertically polarized dipoles; 45 dB requires a vertical separation of 3 wavelengths. The spacing for horizontal separation of vertically polarized antennas is 2.5 wavelengths for 25 dB isolation and 25 wavelengths for 45 dB. To preserve this isolation the feeder cables should also be separated over their routes.

Antenna multi-coupling

A precise method of controlling the RF isolation between mutually sited systems where the frequencies are moderately spaced is for several systems to share one antenna by using multi-coupling techniques (see Section 19.9).

19.8.2 Intermodulation

Any two or more RF signals applied to a non-linear device intermodulate, that is, they combine to form additional frequencies. Table 19.2 lists the combinations for two input frequencies. It must be remembered that the side frequencies produced by modulation of the original carriers will also be present.

19.8.3 Control of intermodulation

A clue to the source of interference will be obtained from examination of the relationship of all frequencies produced on site, and also those produced nearby. That the side frequencies produced by modulation must be

Table 19.2 *Low order intermodulation products*

2nd	$A+B$	5th	$3A+2B$	7th	$4A+3B$
	$A-B$		$3B+2A$		$4B+3A$
			$3A-2B^*$		$4A-3B^*$
3rd	$2A+B$		$3B-2A^*$		$4B-3A^*$
	$2A-B^*$		$4B+A$		$5A+2B$
	$2B+A$		$4A+B$		$5B+2A$
	$2B-A^*$		$4A-B$		$5A-2B$
			$4B-A$		$5B-2A$
4th	$2A+2B$				$6A+B$
	$2A-2B$	6th	$5A+B$		$6B+A$
	$2B+2A$		$5B+A$		$6A-B$
	$2B-2A$		$5A-B$		$6B-A$
	$3A+B$		$5B-A$		
	$3A-B$		$4B+2A$		
	$3A+A$		$4B+2B$		
	$3B-A$		$4B-2A$		
			$4B-2B$		
			$3A+3B$		
			$3A-3B$		
			$3B-3A$		

*indicates in-band products. No eighth-order products fall in-band but ninth-order in-band products are produced by:

$$5A-4B$$
$$5B-4A$$

considered was proved in an intermodulation situation where the interference was only received when a nearby band III television transmitter radiated a peak white signal.

Corroded joints in metalwork close to the site, receiver RF stages subjected to very large signals and transmitter output stages are all non-linear devices and can create intermodulation.

Regular maintenance of the metalwork around the site reduces the chances of intermodulation from this source, but it is difficult to see and eliminate all corrosion on an inspection. Should interference from this cause be suspected after elimination of the other sources, a successful method of locating the offending joint requires a receiver tuned to the interfering signal and fitted with a whip antenna. While receiving the interference, a very sharp null in signal strength occurs when the tip of the whip points directly at the source of the signal.

Receiver RF amplifiers are usually designed to cope with small signals. When a strong signal radiated from a nearby transmitter is received the stage may well be overloaded and driven into non-linearity, or even into a blocked state where the receiver is effectively dead and may take some seconds to recover after the signal is removed. In the non-linear state the receiver is in an ideal condition to create intermodulation. If this situation is suspected a method of identifying it is to install a variable RF attenuator in the antenna feed to the suspect receiver. With zero attenuation the interference will be received but, if the receiver is the cause, increasing attenuation will produce little reduction of the interference until a point is reached where the interference suddenly disappears – the receiver at this point has re-entered a linear state. If the interference reduces gradually to zero with increased attenuation, the source is elsewhere.

Most amplifiers are designed to be linear, that is, the output signal level will follow that of the input signal. However, with a sufficiently high input signal overloading occurs, the amplifier becomes non-linear and compression of the signal results (Figure 19.2).

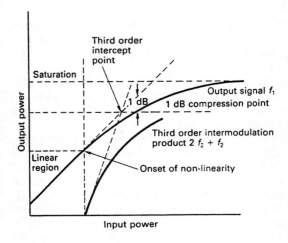

Fig. 19.2 *Production of intermodulation*

The point where 1 dB of compression occurs is a commonly referred to amplifier parameter. If the input level is increased further the gain of the amplifier is reduced until saturation is reached when the output level can no longer increase.

At the onset of non-linearity harmonics and inter-modulation, if any other frequencies are present, are produced. The strength of these rises rapidly with an increase of input, and the point where an extension of the almost linear portion of their curve crosses an extension of the linear gain line of the amplifier is the third-order intercept – usually it is only the third-order products which are considered for this purpose.

The class C operated output stages of transmitters are by design non-linear. Strong signals from a neigh-bouring transmitter or antenna applied via the antenna feed will mix with the transmitted frequency to create intermodulation products. Increased isolation between antennas and feeders may be the simplest remedy. Alternatively, a circulator or filters may be connected to the output of the transmitter, and the possibility of direct radiation from transmitters must not be ignored.

A circulator is a uni-directional device with either three or four ports as shown in Figure 19.3(a). A signal entering at port 1 will leave at port 2 but not at ports 3 or 4. Similarly, a signal entering at port 2 will leave at port 3 but not ports 1 or 4. Circulators are used for combining the ouputs of transmitters for application to a single antenna but if a three-port circulator is connected as in Figure 19.3(b) it forms an isolator. Typical isolation over a 1.5% bandwidth is 20–30 dB with an insertion loss of 0.7 dB.

19.9 Antenna multi-coupling

The sharing of antennas and feeders between a num-ber of co-sited services is not only a precise method of controlling frequency isolation, it is an economic solution to the problems of antenna and tower man-agement. An antenna may be shared between a number of receivers or transmitters, or shared simultaneously by both receivers and transmitters. Receiver sharing requires a splitter, i.e. a filter, to separate the frequen-cies for each receiver and an amplifier to make up the

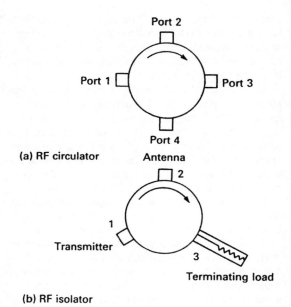

(a) RF circulator

(b) RF isolator

Fig. 19.3 *RF circulator and isolator*

filter losses. Transmitters can share an antenna through circulator/isolators and filters. To obtain the required selectivity, stability and power handling capability the filters are often solidly constructed cavity resonators with an insertion loss of 1 dB and a bandwidth of 0.1% to the −20 dB points. Figure 19.4 shows the layout of a transmitter/receiver sharing system.

19.10 Emergency power supplies

The installation of equipment to provide power for all the equipment on a site during a failure of the public supplies has declined in recent years. But although lengthy power failures seem to be a thing of the past, they could recur, and electricity supplies do still fail for short periods. To provide a complete supply for large station necessitates a standby generator. On smaller stations batteries may be capable of supplying the demands of the electronic equipment and emergency

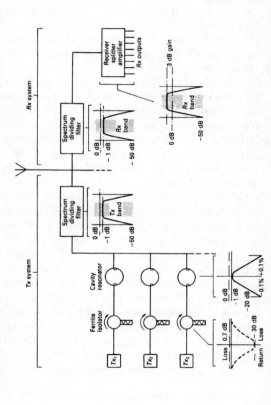

Fig. 19.4 Typical sub-band transmitter/receiver sharing (© Crown Copyright 1991, Radiocommunications Agency)

lighting for one or two hours. The trend on multi-user shared sites appears to be for the individual user to provide the emergency power, derived from batteries, for his own equipment. Factors to consider are:

1. Standby motor generators are expensive. They need housing, ventilation and fuel storage. They take time to start up, 45 seconds being typical.
2. Uninterruptible power supplies (UPS). Where no interruptions can be tolerated the most efficient supply is provided by batteries alone but the equipment must be designed to operate directly from a DC, low voltage supply.

   To supply power at 220/240 V AC an inverter is the accepted method. Double inverters permanently charge a battery from the AC supply and the battery drives an oscillator to reproduce the AC supply voltage, a very inefficient process. Also, should any item of the inverter break down the supply is lost. This is overcome in some inverters by a switch which transfers the load to the public supply when a failure of the inverter occurs. Single inverters charge a battery and supply the equipment directly from the public supply until a failure occurs. When this happens the supply changes over to the battery and oscillator. Single inverters are more efficient but there is a short break in supply while the change-over occurs. Adequate stabilization and filtering to avoid disturbances on the supply must be provided. One possible hazard with all inverter supplies is that where more than one inverter is used, the outputs of the inverters are unlikely to be in phase and a 480 V difference could exist between the supplies for two items of equipment.

For all battery-operated supplies correct charging of the battery is crucial. Nickel–cadmium batteries require constant-current charging with a fall to the trickle rate when fully charged and sophisticated circuitry is needed to identify the precise point of full charge (see Section 21.3.2).

Lead acid batteries produce hydrogen if overcharged so their storage must be designed so that no danger, either to personnel or of explosion, exists. For reliability, natural ventilation is preferred to forced. Recommendations for the accommodation of lead acid batteries other than sealed are given in BS 6133: 1985. Batteries which gas also need more frequent maintenance. Because of the reduction in performance of batteries at low temperatures the ambient temperature of the battery storage room should not fall below 4°C, and to reduce water evaporation lead acid batteries should not be run continuously above 40°C.

The period of time over which a lead acid battery must be fully recharged determines the capacity of the battery more than the period of time over which it must supply power. Because of the need to reduce the charge rate – to avoid gassing – to a very low level as the battery approaches full charge, the last 15% of capacity takes a very long time to acquire (see Section 21.3).

19.11 Approval and certification

No base radio station may be operated prior to inspection of the installation by the Radio Investigation Service of the Radiocommunications Agency and the issue of their certificate of approval. The installation may not subsequently be modified without re-approval and certification.

References

Belcher R. *et al.* (1989). *Newnes Mobile Radio Servicing Handbook*. Butterworth-Heinemann, Oxford.

BS 6651: 1985. *Code of Practice for the Protection of Structures against Lightning*.

BS 6701: Part 1:1990. *Code of Practice for installation of apparatus intended for connection to certain telecommunications systems*.

BS 7430: 1991. *Code of Practice for Earthing. Earthing of Telecommunications Installations*. International Telecommunications Union.

MPT 1331. *Code of Practice For Radio Site Engineering*.

MPT 1351. *Code of Practice For Repeater Operation at Communal Sites*.

MPT 1367. *Guidance Note to Legal Requirements covering the Installation and use of Radio Apparatus and other Apparatus Generating Radio Frequency Emissions.*

MPT 1368. *Code of Practice for the inspection of a Land Mobile Radio System for conformity with the Wireless Telegraphy Act Licence and Performance Specification.*

20

INSTRUMENTATION

20.1 Accuracy, resolution and stability

All measurements are subject to error and two instruments applied to the same piece of equipment under test may give a different answer. Tolerances must therefore be accepted. The errors arise from the following sources:

1. Human error, e.g. precision in reading a scale, use of incorrect instrument or range setting for the purpose.
2. The accuracy to which the instrument is able to display the result of a measurement or, in the case of a generator, the frequency or output level.
3. Accuracy of calibration.
4. Tolerances in the components used in the construction of the instrument. Variations in the load applied to the instrument.
5. Variations caused by long-term drift in the values of components.
6. Variations due to temperature and supply voltage fluctuations, and the warm-up time required by some instruments.
7. An effect on the circuit under test by the connection of the instrument.

There is an important difference between the accuracy and the resolution of an instrument. The accuracy is a statement of the maximum errors which may occur from the causes in groups 3 to 6 above. In instrument specifications stability, groups 5 and 6, is usually quoted separately.

The accuracy of analogue measuring instruments is normally quoted as a percentage of full scale deflection (FSD). This is the accuracy of the instrument movement and components plus the scale calibration. The scale graduations, though, may not permit the user to determine the reading to the accuracy of the instrument perhaps because the graduations are cramped or parallax reading error occurs. These factors decide the resolution or precision of reading.

The accuracy of instruments with digital displays is usually quoted as a percentage of the reading plus or minus one count or one digit. While digital instruments are generally more accurate than their analogue counterparts, the fact that the least significant figure may be in error affects the resolution. Figure 20.1 shows how this can arise. Most digital instruments use a gating process to switch the input to the measuring circuitry for the appropriate period of time. The gating time itself may vary and affect the accuracy, but even with a perfectly accurate and stable gating time, the phase of the input signal at the time of switching affects the number of pulses passing through the gate and thus the resolution.

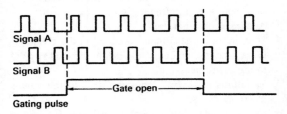

Fig. 20.1 *Gating error introduced by signal phasing*

Performance figures taken from a number of manufacturer's catalogues are listed below. They show only the more important features of the specifications and are typical of those for high quality instruments used in radio work.

20.2 Audio instruments

20.2.1 Output power meters

Those for radio applications measure RMS audio power and are usually calibrated in watts and dBW or dBm. They usually contain a dummy load resistor of adjustable impedance as a substitute for the receiver loudspeaker when in use. Performance:

Range of measurement	300 µW to 30 W
Frequency range	20 Hz to 35 kHz
Input impedance	Switchable in steps between 2.5 Ω and 20 kΩ
Input configuration	Balanced or unbalanced
Accuracy	±7% FSD ±10 µW, from 100 Hz to 10 kHz and from 5 to 35°C

20.2.2 Distortion factor and Sinad meters

These may be separate instruments or combined, and may also be incorporated within an audio output power meter. Both operate on the same principle. An audio input amplifier, with automatic gain control to give a fixed reference level at its output, applies the output from the unit under test, first to a notch filter to remove the test frequency of 1000 Hz, then to an AC voltmeter. The voltmeter is calibrated in percentage distortion or dB Sinad.

In testing a receiver Sinad ratio, the instrument is connected across the loudspeaker or equivalent dummy load. An RF signal generator connected to the antenna socket is adjusted to apply a moderately high level signal modulated with a 1 kHz tone – which must be pure and match the notch filter in the Sinad meter – at 60% of the system peak modulation. To be correct, the receiver volume control is adjusted to give the rated audio output but if only Sinad sensitivity is being measured a lower level is acceptable. The Sinad meter will now display the combined level of all the frequencies except 1000 Hz present in the receiver's output. These are the noise and distortion products. With the high level input applied these will be low and the meter deflection will be small indicating a high Sinad ratio. The RF input level is now reduced until the meter reads 12 dB (the standard reference) and the RF level, µV or dBm, is the Sinad sensitivity for the receiver.

The oscillator to produce the test tone may be included in the meter and filters may be provided to comply with various weightings.

Distortion factor meter performance:

Distortion range	0.01 to 30% in six ranges
Fundamental tone frequencies	400 Hz ±5% and 1 kHz ±5%
Bandwidth to upper −3 dB point	10 kHz
Input level	0.1 V RMS to 100 V RMS
Input impedance	100 kHz (distortion and Sinad meters are usually connected across the loudspeaker as a voltmeter)
Measurement accuracy for	±2% of reading, ±3% FSD each range

Sinad meter performance:

Sinad scale	0 to 24 dB, linear
Fundamental null frequency	1 kHz ±15 Hz
Rejection of null frequency	>60 dB
Input level	0.01 V to 3 V automatic
Input impedance	100 k Ω
Filter	Flat or CCITT P53A (psophometric). C message weighting – optional
Measurement accuracy	±1 dB

Internal test tone generator:

Frequency	1 kHz ±1 Hz
Level	0–3 V RMS
Distortion	<0.05%
Output impedance	600 Ω

20.2.3 Audio signal generators

The performance specification details listed are for a general purpose instrument. The instrument also provides an output for synchronizing to an external standard and a TTL compatible output. A high power version capable of delivering 3 watts into a 3 Ω load in the audio band is available.

Frequency range	10 Hz to 1 MHz in 5 decades
Scale accuracy	Within ±3%. Typically ±1% up to 100 kHz. To within ±4% 100 kHz to 1 MHz
Harmonic distortion (at 12 V pk–pk into 600 Ω)	Less than 0.2% 100 Hz to 20 kHz (typically 0.02% at 1.2 kHz) Less than 0.5% 10 Hz to 100 Hz and 20 kHz to 60 kHz Less than 0.9% 60 kHz to 100 kHz Less than 5% 100 kHz to 1 MHz
Rise and fall times (square waves into 600 Ω)	Less than 200 ns. Typically 160 ns
Output level into 600 Ω	From approx. 1 mV to 12 V peak to peak
Frequency response (output flatness vs. frequency)	Less than 2% variation 10 Hz to 100 kHz Less than 10% variation 100 kHz to 1 MHz
Output impedance	60 Ω approximately
Load impedance	600 Ω recommended. Will drive lower

20.3 Radio frequency instruments

20.3.1 RF power meters

Direct reading RF power meters either contain a non-reactive load or use an external load and may be calibrated in watts or dBm. RF calorimeters convert the RF energy into heat and measure the temperature of the heated element. At low powers, 'dry' calorimeters are used but their long thermal time constant inhibits their use at high power levels. To measure high powers 'flow' calorimeters, where a fluid flows around a closed system and the output temperature of the fluid is measured, are used. Power can be determined from:

$$P = F(T_{out} - T_{in})c(T)$$

where

P = power
F = mass flow rate of the fluid
$c(T)$ = the fluid's specific heat
T_{in} = temperature of the fluid entering the load
T_{out} = temperature of the fluid after being heated by the load

Thruline type instruments require an external load which may be either an antenna system or a load resistor. This type of meter reads forward and reflected power and, in some instruments, VSWR directly. Performance:

Direct reading absorption wattmeter:

Frequency range	2 MHz to 2.3 GHz depending on plug-in element selected
Power range	Selectable by choice of element between 1 W to 1000 W (versions up to 2.5 kW available)
VSWR	1.1 max. DC to 1 GHz
	1.25 max. 1 GHz to 2.3 GHz
Input impedance	50 Ω
Accuracy	±5% FSD to 1 GHz
	±8% FSD 1 GHz to 2.3 GHz

Flow calorimeter:

Frequency range	DC to 2.5 GHz
Power range	10 to 200 W
Input VSWR	DC to 1 GHz; 1.10 max. 1 to 2.5 GHz; 1.25 max.
Input impedance	50 Ω
Accuracy	10 W to 25 W; ±3% of reading
	25 W to 200 W; ±1.25% of reading
Response time	Less than 1 minute maximum to reach 97% of final reading

Portable Thruline instrument:

Frequency range	0.45 MHz to 2.3 GHz depending on choice of plug-in element
Power range	Forward and reflected power; 100 mW to 10 kW RMS depending on element used
Insertion VSWR	1.05:1 max. to 1.0 GHz
Accuracy	±5% of FSD

20.3.2 RF signal generators

The range of instruments designed for use on specific systems, e.g. cellular and digitally modulated, is so wide that manufacturers' catalogues must be consulted for each application. In addition to the accuracy of the carrier frequency and output level, RF leakage, spectral purity and modulation noise levels are important. The output level may be calibrated in μ V, dBV or dBm, and may refer to either an unterminated instrument (p.d.) or terminated in a load equal to the output impedance of the generator (e.m.f.). If the instrument is calibrated in p.d., the output voltage must be halved when the instrument is terminated in an equal impedance. The performance figures quoted below are typical for a current standard generator:

Frequency range	10 kHz to 5.4 GHz in switchable ranges
RF carrier accuracy	Resolution: 0.1 Hz
	Temperature stability: better than ± 5 in 10^8 over the operating range of 0 to 50°C
	Ageing rate: better than 2 in 10^7 per year
	Warm-up time: within 2 in 10^7 of final frequency within 10 minutes from switch on at 20°C ambient
RF output	Range: -144 dBm to $+13$ dBm
	Resolution: 0.1 dB
	Impedance: 50 Ω
	Accuracy at 22°C $\pm 5\%$
	VSWR: for outputs <0 dBm, <1.25:1 to 2.2 GHz, <1.4:1 to 2.7 GHz, <1.5:1 to 5.4 GHz.
	Output protection: reverse power of 50 W from a source VSWR of up to 5:1 (important when testing transmitter/receivers)
Spectral purity (at levels up to $+7$ dBm)	Harmonics: better than -30 dBc to 1 GHz, better than -27 dBc above 1 GHz

	Subharmonics: better than −90 dBc to 1.35 GHz, better than −40 dBc to 2.3 GHz, better than −30 dBc to 2.7 GHz
	Non-harmonics: better than −70 dBc at offsets from the carrier of 3 kHz or greater
Noise	Residual FM (FM off): <7 Hz RMS deviation in a 300 Hz to 3.4 kHz unweighted bandwidth at 470 MHz
	SSB phase noise: < −116 dBc/Hz (typically −122 dBc/Hz) at an offset of 20 kHz from a carrier frequency of 470 MHz
	FM on AM: typically <100 Hz for 30% AM depth at a modulation frequency of 1 kHz and a carrier frequency of 500 MHz
	Phase modulation on AM: Typically <0.1 radians at a carrier frequency of 500 MHz for 30% AM depth for modulation rates up to 10 kHz
RF leakage	Less than 0.5 µV p.d. at the carrier frequency in a two turn 25 mm loop 25 mm or more from any part of the case
Modulation modes	Single: FM, wideband FM, phase, AM or optional pulse
	Dual: two independent channels of differing modulation type (e.g. AM with FM)
	Composite: two independent channels of the same modulation type (e.g. FM1 with FM2)
	Dual composite: a combination of dual and composite modes providing four independent channels

	(e.g. AM1 with AM2 and FM1 and FM2)
Frequency modulation	Deviation: peak deviation from 0 to 1 MHz for carrier frequencies up to 21.09375 MHz, peak deviation from 0 to 1% of carrier frequency above 21.09375 MHz
	Displayed resolution: 1 Hz or 1 least significant digit, whichever is the greater
	Accuracy at 1 kHz: $\pm5\%$ of indication ±10 Hz excluding residual FM Bandwidth (1 dB): DC to 300 kHz (DC coupled), 10 Hz to 300 kHz (AC coupled)
	Distortion: using external modulation without automatic level control (ALC), <3% at maximum deviation for modulation frequencies up to 20 kHz
	Less than 0.3% at 10% of max. deviation for modulation frequencies up to 20 kHz
Wideband FM	Deviation and accuracy: as for FM Bandwidth (3 dB): typically 10 MHz (DC or AC coupled)
Phase modulation	Deviation: 0 to 10 radians
	Resolution: 0.01 radians
	Accuracy at 1 kHz: $\pm5\%$ of indicated deviation excluding residual phase modulation Bandwidth (3 dB): 100 Hz to 10 kHz Distortion: <3% at maximum deviation at 1 kHz modulation rate
Amplitude modulation	For carrier frequencies up to 1 GHz Range: 0 to 99.9%
	Resolution: 0.1%
	Accuracy: $\pm4\%$ of setting $\pm1\%$ Bandwidth (1 dB): with modulation ALC off,

DC to 30 kHz (DC coupled),
10 Hz to 30 kHz (AC coupled)
Distortion: for a modulation
rate of 1 kHz, <1% THD for
depths up to 30%, <3% THD
for depths up to 80%

The 50 ohm output impedance of many signal generators matches the input impedance of most VHF and
UHF receivers directly. To simulate the impedance of
antennas at HF and below a dummy antenna is usually inserted between the generator and the receiver
under test. Figure 20.2 shows the circuit of a standard
dummy antenna.

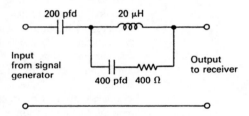

Fig. 20.2 *Standard dummy antenna*

20.3.3 Frequency counters

Instruments are necessary to measure not only the
precise frequencies of transmitters and receiver local
oscillators but also those of low frequency signalling
tones. Consequently they have a wide operating range.

The output of a transmitter must never be connected
directly into a counter. The signal should either be
obtained off-air or through a 'sniffer', appropriate
for the frequency range, which siphons off a fragment
of the transmitter output. It is also essential to ensure
that the frequency displayed is the fundamental and
not a harmonic or some other spurious frequency.
Performance:

Frequency range	10 Hz to 1.2 GHz (other instruments to 26 GHz)
Resolution	From 0.1 Hz to 1 kHz depending on range selected

Accuracy	Internal oscillator: ±1 ppm from 0°C to +55°C
	Ageing: ±1 ppm per year
Gate-time	0.1 s, 1.0 s, 10 s

20.3.4 Modulation meters

The remarks in the above section relating to the connection of a transmitter and the selection of the fundamental frequency apply equally to modulation meters. Performance:

Frequency range	1.5 Mhz to 2.0 GHz with fully automatic tuning
Sensitivity	From −30 dBm (7 mV) to −10 dBm (70 mV) depending on RF range with fully automatic level setting Absolute safe maximum input: +30 dBm (7 V) intermittent
Input impedance	50 Ω
FM measurement	
Maximum deviation	Carrier frequency <5 MHz, 20 kHz (mod. frequency 50 Hz to 10 kHz) Carrier frequency <8 MHz, 50 kHz (mod. frequency 50 Hz to 15 kHz) Carrier frequency >8 MHz, 100 kHz (mod. frequency 25 Hz to 30 kHz)
Deviation range	Automatic, 0–10 kHz and 10–100 kHz
Resolution	10 Hz, 0–9.99 kHz deviation 100 Hz, 10–99 kHz deviation
Accuracy	±2% of reading ±1 changing digit at 1 kHz mod. frequency for deviations above 3 kHz ±2%–3% of reading ±1 changing digit at 1 kHz mod. frequency for deviations less than 3 kHz
Residual FM	Less than 100 Hz at 500 MHz decreasing with frequency (300 Hz-3 kHz filter selected)
AM rejection	Additional deviation error <100 Hz at 50% AM (mod. frequency <3 kHz)

AM measurement

Maximum modulation	0.1%–99.8%. Range selection fully automatic
Modulation frequency	50 Hz to 15 kHz for carrier frequencies below 6 MHz 25 Hz to 30 kHz for carrier frequencies above 6 MHz
Accuracy	±3% of reading ±1 changing digit at 1 kHz mod. frequency for depth above 8% and below 92%
Residual AM	Less than 0.15% with signal input greater than −10 dBm (300 Hz-3 kHz filter selected)

PM measurement

Maximum deviation	Carrier frequency <5 MHz, 1 radian (mod. frequency 50 Hz to 10 kHz) Carrier frequency <8 MHz, 2.5 radians (mod. frequency 50 Hz to 15 kHz) Carrier frequency >8 MHz, 5 radians (mod. frequency 25 Hz to 30 kHz)
Accuracy	±3% of reading ±1 changing digit at 1 kHz mod. frequency for deviations above 0.15 radians ±2–3% of reading at 1 kHz mod. frequency for deviations less than 0.15 radians
Residual PM	As FM measurement
Demodulation modes	Peak positive (P+) Peak negative (P−) Peak average ((P−) + (P+))/2
Filter bandwidths (3 dB points)	300 Hz-3 kHz 50 Hz-15 kHz 50 Hz-30 kHz

20.3.5 Spectrum analysers

Today's tight regulation of spectrum usage makes a spectrum analyser an essential tool for the precise alignment of radio equipment and for the measurement of modulation products and the noise content of signals.

A spectrum analyser is essentially a superheterodyne receiver with an adjustable IF bandwidth which sweeps across a portion of the frequency spectrum in synchronism with the horizontal trace of a cathode ray tube display. A signal at any frequency within the swept band will, while it remains within the IF passband, appear as a vertical displacement of the display trace. The design of the filter is crucial but must be a compromise. An analyser using multiple filters permits fast measurement but low resolution; a single filter offers high resolution but slower response time. Also, the shape factor of a wide filter inhibits the display of low level signals close to the centre frequency of the filter and a narrow filter, while permitting the display of these signals, may fail to display transients. The bandwidth determines the sweep speed, narrower bandwidths requiring slower speeds.

The possibility exists of spurious signals, e.g. intermodulation, being produced within the analyser with large input signals. A method of reducing the possibility is to insert a circulator and a notch filter between the unit under test and the analyser input to reduce the fundamental frequency amplitude. The circulator must be installed between the filter and the unit under test to absorb the reverse power produced by the impedance variations of the filter.

Although spectrum analysers operating well into the microwave region are available, only the more important features of a VHF instrument are detailed below:

Measurement range	100 Hz to 400 MHz in 1 Hz steps
Frequency span (of display)	10 Hz/div. to 20 MHz/div.
Display (vertical range)	From 5 dB-100 dB full scale in 5 ranges
Display (fidelity)	dB/div. <0.3% departure from true logarithmic relationship anywhere over the top 80 dB of display, typically <0.05 dB per dB. Linearity, better than ±2% FSD
Accuracy (of internal	Stability: better than ±2 parts in

frequency standard)	10^6 from 0 to 50°C
	Ageing: better than 1 part in 10^6/ year
Resolution	Twelve filters with 3 dB bandwidths of 3 Hz to 1 MHz in a 1, 3, 10 sequence
Amplitude measurement range	From +27 dBm down to the noise floor. Overload protected to +47 dBm (50 W)

Many instruments are programmable and suitable for incorporation in automatic testing (ATE) systems, and instruments with similar features to those described are also offered in a combination test set form.

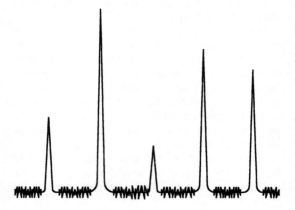

Fig. 20.3 *Amplitude-vs-frequency spectrum analyser display*

20.3.6 Network analysers

A network analysers examines *incident, reflected* and *transmitted* signals through a circuit or device, and displays the magnitude and phase of these signals. A spectrum analyser, on the other hand, measures only one channel, and displays magnitude and frequency.

Figure 20.4 shows an optical analogy to the netwok analyser. A 'device' of different optical density than ambient is in the path of an incident ray of ˏlight (R). When the light hits the surface, part of it is reflected

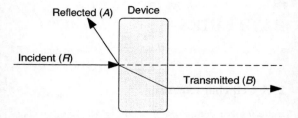

Fig. 20.4 *Optical analogy to network analyser*

(A) and part is transmitted through the 'devices' (even though refracted a bit).

A network analyser consists of a three-channel RF receiver and a display. The incident signal is considered the *reference signal*, so is designated the R-channel. The other two channels receive the reflected signal on the A-channel, and the transmitted signal on the B-channel.

The uses of scaler and vector network analysers differ from the uses of spectrum analysers. The spectrum analyser measures external signals of unknown frequency and modulation type. Even when a tracking generator is added, to allow the spectrum analyser to perform stimulus-response tests, the spectrum analyser cannot do the job of the network analyser. The network analyser, by contrast, contains a known signal source, and is capable of sweeping a range of frequencies and power output levels. It can also perform ratio measurements.

References

Belcher, R. *et al.* (1989). *Newnes Mobile Radio Servicing Handbook*. Butterworth-Heinemann, Oxford

Bird Electronic Corp., Catalog GC-92.

Edis, E. A. and Varrall, J. E. (1992) *Newnes Telecommunications Pocket Book*. Butterworth-Heinemann, Oxford.

Farnell Instruments Ltd, Catalogue, 1993.

Marconi Instruments Ltd, Test and Measurement Instrument Systems 1993/94.

RTT Systems Ltd, Catalogue, 1993.

21
BATTERIES

21.1 Cell characteristics

Batteries are composed of cells which exhibit characteristics peculiar to their chemical constituents and construction. Common features, important to users, are their ability to store energy within a small space and with the least weight, and to release it at an adequate rate for the purpose under consideration.

21.1.1 Capacity

The amount of energy a battery can store is measured in ampere hours (Ah) at a specified discharge rate. For large cells this is usually the 10 hour rate but American practice, which is now almost universal, at least for smaller cells, is to use the 1 hour rate. It is important, therefore, to be certain which rate is referred to. The capacity reduces as the rate of discharge increases. Thus a battery of 60 ampere hours capacity at the 10 hour rate will provide 6 amps for ten hours before reaching the point at which it is considered to be discharged. If a current of 12 amps is taken, the battery will become discharged in less than 5 hours, and if the current is 3 amps, it will last longer than 20 hours. The rate of discharge is often referred to in terms of the C-rate which may be expressed in several ways. 1C, C or C_1 are numercially the same as the rated capacity, e.g. a 500 mA Nicad cell supplying 500 mA, which may be expressed as C, 1C or C_1 continuously will be discharged in approximately 1 hour. If the cell supplies current at a reduced rate, 0.5 C, 0.5 C_1, or C/2, i.e. 250 mA, it will last approximately 125 minutes. The subscript, e.g. C_5, indicates the hourly discharge rate.

The terminal voltage at which a cell is considered discharged also varies with the discharge rate; a lead acid cell discharged at 1C is considered to be discharged when the terminal voltage falls to 1.75 V. At C_{10} the cell is considered to be discharged at 1.85 V.

The capacity of a battery or cell may also be specified at a given ambient temperature, usually 20°C. Lower temperatures reduce the effective capacity and maximum current off-take, higher temperatures increase them slightly.

For radio use battery duration may be quoted in terms of standby, receive and talk time. The duty cycle obviously varies from user to user but a useful standard for a radio-telephone is 90% standby, 5% receive and 5% transmit. On an open channel PMR system, 80%/15%/5% is more typical. Measuring the current drains during these activities enables the battery requirements to be calculated.

21.1.2 Internal resistance

The maximum instantaneous current which a battery can deliver is determined by its internal resistance. In this respect a battery behaves like any other generator (see Section 3.2), where increasing load currents produce an increasing voltage drop across the internal resistance. The internal resistance of a battery is seldom specified, but for a battery in good condition it is extremely low (one quoted figure is $15\,m\Omega$ for a fully charged 500 mAh Nicad cell) and the voltage drop in the connecting leads will govern the maximum withdrawable current. Equally, the resistance of the meter used will affect the measurement of charge or discharge currents. If using an analogue meter, the older low resistance types are preferable (I keep a model 40 Avometer, $0.03\ \Omega$ int. res. on 12 amp range, for the purpose). Low current-rating fuses also present a resistance higher than that of a battery ($4\ \Omega$ measured for a 250 mA fuse, $0.5\ \Omega$ for a 1 A) and, probably, the connecting leads.

21.2.3 Power:weight and volume ratios

It is the battery that now limits the size to which radio equipment can be reduced. Recent developments have increased the power to weight and power to volume ratios which are possible. Typical ratios are referred to in the sections dealing with individual battery types.

21.1.4 Recharging conditions

The initial charge rate is the current flowing through a discharged battery to replace the charge in a specified time. Unless supplied from a constant current charger the current will fall as the battery voltage rises, but as full charge is approached the charge rate is usually reduced to a trickle or finishing charge rate.

Trickle charging maintains the cells in a fully charged condition by passing a very small current through them sufficient merely to replace any self-discharge losses through leakage.

Finishing charge is a rate to which the charging current is reduced when a battery reaches about 85% of its full capacity. It is a rate at which gassing is unlikely to occur.

Float charging maintains the cell voltage at its nominal while it is supplying continuous and variable loads.

21.2 Non-rechargeable, primary batteries

While rechargeable batteries are the obvious choice for use in equipment which is in use continuously, disposable types have economic and logistic advantages for some applications – the batteries are more expensive long term but no charger is needed. Also primary batteries have up to 4 times the available capacity of their nickel cadmium equivalents with less weight. Details of disposable batteries with some standard nickel-cadmium rechargeable equivalents are given in Table 21.1.

Lithium primary batteries are now available and have the following qualities:

1. A high cell voltage of 3.6 V on load. The voltage is constant – after an initial fall when load is first applied – until discharged when a rapid fall occurs.
2. An operating temperature range of 55°C to +75°C is possible.
3. Energy densities up to 630 mWh/g. A standard AA size cell has a power/weight ratio of 340 mWh/g.
4. Long shelf life. Ten years at room temperature for a 10% fall in capacity is envisaged.

Table 21.1 *Cells and batteries*

USA size	Nominal voltage (V)	Type†	IEC equivalent	Length (or diameter)	Width	Height	Contacts	Current (mA)	Weight (g)
Zinc carbon									
N	1.5	D23	R1	12	–	00.1	Cap and base	1–5	7
AAA		HP16	R03	10.5	–	45	Cap and base	0–10 000	8.5
AA		HP7	R6	14.5	–	50.5	Cap and base	0–75	16.5
AA		C7	R6	14.5	–	50.5	Cap and base	0–75	16.5
C		SP11	R14	26.2	–	50	Cap and base	20–60	45
C		HP11	R14	26.2	–	50	Cap and base	0–1000	45
C		C11	R14	26.2	–	50	Cap and base	0–5	45
D		SP2	R20	34.2	–	61.8	Cap and base	25–100	90
D		HP2	R20	34.2	–	61.8	Cap and base	0–2000	90
	4.5	AD28	3R25	101.6	34.9	105	Socket	30–300	453.6
		1289	3R12	62	22	67	Flat springs	0–300	113
	6.0	PP8	4-F100-4	65.1	51.5	200.8	Press studs	20–151	1100

(continued overleaf)

Table 21.1 (continued)

USA size	Nominal voltage (V)	Type†	IEC equivalent	Maximum dimensions (mm)			Contacts	Current (mA)	Weight (g)
				Length (or diameter)	Width	Height			
		P1996	4-R25	67	67	102	Spiral springs	30–300	581
		991		135.7	72.2	125.4	Two screws	30–500	1470
		PP3-P	6-F22	26.5	17.5	48.5	Press studs	0–50	39
	9.0	PP3-C	6-F22	26.5	17.5	48.5	Press studs	0–50	39
		PP3	6-F22	26.5	17.5	48.5	Press studs	0–10	38
		PP4	6-F20	26.5	–	50	Press studs	0–10	51
		PP6	6-F50-2	36	34.5	70	Press studs	2.5–15	142
		PP7	6-F90	46	45	61.9	Press studs	5–20	198
		PP9	6-F100	66	52	81	Press studs	5–50	425
		PP10	6-F100-3	66	52	225	Socket	15–150	1250
	15.0	B154	10-F15	16	15	35	End contacts	0.1–0.5	14.2
		B121	10-F20	27	16	37	End contacts	0.1–0.1	21
	22.5	B155	15-F15	16	15	51	End contacts	0.1–0.5	20
		B122	15-F20	27	16	51	End contacts	0.1–1.0	32

Manganese alkaline	ED	1.5	MN1300*	LR20	34.2	—	61.5	Cap and base	10.00†	125
	C		MN1400*	LR14	26.2	—	50	Cap and base	5.50†	65
	AA		MN1500*	LR6	14.5	—	50.5	Cap and base	1.80†	23
	AAA		MN2400*	LR03	10.5	—	44.5	Cap and base	0.80†	13
	N		MN9100*	LR1	12	—	30.2	Cap and base	0.65†	9.6
Mercuric oxide		1.35/1.4	RM675H	NR07	11.6	—	5.4	Cap and base (button)	0.21†	2.6
			RM625N	MR9	15.6	—	6.2	Cap and base (button)	0.25†	4.3
			RM575H	NR08	11.6	—	3.5	Cap and base (button)	0.12†	1.4
			RM1H	NR50	16.4	—	16.8	Cap and base (button)	1.00†	12.0
Silver oxide		1.5	10L14	5R44	11.56	—	5.33	Cap and base (button)	0.13†	2.2
			10L124	5R43	11.56	—	4.19	Cap and base (button)	0.13†	1.7
			10L123	5R48	7.75	—	5.33	Cap and base (button)	0.08†	1.0
			10L125	5R41	7.75	—	3.58	Cap and base (button)	0.04†	0.8
Nickel cadmium	AA	1.25	NC828	—				Button	0.28†	16.5
	C		NCC50	—	See HP7			Button	0.60†	30.0
			NCC200	—	See HP11			Button	2.00†	78.8

(continued overleaf)

Table 21.1 (continued)

USA size	Nominal voltage (V)	Type‡	IEC equivalent	Maximum dimensions (mm)			Contacts	Current (mA)	Weight (g)
				Length (or diameter)	Width	Height			
D		NCC400	–	See HP2			Button	4.00†	170.0
	10.0	NC828/8	–				Button stack	0.28†	125.0
	12.0	10/2250K	–				Button stack	0.225†	135.0
	9.0	TR7/8	(DEAC)	See PP3			Press studs	0.07†	45.0
AA	1.25	501RS	(DEAC)	See HP7			Press studs	0.50†	30.0
C		RS1.8	(DEAC)	See HP11			Press studs	1.80†	65.0
D		RS4	(DEAC)	See HP2			Press studs	4.00†	150.0

† Capacity in ampere hours.
‡ BEREC types unless otherwise indicated.
* Also Duracell (Mallory).

They have a disadvantage in that fire, explosion or severe burns may result if the batteries are mistreated. They must not be recharged, disassembled, heated above 100°C, incinerated or the contents exposed to water.

One use of lithium batteries is as a power source for memories but protection against charging is needed. Figure 21.1 shows a typical circuit.

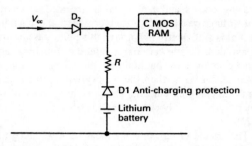

Fig. 21.1 *Typical lithium battery application*

21.3 Rechargeable batteries

21.3.1 Lead acid batteries

Lead acid batteries, whether of free electrolyte or low-maintenance sealed construction, have a nominal terminal voltage on load of 2.0 V per cell. This voltage falls on load in a gradual curve, shown in Figure 21.2, until the discharged voltage of between 1.75 and 1.85 volts per cell is reached.

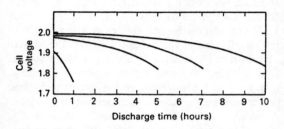

Fig. 21.2 *Typical lead acid cell discharge characteristics*

A discharged battery will, because of its inefficiencies, require a recharge equal to the amperes × hours discharged +11%, e.g. a cell discharged at 5 amps for 10 hours will require a recharge of 55.5 ampere hours with a constant current at the 10 hour rate. Because of the reducing current as full charge is approached a recharge time of 1.4 to 1.5 times the capacity to be restored is more practical. The final on-charge cell voltage can increase to approximately 2.7 volts. Gassing occurs and hydrogen is liberated when the cell voltage reaches 23 V but provided the charging current is sufficiently low above this point gassing will be avoided. This lower charge rate, the 'finishing rate' can be applied by maintaining the charging voltage at about 2.4 volts when the battery will automatically limit the charging current. The specific gravity of the electrolyte in a fully charged cell is between 1.205 and 1.215.

The trickle charge current must be low enough to avoid gassing. A current of 7% of the 10 hour capacity is typical.

Float charging should maintain a cell voltage of approximately 2.25 volts.

The power:weight ratio of lead acid batteries is poor, a small (4 Ah), 6.0 V, sealed lead acid battery having a power:weight ratio of 26 mWh/g.

21.3.2 Nicked cadmium (Nicad) batteries

At present the Nicad is probably the most commonly used rechargeable battery for portable applications. A standard size AA cell has a power:weight ratio of 27 mWh/g. The on-load Nicad cell voltage, after an initial fall from the on-charge voltage of between 1.3 and 1.4 V, remains substantially constant at about 1.2 V until the discharged voltage of 0.9 to 1.1 V is reached. Thereafter the voltage falls rapidly. This is illustrated in Figure 21.3. While the constant voltage is ideal during discharge it poses a problem in that it is difficult to reliably measure the intermediate state of charge which created difficulties with recharging. A Nicad battery which is repeatedly partially discharged and then recharged may, after many cycles, behave as though it were fully discharged when the repeated recharge condition is reached (the memory effect) and, with a fixed time charger, the possibility of over-charging is present. One solution fully discharged all batteries after use to a predetermined level, typically 1.1 V per cell, and then recharged them at a constant current for a fixed period of time. Unfortunately, this procedure shortened the life of the batteries; Figure 21.4 shows the life expectancy of a cell with repeated discharges. The present solution is to charge the batteries automatically to the fully charged state and then reduce the current to the trickle charge level. Batteries which are subjected to repeated partial discharge may then be occasionally fully discharged to obviate the memory effect.

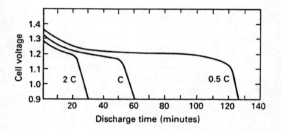

Fig. 21.3 *Typical Nicad cell discharge characteristics*

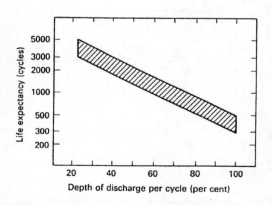

Fig. 21.4 *Nicad cell: effect of repeated discharge vs. cell life*

Constant current, automatic charging is recommended. Chargers vary in complexity, some detecting the end-of-charge point by sensing a variation of voltage. At end of charge the cell voltage first rises and then falls slightly as in Figure 21.5. More sophisticated chargers also sense the cell case temperature which rises during charge. Batteries are available for standard charging at the ten hour rate where 14 to 15 hours will be required to recharge a fully discharged battery, fast charging at 5 C and rapid charging at C rate.

Cells may be fitted with a re-sealing one-way vent which opens at about 200 psi and closes at about 175 psi to relieve any excess internal pressure caused by a fault or abuse.

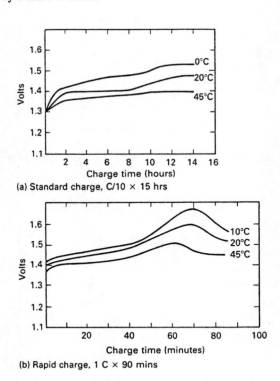

(a) Standard charge, C/10 × 15 hrs

(b) Rapid charge, 1 C × 90 mins

Fig. 21.5 *Nicad cell: discharging characteristics*

22

SATELLITE COMMUNICATIONS

22.1 Earth orbits

Communications satellites are required to illuminate the earth with radio signals and their orbits are chosen according to the size and location of the part of the earth's surface they must light up.

A satellite orbiting the earth is continuously pulled by a centripetal force, in this case gravity, towards the centre of the earth. It is also pulled by centrifugal force to leave its orbit at a tangent. When these opposing forces are equal in magnitude the satellite is in a stable orbit. There is, then, for a given height (the radius of the path minus the radius of the earth, 6378 km), a velocity at which the conditions for stable orbit apply, and which determines the orbiting time.

22.1.1 Geostationary orbits

Satellites relay information from ground stations, either fixed or mobile, or between satellites. It is an advantage for some purposes, therefore, to use satellites with an orbit time identical to that of the earth so that no tracking from the ground stations is needed. Geostationary satellites have the same angular velocity as the earth making them appear to be stationary. Their height is 35 788 kms and four such satellites cover the earth from latitude 81.3°N to 81.3°S as in Figure 22.1.

The disadvantages of geostationary satellites are that they are in a high earth orbit (HEO) resulting in a signal delay of 240 ms for the complete go and return path. Also they are in an equatorial orbit so that signals to the higher latitudes travel at a shallow angle to the earth's surface rendering them unsuitable for mobile use where communications must be achieved at street level in cities.

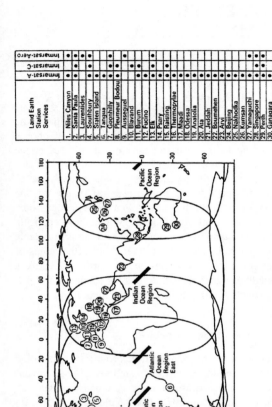

Fig. 22.1 *Immarsat coverage and GES locations*

Land Earth Station Services	Inmarsat-A	Inmarsat-C	Inmarsat-Aero
1. Niles Canyon	●	●	●
2. Santa Paula	●	●	●
3. Laurentides	●	●	●
4. Southbury	●		●
5. Staten Island	●	●	
6. Tangua	●		●
7. Goonhilly	●	●	●
8. Pleumeur Bodou	●		
9. Aussaguel	●	●	
10. Blavand	●	●	
11. Burum	●	●	●
12. Fucino	●	●	
13. Elk	●		
14. Psary	●		
15. Raisting	●	●	
16. Thermopylae	●		
17. Maadi	●		
18. Odessa	●		
19. Anatolia	●		
20. Ata	●		
21. Jeddah	●		
22. Boumehen	●		
23. Arvi	●		
24. Beijing	●		
25. Nakhodka	●		
26. Kumsan	●		
27. Yamaguchi	●	●	●
28. Singapore	●	●	●
29. Perth	●	●	●
30. Ganagara	●		●

As each satellite covers a large portion of the earth, the design of their antennas to permit repeated frequency re-use is important and antennas with small, steerable footprints have been developed for this purpose. An advantage, in addition to the lack of tracking, is that the shadowing by the earth is minimal so that solar power cells receive almost continuous illumination.

22.1.2 Elliptical orbits

A polar orbit, where the satellite follows a North/South track, provides the opportunity to survey the earth in a series of strips. The satellites used for this purpose are in low each orbit (LEO) and consequently have a high velocity. Their orbit time is approximately 1.5 hours and between successive orbits the earth has rotated 22.5°. Sixteen orbits are therefore needed to scan the earth's surface. Until recently polar orbits were used only for optical surveillance but now several projects for radio communication are either in the early stages of installation or under development. These use a number of satellites so that for mobile communications, for instance, there is a satellite continuously in view. Tracking of these extremely fast satellites by the ground stations, which may themselves be moving, along with Doppler effect has been a major obstacle which is now being overcome.

22.2 Communications by satellite link

Satellites are radio links and receive signals from the ground and other satellites which they must re-transmit. The signals from ground stations are comparatively weak and require high power amplification for onward transmission. As the satellite's receive and transmit antennas must be close together, the possibility of RF instability prohibits on-frequency repetition. The up-link, from ground station to satellite, must therefore be converted to another before re-transmission. The up-link frequency is normally higher than the down-link and the frequency converter is referred to as a 'down-converter'.

The frequency bands used for communications purposes are listed in Table 22.1

Table 22.1 *Communications satellite frequencies*

Frequency band (GHz)	Link	European telecom links	International telecom links
1.5–1.6	Down	Mobile	
1.6–1.7	Up	Mobile	
3.4–4.2	Down	Fixed	Fixed
4.5–4.8	Down	Fixed	Fixed
5.9–7.0	Up	Fixed	Fixed
10.7–11.7	Down	Fixed (+ Non-DBS television)	
11.7–12.5	Down	Fixed (+ television)	
12.5–12.75	Down	Fixed (private links)	
12.75–13.75	Up	Fixed (private links)	
14.0–14.8	Up	Fixed	
17.3–18.3	Up	Fixed	
17.7–20.2	Down	Fixed	
20.2–21.2	Down	Mobile	
27.0–30.0	Up	Fixed	
30.0–31.0	Up	Mobile	
22.5–23.55		Allocated for intersatellite links	
32.0–33.0		Allocated for intersatellite links	
54.25–58.2		Allocated for intersatellite links	
59.0–64.0		Allocated for intersatellite links	

Satellite television

The broadcasting of television may be via either communications satellites or Direct Broadcasting by Satellite (DBS) satellites. The positions of non-DBS satellites relative to the UK are shown in Figure 22.2 and Table 22.3 lists the European channels. Figure 22.3 shows the world allocations of DBS satellites and Tables 22.3 and 22.4 list the channel frequencies and national allocations. The frequency plan for the Astra satellite is in Table 22.5.

Table 22.2 *European satellite television channels broadcast via communications satellites*

Channel	Frequency (GHz)	Polarization	Audio sub-carrier	Video standard	Satellite
3Sat	11.175	H	6.65	PAL	Eutelsat 1 F1
Anglovision	11.515	V	6.6	PAL	Intelsat VA-F11
Arts Channel	11.135	H	6.6	PAL	Intelsat VA-F11
BBC1/2	11.175	H	6.65	PAL	Intelsat VA-F11

Channel	Frequency (GHz)	Polarization	Audio sub-carrier	Video standard	Satellite
CanalJ	12.564	V	5.8	PAL	Telecom 1 F2
Children's Channel	11.015	H	6.6	PAL	Intelsat VA-F11
CNN	11.155	V	6.65	PAL	Intelsat VA-F11
Filmnet	11.140	V	6.6	PAL	Eutelsat 1 F1
Moscow 1	3.675	–	7.0	SECAM	Gorizont-12
Moscow 2	3.	–	7.0	SECAM	Gorizont-7
Infofilm & Video	11.015	H	6.6	PAL	Intelsat F2
La Cinq	12.606	V	5.8	SECAM	Telecom 1 F2
Lifestyle	11.135	H	6.6	PAL	Intelsat VA-F11
M6	12.648	V	5.8	SECAM	Telecom 1 F2
MTV	10.975	H	6.65	PAL	Intelsat VA-F11
Norsk Rikskringkasti	11.644	H	Digital	C-MAC	Eutelsat 1 F2
Premiere	11.015	H	6.6	PAL	Intelsat VA-F11
RAI-Uno	11.007	H	6.6	PAL	Eutelsat 1 F1
RTL-Plus	11.091	V	6.65	PAL	Eutelsat 1 F1
SAT1	11.507	V	6.65	PAL	Eutelsat 1 F1
Satellite Information	11.575	H	Digital	B-MAC	Intelsat VA-F11
Screensport	11.135	H	6.6	PAL	Intelsat VA-F11
Skychannel	11.650	H	6.65	PAL	Eutelsat 1 F1
Superchannel	10.674	V	6.65	PAL	Eutelsat 1 F1
SVT-2	11.178	H	Digital	C-MAC	Intelsat F2
SVT-2	11.133	H	Digital	C-MAC	Intelsat F2
Teleclub	11.987	V	6.5	PAL	Eutelsat 1 F1
TV5	11.472	H	6.65	PAL	Eutelsat 1 F1
Worldnet	11.512	H	6.65	PAL	Eutelsat 1 F1
Worldnet	11.591	H	6.6	SECAM	Eutelsat 1 F1
Worldnet	12.732	V	5.8	NTSC	Telecom 1 F2

Table 22.3 *DBS television channels*

Channel number	Frequency (GHz)		
1	11.72748	21	12.11108
2	11.74666	22	12.13026
3	11.76584	23	12.14944

(continued overleaf)

Table 22.3 *(continued)*

Channel number	Frequency (GHz)		
4	11.78502	24	12.16862
5	11.80420	25	12.18780
6	11.82338	26	12.20698
7	11.84256	27	12.22616
8	11.86174	28	12.24534
9	11.88092	29	12.26452
10	11.90010	30	12.28370
11	11.91928	31	12.30288
12	11.93846	32	12.32206
13	11.95764	33	12.34124
14	11.97682	34	12.36042
15	11.99600	35	12.37960
16	12.01518	36	12.39878
17	12.03436	37	12.41796
18	12.05354	38	12.43714
19	12.07272	39	12.45632
20	12.09190	40	12.47550

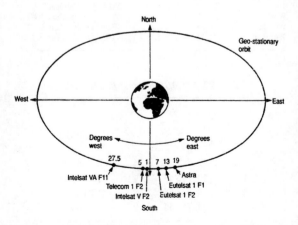

Fig. 22.2 *Non-DBS satellite television: positions of the main non-DBS satellites relevant to the UK*

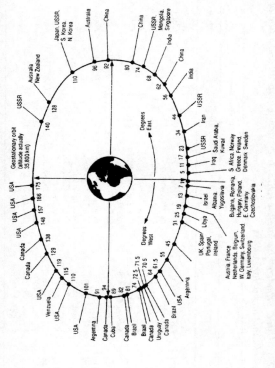

Fig. 22.3 *World allocations of DBS satellite positions*

Table 22.4 *Direct broadcast by satellite (DBS) television. European allocation of satellite positions, channel and polarizations*

Band		Orbital position		
	5° E	19° W	31° W	37° W
11.7–12.1 GHz RH polarized	*Turkey:* CH. 1, 5, 9, 13, 17 *Greece:* CH. 3, 7 11, 15, 19	*France:* CH. 1, 5, 9, 13, 17 *Luxembourg:* CH. 3, 7, 11, 15, 19	*Eire:* CH. 2, 6, 10, 14, 18 *UK:* CH. 4, 8, 12, 16, 20	*San Marino:* CH. 1, 5, 9, 13, 17 *Lichtenstein:* CH. 3, 7, 11, 15, 19
11.7–12.1 GHz LH polarized	*Finland:* CH. 2, 6, 10 *Norway:* CH. 14, 18 *Sweden* CH. 4, 8 *Denmark:* CH. 12, 16, 20	*Germany:* CH. 2, 6, 10, 14, 18 *Austria:* CH. 4, 8, 12, 16, 20	*Portugal:* CH. 3, 7, 11, 15, 19	*Andorra:* CH. 3, 8, 12, 16, 20
12.1–12.5 GHz RH polarized	*Cyprus:* CH. 21, 25, 29, 33, 37 *Iceland, etc.:* CH. 23, 27, 31, 35, 39	*Belgium:* CH. 21, 25, 29, 33, 37 *Netherlands:* CH. 23, 27, 31, 35, 39	*Iceland:* CH. 21, 25, 29, 33, 37 *Spain:* CH. 23, 27, 31, 35, 39	*Monaco:* CH. 21, 25, 29, 33, 37 *Vatican* CH. 23, 27, 31, 35, 39
12.1–12.5 GHz LH polarized	*Nordic group:** CH. 22, 24, 26, 28 30, 32, 36, 40 *Sweden:* CH. 34 *Norway:* CH. 38	*Switzerland:* CH. 22, 26, 30, 34, 38 *Italy:* CH. 24, 28, 32, 36, 40		

*Wide beam channels: Denmark, Finland, Norway, Sweden.

Table 22.5 Astra television channels

Channel	Astra 1-A (GHz)	Channel	Astra 1-B (GHz)	Channel	Astra 1-C (GHz)	Channel	Astra 1-D (GHz)
1	11.21425	17	11.46425	33	10.96425	49	10.71425
2	11.22900	18	11.47900	34	10.97900	50	10.72900
3	11.24375	19	11.49375	35	10.99375	51	10.74375
4	11.25850	20	11.50850	36	10.00850	52	10.75850
5	11.27325	21	11.52325	37	10.02325	53	10.77325
6	11.28800	22	11.53800	38	10.03800	54	10.78800
7	11.30275	23	11.55275	39	10.05275	55	10.80275
8	11.31750	24	11.56750	40	10.06750	56	10.81750
9	11.33225	25	11.58225	41	10.08225	57	10.83225
10	11.34700	26	11.59700	42	10.09700	58	10.84700
11	11.36175	27	11.61175	43	10.11175	59	10.86175
12	11.37650	28	11.62650	44	10.12650	60	10.87650
13	11.39125	29	11.64125	45	10.14125	61	10.89125
14	11.40600	30	11.65600	46	10.15600	62	10.90600
15	11.42075	31	11.67075	47	10.17075	63	10.92075
16	11.43550	32	11.68550	48	10.18550	64	10.83550

22.3 Proposed satellite television formats

Most current European satellite television programmes (non-DBS) are broadcast as fairly standard PAL signals, FM modulated into the satellite channel. DBS transmissions, and those from Astra satellite, will probably be of a multiplexed analogue component (MAC) format. In MAC, data corresponding to sound tracks and subtitles etc., an analogue signal corresponding to chrominance and an analogue signal corresponding to luminance are transmitted separately in each broadcast line of the picture.

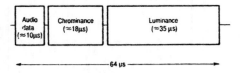

In order to achieve multiplexing of the three parts of the format, time compression of chrominance and luminance signals occurs before transmission, and they must be reconstituted at the receiver. Two main variations of the MAC format have been selected for European broadcasters, D-MAC and D2-MAC. They differ basically in the number of data channels, and hence the overall bandwidth required. Both modulate video signals in FM, and data signals in duobinary FM.

22.4 Global positioning system (GPS)

This system uses an American government satellite network comprised of 24 satellites in three circular orbits at a height of 20 000 km and with twelve-hour periods. At least six satellites should be visible from any point around the earth at any one time.

Each satellite transmits continuously updated information about its orbit on two frequencies, 1227 MHz and 1575 MHz. One radio channel carries two pseudo-random codes, one very long and the other very short. The second channel is modulated only with the short code. The codes enable the satellite to be positively identified, and the distance from a receiver on, or close to, the earth to be calculated.

A receiver's position, in three-dimensional space, is identified by measuring and calculating the distance from three satellites. The short codes provide the initial fix but increased precision is obtained from the long codes. There is an error due to time variations arising from various sources of which satellite speed is one, but taking a measurement from a fourth satellite enables a correction factor to be applied.

Receivers which utilize only the short PN codes provide a resolution accurate to about 100 m. Those which can process the long codes provide a fix accurate to about 45 m.

References

Lewis, G. E. (1988). *Communication Services via Satellite*. BSP Professional Books, Oxford.

Long, M. *The 1993/1994 World Satellite Annual*. Mark Long Enterprises Inc., Ft Lauderdale, USA.

23

CONNECTORS AND INTERFACES

23.1 Audio and video connectors

Audio connectors

The DIN standards devised by the German Industrial Standards Board are widely used for the connection of audio equipment. The connectors are shown below. The 3-way and 5-way 45 are the most common and connections for those are listed.

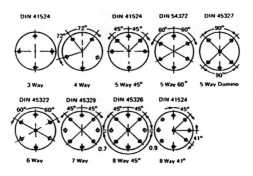

	Mono	Stereo
Microphone	Input 1	Input LH 1
	O V 2	Input RH 4
		O V 2
	Pin 3 available for polarizing voltage	Pins 3 and 5 available for polarizing voltage
Tape recorder inputs and monitor outputs	Input 1	Input LH 1
	O V 2	Input R4 4
	Output 3	O V 2
		Output LH 3
		Output RH 5

	Mono	Stereo
Tape recorder replay output	Output, low Z 1	Output LH, low Z1
	O V 2	Output RH, low Z4
	Output, high Z3	O V 2
		Output LH, high Z3
		Output RH, high Z5
Amplifiers	Output to tape 1	Output LH 1
	O V 2	Output RH 1
	Input from tape 3	O V 2
		Input LH 3
		Input RH 5

Variations on the above exist between different manufacturers.

Videorecorder/televisions/camera connectors

Standard pin configurations for videorecorders, televisions and videocameras are shown below. Many follow standard DIN connector pinouts, but videocamera and SCART connectors differ significantly.

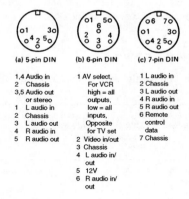

(a) 5-pin DIN

1,4 Audio in
2 Chassis
3,5 Audio out
or stereo
1 L audio in
2 Chassis
3 L audio out
4 R audio in
5 R audio out

(b) 6-pin DIN

1 AV select,
For VCR
high = all
outputs,
low = all
inputs,
Opposite
for TV set
2 Video in/out
3 Chassis
4 L audio in/
out
5 12V
6 R audio in/
out

(c) 7-pin DIN

1 L audio in
2 Chassis
3 L audio out
4 R audio in
5 R audio out
6 Remote
control
data
7 Chassis

SCART (BS 6552)

SCART connectors, also know as Peritelevision or Euroconnector connectors, feature two control systems which allow remote control over the television's or videorecorder's functions.

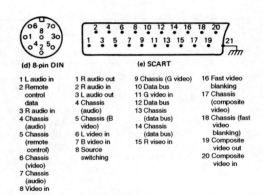

(d) 8-pin DIN

1 L audio in
2 Remote control data
3 R audio in
4 Chassis (audio)
5 Chassis (remote control)
6 Chassis (video)
7 Chassis (audio)
8 Video in

(e) SCART

1 R audio out	9 Chassis (G video)
2 R audio in	10 Data bus
3 L audio out	11 G video in
4 Chassis (audio)	12 Data bus
5 Chassis (B video)	13 Chassis (data bus)
6 L video in	14 Chassis (data bus)
7 B video in	15 R viseo in
8 Source switching	16 Fast video blanking
	17 Chassis (composite video)
	18 Chassis (fast video blanking)
	19 Composite video out
	20 Composite video in

The simplest is a source switching input (pin 8) in which an external source (videorecorder, computer, etc.) can, by issuing a 12 volt signal, cause the television to switch to baseband inputs.

A more complex control system, called *domestic data bus* (D^2B), is given through pins 10 and 12, in which serial data can be passed between controlling microprocessors in the television and external equipment. No standard yet exists for D^2B.

23.2 Co-axial connector

The most commonly used connectors for RF cables are:

Type	Impedance, Z_0 (ohms)	Max. VSWR to (frequency, GHz)	Maximum proof RF voltage	Notes
N	50 or 75	1.30 (4)	1.5 kV (5 MHz at sea level)	Screw together
BNC	50 or 75	1.01 (1) 1.30 (4)	1 kV (5 MHz at sea level)	Miniature Bayonet fitting
TNC	50 or 75	As BNC	As BNC	Miniature, Robust screw together
SMB	50	1.46 (4)	1.5 kV (At sea level. Cable dependent)	Sub-miniature Snap together

Type	Impedance, Z_0 (ohms)	Max. VSWR to (frequency, GHz)	Maximum proof RF voltage	Notes
SMC	50	1.41 (4) 1.69 (10)	As SMB	Sub-miniature Screw together
SMD	50	As SMB	As SMB	Sub-miniature Push together
BT 43	75			Developed from SMB range for use in telecomms and data transmission
7–16	50	1.3 (5 GHz)	2.7 (connector)	Suitable for medium to high power applications in the cellular and broadcast industries. Screw together
PL259/ SO239/ UHF 83	50	–	500 (pk)	Non-constant imp. High VSWR makes it unsuitable for use above 144 MHz and for extending RF cables. Very robust. Screw together
C	75	–	–	Bayonet fitting
F	50	–	–	American CCTV connector used on some 144 MHz hand-portable transceivers. Plugs use inner conductor of cable as centre pin
Belling Lee	50	–	–	British TV antenna connector. Aluminium versions may corrode when used outdoors
GR	50	1 MHz max. frequency	–	Constant imp. sexless connector
Phono	–	–	–	American connector for audio use

Examples of assembly instructions for co-axial RF connectors (by kind permission of M/ACOM Greenpar Ltd.)

 measuring instrument – a rule is shown, but better results are obtained by using a Vernier gauge

 stout trimming blade, suitable for cutting copper wire braid

 crimping tool

 soldering iron

 side cutters, also for trimming braid

 spanner, of the relevant size for the connector

 small screwdriver

 hacksaw, sometimes appropriate for semi-rigid cable, although for repetitive operations a power trimmer should be considered

Assembly instructions Type N

Cable types:

50 ohm: PSF1/4M (BBC), RG 8A/U, RG 213/U, URM 67

75 ohm: RG 11A/U, RG 63B/U, RG 114A/U, URM 64

1 Side clamp nut and plain gasket over cable and trim outer sheath from cable as shown.

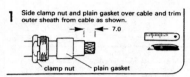

2 Fold back braid and push ferrule over dielectric to trap braid between outer sheath and ferrule. Trim off surplus braid.

3 Trim back dielectric and check the length of the protruding centre conductor.

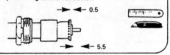

4 Tin centre conductor, then slide rear insulator over dielectric, to butt against ferrule.

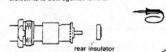

5 Fit contact (male for plugs, female for jacks) onto centre conductor. Hold cable and contact tightly together and solder

6 Slide plain gasket and clamp nut up to ferrule trapping braid. Fit front insulator over contact to butt against rear insulator and press sub-assembly into body as far as possible.

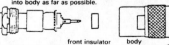

7 Engage and tighten clamp nut.

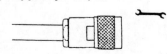

1 Slide crimp sleeve over cable

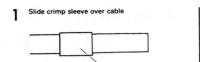

crimp sleeve

2 Trim back outer sheath and braid to dimensions shown. Any foils adjacent to the dielectric should be left in place but trimmed 1.5 mm from the face of the dielectric

18.2

4.7

10.3

3 Fit contact (male for plugs, female for jacks) over centre conductor to butt against dielectric and crimp.

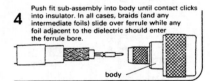

male contact

4 Push fit sub-assembly into body until contact clicks into insulator. In all cases, braids (and any intermediate foils) slide over ferrule while any foil adjacent to the dielectric should enter the ferrule bore.

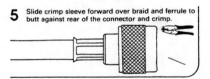

body

5 Slide crimp sleeve forward over braid and ferrule to butt against rear of the connector and crimp.

290

Assembly instructions Type BNC
Cable types:
RG 58C/U, RG 141A/U, URM 43, URM 76

1 Slide clamp nut and plain gasket over cable and trim outer sheath from cable, as shown.

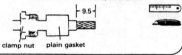

clamp nut   plain gasket

2 Fold back braid and push ferrule over dielectric to trap braid between outer sheath and ferrule. Trim off surplus braid.

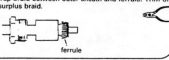

ferrule

3 Trim back dielectric and check the length of the protruding centre conductor.

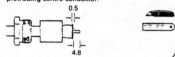

4 Tin centre conductor, then slide rear insulator over dielectric, to butt against ferrule.

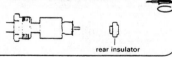

rear insulator

5 Fit contact (male for plugs, female for jacks) onto centre conductor, with collar pressed into recess in rear insulator. Hold cable and contact tightly together, and solder.

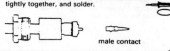

male contact

6 Slide plain gasket and clamp nut up to ferrule, trapping braid. Fit front insulator over contact to butt against rear insulator and press sub-assembly into body as far as possible.

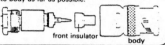

front insulator   body

7 Engage and tighten clamp nut

1 Slide metal crimp sleeve over cable, trim outer sheath from cable as shown.

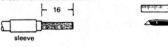

sleeve

2 Trim back braid and dielectric to the dimensions shown.

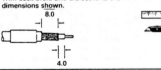

3 Fit contact over centre conductor to butt against the dielectric, then crimp.

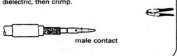

male contact

4 Press sub-assembly into body, until contact clicks into place and ensuring that the knurled ferrule is inserted between the dielectric and braid.

5 Slide the sleeve along the cable, until it butts against the body sub-assembly. Crimp, using the tool listed below.

Note: a plug is shown, but these instructions are relevant to both plugs and jacks. The shape of contacts and insulators may also vary from the drawings shown.

292

Assembly instructions Types SMB/SMC/SMD
Cable types:
  TM 3306, RG174A/U, RG188A/U, RG316/U
  TM 3263, RG178B/U, RG196A/U, URM 110

1 Slide clamp nut, a washer, a gasket and the other washer over the cable trim outer sheath to dimension shown.

2 Fold back braid. Push ferrule over dielectric to trap braid between outer sheath and ferrule.

3 Trim off surplus braid. Trim dielectric flush with ferrule and check length of centre conductor. Tin centre conductor.

4 Side rear insulator over centre conductor until it butts against ferrule.

5 Fit contact onto centre conductor until it butts against rear insulator. Hold cable and contact tightly together and solder.

6 Fit front insulator over contact until it butts against internal shoulder.

7 Press sub-assembly into body as far as possible. Engage and tighten clamp nut.

1 Place crimp sleeve over sheat and trim cable to the dimensions shown. Ensure that the centre conductor is not damaged.

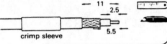

2 Place contact on centre conductor and crimp.

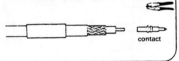

3 Push contact into body sub-assembly. Ensure the contact 'clicks' into rear insulator with the ferrule between the braid and the dielectric.

4 Slide crimp sleeve over the braid and crimp.

23.3 Interfaces

23.3.1 Connectors and connections

Data interchange by modems

When transmitting and receiving data across telephone or other circuits, the equipment which actually generates and uses the data (e.g. a computer or VDU terminal) is known as *data terminating equipment* (DTE). The equipment which terminates the telephone line and converts the basic data signals into signals which can be transmitted is known as *data circuit-terminating equipment* (DCE). As far as the user is concerned the interface between DTE and DCE is the most important. CCITT recommendation V24 defines the signal interchanges and functions between DTE and DCE; these are commonly known as the 100 series interchanges circuits:

Interchange circuit		Data		Control		Timing	
Number	Name	From DCE	To DCE	From DCE	To DCE	From DCE	To DCE
101	Protective ground or earth						
102	Signal ground or common return						
103	Transmitted data		•				
104	Received data	•					
105	Request to send				•		
106	Ready for sending			•			
107	Data set ready			•			
108/1	Connect data set to line				•		
108/2	Data terminal ready				•		
109	Data channel received line signal detector			•			
110	Signal quality detector			•			
111	Data signalling rate selector (DTE)				•		
112	Data signalling rate selector (DCE)			•			
113	Transmitter signal element timing (DTE)						•
114	Transmitter signal element timing (DCE)					•	
115	Receiver signal element timing (DCE)					•	

(*continued overleaf*)

Interchange circuit		Data		Control		Timing	
Number	Name	From DCE	To DCE	From DCE	To DCE	From DCE	To DCE
116	Select stand by				•		
117	Standby indicator			•			
118	Transmitted backward channel data		•				
119	Received backward channel data	•					
120	Transmit backward channel line signal				•		
121	Backward channel ready			•			
122	Backward channel received line signal detector			•			
123	Backward channel single quality detector			•			
124	select frequency groups				•		
125	Calling indicator			•			
126	Select transmit frequency				•		
127	Select receive frequency				•		
128	Receiver signal element timing (DTE)						•
129	Request to receive				•		
130	Transmit backward tone				•		
131	Received character timing					•	
132	Return to non-data mode				•		
133	Ready for receiving				•		
134	Received data present			•			
191	Transmitted voice answer				•		
192	Received voice answer			•			

Modem connector pin numbers

The connectors used with 100 series interchange circuits and its pin assignments are defined by international standard ISO 2110 and are (for modems following the CCITT recommendations V21, V23, V26, V26bis, V27 and V27bis) as follows:

	Interchange circuit numbers		
Pin number	V21	V23	V26/V27
1	*1	*1	*1
2	103	103	103
3	104	104	104
4	105	105	105
5	106	106	106
6	107	107	107
7	102	102	102
8	109	109	109
9	*N	*N	*N
10	*N	*N	*N
11	126	*N	*N
12	*F	122	122
13	*F	121	121
14	*F	118	118
15	*F	*2	114
16	*F	119	119
17	*F	*2	115
18	141	141	141
19	*F	120	120
20	108/1-2	108/1-2	108/1-2
21	140	140	140
22	125	125	125
23	*N	111	111
24	*N	*N	113
25	142	142	142

Notes:

*1 Pin 1 is assigned for connecting the shields between tandem sections of shielded cables. It may be connected to protective ground or signal ground.
*F Reserved for future use.
*N Reserved for national use.

Automatic calling

A similar series of interchange circuits is defined in CCITT recommendation V25 for automatic calling answering between modems over the telephone network. This is the 200 series interchange circuits:

Interchange circuit

Number	Name	From DCE	To DCE
201	Signal ground	•	•
202	Call request		•
203	Data line occupied	•	
204	Distant station connected	•	
205	Abandon call	•	
206	Digit signal (2^0)		•
207	Digit signal (2^1)		•
208	Digit signal (2^2)		•
209	Digit signal (2^3)		•
210	Present next digit	•	
211	Digit present		•
213	Power indication	•	

RS 232C

The EIA equivalent of CCITT V24 interface is the RS 232C specification, which similarly defines the electrical interface between DTE and DCE. Although the two have different designations, they are to all practical purposes equivalent. The RS 232C interchange circuits are:

Mnemonic	Name	Data		Control		Timing	
		From DCE	To DCE	From DCE	To DCE	From DCE	To DCE
AA	Protective ground						
AB	Signal ground/common return						
BA	Transmitted data		•				
BB	Received data	•					
CA	Request to send				•		
CB	Clear to send			•			
CC	Data set ready			•			
CD	Data terminal ready				•		
CE	Ring indicator			•			
CF	Received line signal detector			•			
CG	Signal quality detector			•			
CH	Data signal rate selector (DTE)				•		
CI	Data signal rate selector (DCE)			•			
DA	Transmitter signal element timing (DTE)						•
DB	Transmitter signal element timing (DCE)					•	
DD	Receiver signal element timing (DCE)					•	
SBA	Secondary transmitted data		•				
SBB	Secondary received data	•					
SCA	Secondary request to send				•		
SCB	Secondary clear to send			•			
SCF	Secondary received line signal detector			•			

RS 449

The EIA RS 232C standard, although the most common, is by no means perfect. One of its main limitations is the maximum data rate −18.2 K baud. Various improved interchange circuits (RS 422, RS 423) have been developed. The RS 449 standard is capable of very fast data rates (up to 2 Mbaud):

		Data		Control		Timing	
Interchange circuit		From DCE	To DCE	From DCE	To DCE	From DCE	To DCE
Mnemonic	*Name*						
SG	Single ground						
SC	Send common						
RC	Receive common						
IS	Terminal in service				•		
IC	Incoming call			•			
TR	Terminal ready				•		
DM	Data mode			•			
SD	Send data		•				
RD	Receive data	•					
TT	Terminal timing						•
ST	Send timing					•	
RT	Receive timing					•	
RS	Request to send				•		
CS	Clear to send			•			
RR	Receiver ready			•			
SQ	Signal quality			•			
NS	News signal				•		
SF	Select frequency				•		
SR	Signalling rate selector				•		
SI	Signalling rate indicator			•			
SSD	Secondary send data		•				
SRD	Secondary receive data	•					
SRS	Secondary request to send				•		
SCS	Secondary clear to send			•			
SRR	Secondary receiver ready			•			
LL	Local loopback				•		
RL	Remote loopback				•		
TM	Test mode			•			
SS	Select standby				•		
SB	Standby indicator			•			

Primary channel: SD, RD, TT, ST, RT, RS, CS, RR, SQ, NS, SF, SR, SI

Secondary channel: SSD, SRD, SRS, SCS, SRR

Centronics interface

Most personal computers use the Centronics parallel data transfer to a printer. The pin connections of the connector, abbreviations and signal descriptions are shown.

All signals are standard TTL, although not all signals necessarily exist in any given interface.

Pin number	Abbreviation	Signal description
1	STROBE	Strobe
2	DATA1	Data line 1
3	DATA2	Data line 2
4	DATA3	Data line 3
5	DATA4	Data line 4
6	DATA5	Data line 5
7	Data6	Data line 6
8	Data7	Data line 7
9	Data8	Data line 8
10	ACKNLG	Acknowledge data
11	BUSY	Busy
12	PE	Paper end
13	SLCT	Select printer
14	AUTO FEED XT	Automatic line feed at end of line
15	NC	No connection
16	0V	Logic ground
17	CHASSIS GND	Printer chassis (not necessarily the same as logic ground)
18	NC	No connection
19 to 30	GND	Single ground
31	INIT	Initialise
32	ERROR	Error
33	GND	Signal ground
34	NC	No connection
35	Logic 1	Logic 1
36	SLCT IN	Select input to printer

RS 232C/RS 449/V24 comparison

RS 232 C	RS 232C description	RS 449	RS 449 description	V24	V24 description
AB	Signal ground	SG	Signal ground	102	Signal ground
		SC	Send common	102a	DTE common
		RC	Receive common	102b	DCE common
		IS	Terminal in service		
CE	Ring indicator	IC	Incoming call	125	Calling indicator
CD	Data terminal ready	TR	Terminal ready	108/2	Data terminal ready
CC	Data set ready	DM	Data mode	107	Data set ready
BA	Transmitted data	SD	Send data	103	Transmitted data
BB	Receive data	RD	Receive data	104	Received data
DA	Transmitter signal element Timing (DTE source)	TT	Terminal timing	113	Transmitter signal element Timing (DTE source)
DB	Transmitter signal element Timing (DCE source)	ST	Send timing	114	Transmitter signal element Timing (DCE source)

DD	Receiver signal element timing	RT	Receive timing	115	Receiver signal element timing (DCE source)
CA	Request to send	RS	Request to send	105	Request to send
CB	Clear to send	CS	Clear to send	106	Ready for sending
CF	Received line single detector	RR	Receiver ready	109	Data channel received line signal detector
CG	Signal quality detector	SQ	Signal quality	110	Data signal quality detector
		NS	New signal		
		SF	Select frequency	126	Select transmit frequency

(continued overleaf)

(*continued*)

RS 232 C		RS 449		V24	
CH	Data signal rate selector (DTE source)	SR	Signalling rate selector	111	Data signalling rate selector (DTE source)
CI	Data signal rate selector (DCE source)	SI	Signalling rate indicator	112	Data signalling rate selector (DCE source)
SBA	Secondary transmitted data	SSD	Secondary send data	118	Transmitted backward channel data
SBB	Secondary received data	SRD	Secondary receive data	119	Received backward channel data
SGA	Secondary request to send	SRS	Secondary request to send	120	Transmit backward channel line signal
SCB	Secondary clear to send	SCS	Secondary clear to send	121	Backward channel ready

SCF	Secondary received line signal detector			
		SRR	Secondary receiver ready	122	Backward channel received line signal detector
		LL	Local loopback	141	Local loopback
		RL	Remote loopback	140	Remote loopback
		TM	Test mode	143	Test indicator
		SS	Select standby	116	Select standby
		SB	Standby indicator	117	Standby indicator

Reference

M/ACOM Greenpar Ltd, catalogue 1993.

BROADCASTING

24.1 Standard frequency and time transmissions

Frequency (mHz)	Wavelength (m)	Code	Station location	Country	Power (kW)
60 kHz	5000	MSF	Rugby	England	–
–	–	WWVB	Colorado	USA	–
75 kHz	4000	HBG	–	Switzerland	–
77.5 kHz	3871	DCF77	Mainflingen	DDR	–
1.5	200	HD210A	Guayaquil	Ecuador	–
2.5	120	MSF	Rugby	England	0.5
–	–	WWV	Fort Collins	USA	2.5
–	–	WWVH	Kekaha	Hawaii	5
–	–	ZLF	Wellington	New Zealand	–
–	–	RCH	Tashkent	USSR	1
–	–	JJY	–	Japan	–
–	–	ZUO	Olifantsfontein	South Africa	–
3.33	90.09	CHU	Ottawa	Canada	3
3.81	78.7	HD201A	Guayaquil	Ecuador	–
4.5	66.67	VNG	Victoria	Australia	–
4.996	60.05	RWM	Moscow	USSR	5
5	60	MSF	Rugby	England	0.5
–	–	WWVB	Fort Collins	USA	10
–	–	WWVH	Kekaha	Hawaii	10
–	–	ATA	New Delhi	India	–
–	–	LOL	Buenos Aires	Argentina	2
–	–	IBF	Turin	Italy	5
–	–	RCH	Tashkent	USSR	1
–	–	JJY	–	Japan	–
–	–	ZUO	Olifantsfontein	South Africa	–
5.004	59.95	RID	Irkutsk	USSR	1
6.10	49.2	YVTO	Caracas	Venezuela	–
7.335	40.9	CHU	Ottawa	Canada	10
7.5	40	VNG	Lyndhurst	Australia	5
7.6	39.4	HD210A	Guayaquil	Ecuador	–
8	37.5	JJY	–	Japan	–
8.1675	36.73	LQB9	Buenos Aires	Argentina	5
9996	30.01	RWM	Moscow	USSR	–
10	30	MSF	Rugby	England	0.5
–	–	WWVB	Fort Collins	USA	10
–	–	WWVH	Kekaha	Hawaii	10
–	–	BPM	Xian	China	–
–	–	ATA	New Delhi	India	–

(*continued overleaf*)

Frequency (mHz)	Wavelength (m)	Code	Station location	Country	Power (kW)
–	–	JJY	–	Japan	–
–	–	LOL	Buenos Aires	Argentina	2
–	–	RTA	Novosibirsk	USSR	5
–	–	RCH	Tashkent	USSR	1
10.004	29.99	RID	Irkutsk	USSR	1
12	25	VNG	Lyndhurst	Australia	10
14.67	20.45	CHU	Ottawa	Canada	3
14.996	20.01	RWM	Moscow	USSR	8
15	20	WWVB	Fort Collins	USA	10
–	–	WWVH	Kekaha	Hawaii	10
–	–	LOL	Buenos Aires	Argentina	2
–	–	RTA	Novosibirsk	USSR	5
–	–	BPM	Xian	China	–
–	–	ATA	New Delhi	India	–
–	–	JJY	–	Japan	–
15.004	19.99	RID	Irkutsk	USSR	1
16.384	18.31	–	Allouis	France	2000
15.55	17.09	LQC20	Buenos Aires	Argentina	5
20	15	WWVB	Fort Collins	USA	2.5
100	3	ZUO	Olifantsfontein	South Africa	–

24.2 Standard frequency formats

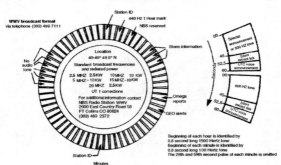

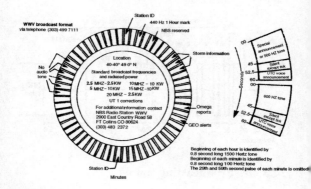

MSF Rugby

Time is inserted in the 60 kHz transmission in two ways, illustrated below.

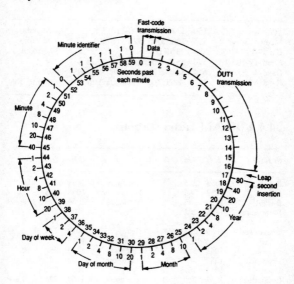

Slow code time and date information is transmitted between the 17th and 59th seconds of the minute-long cycle, in normal BCD coding. *Fast code* time and date BCD coded information is inserted into a 500 ms window in the first second of each minute-long cycle, as illustrated below.

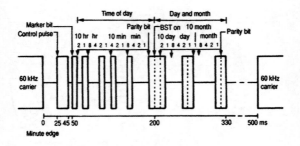

24.3 UK broadcasting bands

Frequency band	Frequency	Use
Long wave	150–285 kHz (2000–1053 m)	AM radio
Medium wave	525–1605 kHz (571–187 m)	AM radio
Band II (VHF)	88–108 MHz	FM radio
Band IV (UHF)	470–582 MHz (channels 21 to 34)	TV
Band V (UHF)	614–854 MHz (channels 39 to 68)	TV
Band VI (SHF)	11.7–12.5 GHz (channels 1 to 40)	satellite TV

24.4 BBC AM radio stations

Radio 1	kHz	m	kW		kHz	m	kW
Barnstaple	1053	285	1	Lisnagarvey	1089	275	10
Barrow	1053	285	1	Londonderry	1053	285	1
Bexhill	1053	285	2	Moorside	1089	275	150
Bournemouth	1485	202	1	Edge			
Brighton	1053	285	2	Postwick	1053	285	10
Brookmans	1089	275	150	Redmoss	1089	275	2
Park				Redruth	1089	275	2
Burghead	1053	285	20	Stagshaw	1053	285	50
Droitwich	1053	285	150	Start Point	1053	285	100
Dundee	1053	285	1	Tywyn	1089	275	1
Enniskillen	1053	285	1	Wallasey	1107	271	0.5
Fareham	1089	275	1	Washford	1089	275	50
Folkestone	1053	285	1	Westerglen	1089	275	50
Hull	1053	285	1	Whitehaven	1089	275	1
Radio 4							
Burghead	198	1515	50	Londonderry	720	417	0.25
Carlisle	1485	202	1	Newcastle	603	498	2
Droitwich	198	1515	500	Plymouth	774	388	1
Enniskillen	774	388	1	Redmoss	1449	207	2

Radio 1	kHz	m	kW		kHz	m	kW
Lisnagarvey	720	417	10	Redruth	756	397	2
London	720	417	0.75	Westerglen	198	1515	50
(Lots Road)							
Radio 5							
Barrow	693	433	1	Lisnagarvey	909	330	10
Bexhill	693	433	1	Londonderry	909	330	1
Bournemouth	909	330	0.25	Moorside Edge	909	330	200
Brighton	693	433	1	Postwick	693	433	10
Brookmans Park	909	330	150	Redmoss	693	433	1
Burghead	693	433	25	Redruth MF	909	330	2
Clevedon	909	330	50	Stagshaw	693	433	50
Droitwich	693	433	150	Start Point	693	433	50
Enniskillen	693	433	1	Tywyn	990	303	1
Exeter	909	330	1	Westerglen	909	330	50
Fareham	909	330	1	Whitehaven	909	330	1
Folkestone	693	433	1				
Radio Scotland							
Burghead	810	370	100	Redmoss	810	370	5
Dumfries	585	513	2	Redmoss			
(R. Solway)				(R. Aberdeen)	990	303	1
				Westerglen	198	370	100
Radio Ulster							
Enniskillen	873	344	1	Londonderry	792	379	1
Lisnagarvey	1341	224	100	(R. Foyle)			
Radio Wales							
Forden	882	340	1	Tywyn	882	340	5
Llandrindod	1125	267	1	Washford	882	340	100
Wells				Wrexham	657	457	2
Penmon	882	340	10	(R. Clwyd)			

24.5 BBC VHF broadcasting

24.5.1 BBC VHF FM radio stations

Notes: **Stereo services:** all services are stereo except where (m) is shown against a frequency.

Polarisation: H indicates horizontal polarisation; M indicates mixed polarisation; V indicates vertical polarisation. All BBC FM transmitters carry RDS information. Main stations shown in roman type. * indicates provisional details. (m) indicates mono, stereo expected.

England, Isle of Man and Channel Islands	Radio 1	Radio 1 and 2	Radio 3	Radio 4	Polarisation	Maximum effective radiated power
Belmont	98.3	88.8	90.9	93.1	M	16 kW
Grantham	97.7	88.1	90.3	92.5	V	35 W
Chatton	99.7*	90.1	92.3	94.5	M	5.6 kW
Berwick-upon-Tweed	98.2	88.6	90.8	93.0	V	20 W
Holme Moss	98.9	89.3	91.5	93.7	M	250 kW
Barnoldswick	99.3	89.7	91.9	94.1	V	20 W
Beecroft Hill	99.4	89.8	92.0	94.2	V	200 W
Cornholme	99.3*	89.7	91.9	94.1	V	20 W
Haslingden	99.5*	89.9	92.1	94.3	V	83 W
Hebden Bridge	98.0*	88.4	90.6	92.8	V	1 kW
Keighley	98.5	88.9	91.1	93.3	V	1 kW
Kendal	98.6*	89.0	91.2	93.4	M	100 W
Luddenden	98.3	88.7	90.9	93.1	V	84 W
Morecambe Bay	99.6	90.0	92.2	94.4	M	10 kW
Olivers Mount	99.5	89.9	92.1	94.3	M	250 W
Pendle Forest	97.8*	90.2	92.6	94.6	M	1 kW
Saddleworth	99.3	89.7	91.9	94.1	V	95 W
Sheffield	99.5	89.9	92.1	94.3	M	320 W
Stanton Moor	99.4*	89.9	92.0	94.2	M	1.2 kW
Todmorden	98.5*	88.9	91.1	93.3	V	100 W
Walsden South	98.0*	88.4	90.6	92.8	V	10 W
Wensleydale	97.9*	88.3	90.5	92.7	H	27 W
Wharfedale	98.0*	88.4	90.6	92.8	M	40 W
Windermere	97.9*	88.3	90.5	92.7	M	64 W
Les Platons (CI)	97.1	89.6	91.1	94.8	M	16 kW
Manningtree	97.7*	88.1	90.3	92.5	M	5 kW
North Hessary Tor	97.7	88.1	90.3	92.5	M	160 kW
*Beacon Hill**	98.4	88.7	90.9	93.1	V	1 kW
Okehampton	98.3*	88.7	90.9	93.1	M	50 W
St. Thomas (Exeter)	98.6*	89.0	91.2	93.4	M	55 W
Oxford	99.1	89.5	91.7	93.9	M	46 kW
*Cirencester Town**	97.7	88.1	90.3	92.5	V	10 W
*Marlborough**	99.7	90.1	92.3	94.5	V	100 W
Peterborough	99.7	90.1	92.3	94.5	M	40 kW
Bow Brickhill	98.2	88.6	90.8	93.0	M	10 kW
Cambridge (Madingley)	98.5*	88.9	91.1	93.3	M	260 W
Northampton	98.5*	88.9	91.1	93.3	M	123 W
Pontop Pike	98.1	88.5	90.7	92.9	M	134 kW
Fenham	99.4*	89.8	92.0	94.2	V	42 W
Newton	99.0	89.4	91.6	93.8	V	100 W
Weardale	99.3	89.7	91.9	94.1	M	100 W
Whitby	99.2	89.6	91.8	94.0	M	40 W
*Woolmoor**	99.6	90.2	92.2	94.4	V	5 kW
Redruth	99.3	89.7	91.9	94.1	M	25 kW
Isles of Scilly	98.4*	88.8	91.0	93.2	M	60 W
Ridge Hill	98.2	88.6	90.8	93.0	M	10 kW
Rowridge	98.2	88.5	90.7	92.9	M	250 kW
Salisbury	99.4	89.8	92.0	94.2	V	20 W

England, Isle of Man and Channel Islands	Radio 1	Radio 1 and 2	Radio 3	Radio 4	Polarisation	Maximum effective radiated power
Ventnor	99.0*	89.4	91.7	93.8	H	20 W
Weymouth	99.6	90.0	92.2	94.4	V	100 W
Sandale	97.7	88.1	90.3	92.5	M	250 kW
Douglas (IOM)	98.0	88.4	90.6	92.8	M	11 kW
Whitehaven	99.3	89.7	91.9	94.1	V	100 W
Sutton Coldfield	97.9	88.3	90.5	92.7	M	250 kW
Buxton	99.6	90.0	92.2	94.4	M	100 W
Chesterfield	98.6	89.0	91.2	93.4	V	400 W
Churchdown Hill	98.6*	89.0	91.2	93.4	M	72 W
Ludlow	99.2*	89.6	91.8	94.0	M	10 W
Swingate	99.1	90.0	92.4	94.4	M	11 kW
Tacolneston	99.3	89.7	91.9	94.1	M	250 kW
Wenvoe (see also Wales)	99.5	89.9	92.1	94.3	M	250 kW
Barnstaple	98.1*	88.5	90.7	92.9	M	1 kW
Bath	98.6*	89.0	91.2	93.4	M	82 W
Calne	97.8	88.2	90.4	92.6	V	32 W
Chalford	98.8	89.2	91.4	93.6	V	100 W
Chippenham	98.4	88.8	91.0	93.2	V	10 W
Combe Martin	98.7	89.1	91.3	93.5	V	4 W
Hutton	99.0	89.4	91.6	93.8	V	40 W
Egford Hill (Frome)	98.7	89.1	91.3	93.5	V	60 W
Ilchester Crescent	98.9*	89.3	91.5	93.7	M	13 kW
Nailsworth	97.8	88.2	90.4	92.6	V	100 W
Westwood	97.9	88.3	90.5	92.7	V	50 W
Winter Hill	98.2	88.6	90.8	93.0	M	4 kW
Darwen	99.1	89.5	91.7	93.9	V	10 W
Wrotham		89.1	91.3	93.5	M	250 kW
	98.8				M	125 kW
*Bexhill**	99.2	88.2	92.2	94.6	V	100 W
Brighton (Whitehawk Hill)	99.7	90.1	92.3	94.5	M	600 W
Caterham	99.3	89.7	91.9	94.1	V	15 W
Folkestone	98.3	88.4	90.6	93.1	V	100 W
Guildford	97.7	88.1	90.3	92.5	M	3 kW
Hastings	97.7	89.6	91.8	94.2	M	500 W
High Wycombe	99.5	89.9	92.1	94.3	M	50 W
Kenley	98.0	88.4	90.6	92.8	V	25 W
Mickleham	99.3	89.7	91.9	94.1	V	25 W
Newhaven	99.3*	89.7	91.9	94.1	M	100 W

Scotland	Radio 1	Radio 1 and 2	Radio 3	Radio 4	Radio Scotland	Polarisation	Maximum effective radiated power
Ashkirk	98.7	89.1	91.3	103.9	93.5	M	50 kW
Eyemouth	99.3	89.7	91.9	104.6	94.1	V	100 W
Innerleithen	99.1*	89.5	91.7	96.1*	93.9	M	20 W
Peebles	98.0	88.4	90.6	95.0	92.8	M	20 W
Black Hill	99.5	89.9	92.1	95.8	94.3	M	250 kW
Bowmore (m)	97.7*	88.1	90.3	95.7	92.5	V	80 W
Campbelltown	98.0	88.4	90.6	95.2	92.8	M	400 W
Girvan	98.5	88.9	91.1	95.3	93.3	V	100 W
Kirkconnel	98.3	88.7	90.9	95.3	93.1	M	40 W
Kirkton Mailer	98.6*	89.0	91.2	104.5*	93.4	M	964 W
Lethanhill	97.9*	88.3	90.5	94.9*	92.7	M	200 W
Lochgilphead		88.3	90.5		92.7	H	10 W
Millburn Muir	97.9	88.3	90.5	104.1	92.7	M	50 W
Port Ellen (m)	98.6*	89.0	91.2	94.7*	93.4	V	65 W
Roseneath	98.8	89.2	91.4	103.8	93.6	M	32 W
Rothesay	98.1*	88.5	90.7	95.1	92.9	M	570 W
South Knapdale	98.9*	89.3	91.5	95.6*	93.7	H	1.1 kW
Strachur	98.2*	88.6	90.8	95.1*	93.0	M	18 W
West Kilbride*	98.7	89.1	91.3	103.5	93.5	V	50 W
Bressay	97.9*	88.3	90.5	94.9*	92.7	M	43 kW
Darvel	99.1	89.5	91.7	104.3	93.9	M	10 kW
Eitshal	99.4	89.8	92.0	95.1	94.2	V	2 kW
Ness of Lewis	97.9	88.3	90.5	96.1	92.7	V	200 W
Ullapool*	97.9	88.3	90.5	96.1	92.7	V	50 W
Forfar	97.9	88.3	90.5	94.9	92.7	M	17 kW
Crieff*	98.9	89.3	91.5	95.3	93.7	V	125 W
Pitlochry	98.8*	89.2	91.4	103.9*	93.6	H	200 W
Rosemount	99.2	89.6	91.8	95.5	94.0	V	32 W
Fort William (m)	98.9*	89.3	91.5	95.9*	93.7	H	1.5 kW
Ardgour*	98.3	88.7	90.9	93.1	95.1	H	50 W
Ballachulish (m)	97.7*	88.1	90.3	94.7*	92.5	H	25 W
Glengorm (m)	99.1*	89.5	91.7	96.1*	93.9	M	5 kW
Kinlochleven (m)	99.3*	89.7	91.9	95.6*	94.1	M	10 W
Mallaig (m)	97.7*	88.1	90.3	94.7*	92.5	H	14 W
Oban (m)	98.5*	88.9	91.1	95.3*	93.3	M	3.6 kW
Keelylang Hill	98.9*	89.3	91.5	95.9*	93.7	M	41 kW
Meldrum	98.3	88.7	90.7	95.3	93.1	M	150 kW
Durris	99.0*	89.4	91.6	95.9*	93.8	M	2.1 kW
Tullich	99.7*	90.1	92.3	104.5*	94.5	M	42 kW
Melvaig	98.7	89.1	91.3	95.7	93.5	M	50 W
Clettraval	98.9	89.3	91.5	95.9	93.7	V	2 kW
Daliburgh*	97.7	88.1	90.3	95.1	92.5	V	1 kW
Penifiler	97.7	88.1	90.3	96.1	92.5	M	19 W
Skriaig	98.1	88.5	90.7	94.8	92.9	M	10 W
Rosmarkie	99.2	89.6	91.8	103.6	94.9	M	20 kW
Grantown	99.4*	89.8	92.0	104.2*	94.6	H	350 W
Kingussie	98.7*	89.1	91.3	95.7*	93.5	H	35 W
Knock More	97.8*	88.2	90.4	94.8*	92.6	M	500 W
Rumster Forest	99.7*	90.1	92.3	104.5*	94.5	M	10 kW
Sandale	97.7	88.1	90.3	92.5	94.7	M	250 kW
Cambret Hill	98.3*	88.7	90.9	95.3*	93.1	H	64 W
Stranraer	99.3*	89.7	91.9	103.6*	94.1	V	31 W

Wales	Radio 1	Radio 1 and 2	Radio 3	Radio 4	R. Cymru	Polarisation	Maximum effective radiated power
Blaenplwyf	98.3	88.7	90.9	104.0	93.1	M	250 kW
Dolgellau (m)	99.7*	90.1	92.3	103.6*	94.5	H	16 W
Ffestiniog	97.7*	88.1	90.3	103.6*	92.5	H	49 W
Llandyfriog	99.7*	90.1	92.3	104.4*	94.5	M	87 W
Machynlleth	99.0*	89.4	91.6	103.6*	93.8	H	60 W
Mynedd Pencarreg	99.3*	89.7	91.9	103.7*	94.1	M	384 W
Haverford West	98.9	89.3	91.5	104.9	93.7	M	20 kW
Llandona	99.4	89.8	92.0	103.6	94.2	M	21 kW
Bettws-y-Coed	97.8*	88.2	90.4	104.9	92.6	H	10 W
Conwy	98.7	89.1	91.3	104.4	93.5	V	50 W
*Deiniolen**	97.7	88.1	90.3	92.5	104.1	V	50 W
Penmaen Rhos	98.0	88.4	90.6	104.6	92.8	V	100 W
Llangollen	98.5	88.9	91.1	93.3	104.3	M	16 kW
Llandinam	99.7*	90.1	92.3	96.1*	94.5	H	20 W
Llanfyllin	98.7	89.1	91.3	93.5	94.7	M	7 W
Llanrhaeadr-ym-Mochnant	99.4	89.8	92.0	94.2	103.8	M	50 W
Long Mountain	99.2*	89.6	91.8	103.6*	94.0	H	24 W
Wenvoe	99.5	89.9	92.1	94.3	96.8	M	250 kW
Aberdare	98.8*	89.2	91.4	104.7*	93.6	M	42 W
Abergavenny	98.3	88.7	90.9	103.5	93.1	M	40 W
Abertillery	98.6	89.0	91.2	93.4	104.3	V	10 W
Blaenavon	98.1	88.5	90.7	92.9	104.0	V	10 W
Brecon	98.5*	88.9	91.1	104.7*	93.3	H	10 W
Carmarthen	98.5	88.9	91.1	95.5	93.3	M	10 W
Carmel	98.0	88.4	90.6	104.6	92.8	M	3 kW
Clyro	99.8	90.2	92.4	94.6	104.9	V	10 W
Croeserw	98.6	89.0	91.2	103.6	93.4	V	10 W
*Cwmafan**	98.1	88.5	90.7	92.9	104.5	V	10 W
Ebbw Vale	98.0	88.4	90.6	92.8	104.6	M	10 W
Kilvey Hill	99.1	89.5	91.7	104.2	93.9	M	925 W
Llandrindod Wells	98.7	89.1	91.3	103.8	93.5	M	2 kW
*Llangeinor**	98.2	88.6	90.8	93.0	104.1	V	10 W
Llanidloes	97.7*	88.1	90.3	104.8*	92.5	H	5 W
*Llyswen**	99.2	89.6	91.8	94.0	104.4	V	10 W
Pennar	99.1	89.5	91.7	93.9	103.7	V	5 W
Pontypool	98.8*	89.2	91.4	104.8*	93.6	M	50 W
Porth	98.0	88.4	90.6	92.8	104.5	V	10 W
Rheola	97.8	88.2	90.4	104.8	92.6	V	10 W
Rhymney	98.9	89.3	91.5	93.7	104.9	V	20 W
Ton Pentre	98.4*	88.8	91.0	104.3*	93.2	M	5 W
Varteg Hill	98.5*	88.9	91.1	103.5*	93.3	M	48 W
Radio Wales							
Blaenavon	–	–	–	–	95.1	V	10 W
Christchurch	–	–	–	–	95.9	M	500 W

315

Northern Ireland	Radio 1	Radio 1 and 2	Radio 3	Radio 4	Radio Ulster	Polarisation	Maximum effective radiated power
Brougher Mountain	99.0*	89.4	91.6	95.6*	93.8	M	9.8 kW
Divis	99.7	90.1	92.3	96.0	94.5	M	250 kW
Cammoney Hill	98.4	88.8	91.0	93.2	95.3	V	20 W
Kilkeel	99.0*	89.4	91.6	103.9*	93.8	H	25 W
Lame	98.7*	89.1	91.3	103.5*	93.5	M	100 W
Rostrevor Forest (m)	97.9*	88.3*	90.5*	103.9*	92.7*	M	32 W
Limavady	99.2*	89.6	91.8	94.0	95.4	M	3.4 kW
Ballycastle	98.4*	88.8	91.0	93.2	95.1	M	100 W
Londonderry	98.3	88.7	90.9		93.1	M	31 kW

24.5.2 BBC VHF test tone transmissions

Transmission starts about 4 minutes after the end of Radio 3 programmes on Mondays and Saturdays.

Time min.	Left channel	Right channel	Purpose
–	250 Hz at zero level	440 Hz at zero level	Identification of left and right channels and setting of reference level
2	900 Hz at +7 dB	900 Hz at +7 dB, antiphase to left channel	Adjustment of phase of regenerated subcarrier (see Note 4) and check of distortion with L-R signal only
6	900 Hz at +7 dB	900 Hz +7 dB, in phase with left channel	Check of distortion with L + R signal only
7	900 Hz at +7 dB	No modulation	Check of L to R cross-talk
8	No modulation	900 Hz at +7 dB	Check of R to L cross-talk
9	Tone sequence at −4 dB: 40 Hz 6·3 kHz 100 Hz 10 kHz 500 Hz 12·5 kHz 1000 Hz 14 kHz This sequence is repeated	No modulation	Check of L-channel frequency response and L to R cross-talk at high and low frequencies
11′40″	No modulation	Tone sequence as for left channel	Check of R-channel frequency response and R to L cross-talk at high and low frequencies
14′20″	No modulation	No modulation	Check of noise level in the presence of pilot
15′20″	End of test transmissions		

Notes:

1. This schedule is subject to variation or cancellation to accord with programme requirements and essential transmission tests.
2. The zero level reference corresponds to 40% of the maximum level of modulation applied to either stereophonic channel before pre-emphasis. All tests are transmitted with pre-emphasis.

3. Periods of tone lasting several minutes are interrupted momentarily at one-minute intervals.
4. With receivers having separate controls of subcarrier phase and crosstalk, the correct order of alignment is to adjust first the subcarrier phase to produce maximum output from either the L or the R channel and then to adjust the crosstalk (or 'separation') control for minimum crosstalk between channels.
5. With receivers in which the only control of crosstalk is by adjustment of subcarrier phase, this adjustment of subcarrier phase, this adjustment should be made on the crosstalk checks.
6. Adjustment of the balance control to produce equal loudness from the L and R loudspeakers is best carried out when listening to the announcements during a stereophonic transmission, which are made from a centre-stage position. If this adjustment is attempted during the tone transmissions, the results may be confused because of the occurrence of standing-wave patterns in the listening room.
7. The outputs of most receivers include significant levels of the 19-kHz tone and its harmonics, which may affect signal-level meters. It is important, therefore, to provide filters with adequate loss at these frequencies if instruments are to be used for the above tests.

24.5.3 Engineering information about broadcast services

Information about all BBC services as well as advice on how best to receive transmissions (including television) can be obtained from:

British Broadcasting Corporation
Engineering Liaison
White City
201 Wood Lane
London W12 7TS

Telephone: 081 752 5040

Transmitter service maps for most main transmitters can also be supplied, but requests for maps should be accompanied by a stamped addressed A4 sized envelope.

Similarly, information about all IBA broadcast services can be obtained from:

Radio:
The Radio Authority
Holbrook House
14 Great Queen Street
Holborn
London WC2B 5DG

Telephone: 071 430 7062

Television:
The Independent Television Commission
Kings Worthy Court
Kings Worthy
Winchester
Hants SO23 7QA

Telephone: 0962 848647

24.6 UK television channels and transmitters

	ITV	Channel 4	BBC 1	BBC 2	Polarisation	Max vision erp(kW)
East of England						
Tacolneston	59	65	62	55	H	250
West Runton	23	29	33	26	V	2
Aldeburgh	23	30	33	26	V	10
Bramford	24	31	21	27	V	0.010
Thetford	23	29	33	26	V	0.02
Little Walsingham	41	47	51	44	V	0.011
Creake	49	42	39	45	V	0.005
Wells next the Sea	50	–	43	–	V	0.09
Burnham	46	–	40	–	V	0.077
Norwich (Central)	49	42	39	45	V	0.034
Bury St. Edmunds	25	32	22	28	V	0.017
Linnet Valley	23	29	33	26	V	0.016
Sudbury	41	47	51	44	H	250
Woodbridge	61	54	58	64	V	0.1
Ipswich (Stoke)	25	32	22	28	V	0.007
Somersham	25	32	22	28	V	0.0025
Wivenhoe Park	61	54	58	64	V	0.011
Felixstowe	60	67	31	63	V	0.0055
Overstrand	41	47	51	44	V	0.063
Sandy Heath	24	21	31	27	H	1000
Northampton						
(Dall. Park)	56	68	66	62	V	0.065
Luton	59	65	55	62	V	0.08
Kings Lynn	52	–	48	–	V	0.34
The Borders and Isle of Man						
Caldbeck	28	32	30	34	H	500
Kendal	61	54	58	64	V	2
Windermere	41	47	51	44	V	0.5
Coniston	24	31	21	27	V	0.09
Hawkshead	23	29	33	26	V	0.061
Whitehaven	43	50	40	46	V	2
Keswick	24	31	21	27	V	0.12
Threlkeld	60	53	57	63	V	0.011
Ainstable	42	49	52	45	V	0.1
Haltwhistle	59	65	55	62	V	2
Gosforth	61	54	58	64	V	0.05
Bassenthwaite	49	42	52	45	V	0.16
Pooley Bridge	46	50	48	40	V	0.013
Moffat	42	49	52	45	V	0.0065
Douglas	48	56	68	66	V	2
Foxdale	23	29	33	26	V	0.008
Glenridding	57	63	60	53	V	0.008
Glencoyne	24	31	21	27	V	0.005
(Glenridding Link)						
Beary Peark	43	50	40	46	V	0.25
Port St. Mary	61	54	58	64	V	0.25
Laxey	61	54	58	64	V	0.025
Langholm	60	53	57	63	V	0.025
Thornhill	60	53	57	63	V	0.5

	ITV	Channel 4	BBC 1	BBC 2	Polarisation	Max vision erp(kW)
Barskeoch Hill	59	65	55	62	V	2
New Galloway	23	29	33	26	V	0.1
Stranraer	60	53	57	63	V	0.25
Portpatrick	61	54	58	64	V	0.006
Cambret Hill	41	47	44	51	H	16
Creetown	61	54	58	64	V	0.032
Kirkcudbright	24	31	21	27	V	0.006
Glenluce	61	54	58	64	V	0.015
St. Bees	61	54	58	64	V	0.012
Workington	61	54	58	64	V	0.01
Bleachgreen	60	53	57	63	V	0.006
Dumfries South	46	50	40	48	V	0.023
Dentdale	60	53	57	63	V	0.052
Union Mills	52	42	39	45	V	0.012
Lowther Valley	46	50	40	48	V	0.026
Pinwherry	25	32	22	28	V	0.056
Ballantrae	61	54	58	64	V	0.0066
Lorton	60	53	57	63	V	0.05
Greystoke	60	53	57	63	V	0.011
Kirkby Stephen	60	53	57	63	V	0.012
Ravenstonedale	60	53	57	63	V	0.011
Orton	43	50	40	46	V	0.031
Sedbergh	43	50	40	46	V	0.5
Grasmere	60	53	57	63	V	0.02
Crosby Ravensworth	60	53	57	63	V	0.006
Crosthwaite	60	53	57	63	V	0.012
Minnigaff	29	23	33	26	V	0.008
Selkirk	59	65	55	62	H	50
Eyemouth	23	29	33	26	V	2
Galashiels	41	47	51	44	V	0.1
Hawick	23	29	33	26	V	0.05
Jedburgh	41	47	51	44	V	0.16
Bonchester Bridge	49	42	39	45	V	0.006
Lauder	25	32	22	28	V	0.011
Peebles	25	32	22	28	V	0.1
Innerleithen	61	54	58	64	V	0.1
Berwick-upon-Tweed	24	31	21	27	V	0.038
Stow	23	29	33	26	V	0.005
Yetholm	41	47	51	44	V	0.006
Midlands (East)						
Waltham	61	54	58	64	H	250
Ashford-in-the-Water	23	29	33	26	V	0.0011
Belper	68	62	66	56	V	0.030
Birchover	49	42	39	45	H	0.012
Bolehill	53	60	63	57	V	0.25
Eastwood	23	29	33	26	V	0.0072
Leicester City	25	32	22	28	V	0.008
Lincoln Central	42	52	39	45	V	0.010
Little Eaton	23	29	33	26	V	0.008
Matlock	24	31	21	27	V	0.017
Nottingham	24	31	21	27	V	2

(continued overleaf)

	ITV	Channel 4	BBC 1	BBC 2	Polarisation	Max vision erp(kW)
Parwich	24	31	21	27	V	0.0031
Stamford	49	42	39	45	V	0.0032
Stanton Moor	59	65	55	62	V	2
Midlands (South)						
Oxford	60	53	57	63	H	500
Ascott-under-						
Wychwood	24	31	21	27	V	0.029
Charlbury	41	47	51	44	V	0.0033
Guiting Power	41	47	51	44	V	0.012
Icombe Hill	25	32	22	28	V	0.11
Kenilworth	60	53	57	63	V	0.010
Over Norton	55	67	65	48	V	0.031
Ridge Hill	25	32	22	28	H	100
Andoversford	59	65	55	62	V	0.056
Eardiston	61	54	58	64	V	0.006
Garth Hill	60	53	57	63	V	0.025
Hereford	41	47	51	44	V	0.040
Hope-under-						
Dinmore	60	53	63	57	V	0.0018
Kington	49	42	39	45	V	0.025
Knucklas	42	49	39	45	V	0.010
New Radnor	41	47	51	44	V	0.125
Peterchurch	60	53	57	63	V	0.076
Presteigne	52	66	48	56	V	0.016
Queslett	61	54	58	64	V	0.0125
Ross-on-Wye	65	59	55	62	V	0.008
St. Briavels	43	50	40	46	V	0.012
Upper Soudley	43	50	40	46	V	0.0017
Midlands (West)						
Sutton Coldfield	43	50	46	40	H	500
Allesley Park	25	32	22	28	V	0.033
Ambergate	25	32	22	28	V	0.030
Ashbourne	25	32	22	28	V	0.25
Brailes	34	59	30	52	V	0.040
Bretch Hill	55	67	65	48	V	0.087
Bridgnorth	56	66	62	68	V	0.016
Brierley Hill	60	53	57	63	V	10
Bromsgrove	24	21	31	27	V	4
Cheadle	56	68	48	66	V	0.024
Fenton	24	21	31	27	V	10
Gravelly Hill	62	68	66	56	H	0.016
Hamstead	24	31	21	27	H	0.0018
Harborne	34	67	30	48	V	0.040
Hartington	56	68	66	48	V	0.033
Ipstones Edge	60	53	57	63	V	0.028
Ironbridge	61	54	58	64	V	0.011
Kiddeminster	61	54	58	64	V	2
Kinver	56	68	66	48	H	0.012
Lark Stoke	23	29	33	26	V	7.60
Leamington Spa	66	68	56	62	V	0.20
Leek	25	32	22	28	V	1

	ITV	Channel 4	BBC 1	BBC 2	Polarisation	Max vision erp(kW)
Long Compton	25	32	22	28	V	0.004
Malvern	66	68	56	62	V	2
Oakamoor	24	31	21	27	V	0.011
Oakeley Mynd	49	42	39	45	V	0.050
Perry Beeches	22	28	25	32	V	0.008
Redditch	25	32	22	28	V	0.0016
Rugeley	56	68	66	48	V	0.008
Tenbury Wells	60	53	57	63	V	0.014
Turves Green	62	68	56	66	V	0.0012
Whittingslow	60	53	57	63	V	0.056
Winchcombe	61	54	58	64	V	0.006
Winshill	56	68	66	48	H	0.006
Woodford Halse	25	32	22	28	V	0.007
The Wrekin	23	29	26	33	H	100
Bucknell	49	42	39	45	V	0.008
Clun	59	65	55	62	V	0.056
Coalbrookdale	47	41	51	44	V	0.0035
Haden Hill	49	42	39	52	V	0.008
Halesowen	61	54	58	64	V	0.012
Hazler Hill	41	47	51	44	V	0.025
Ludlow	42	49	39	45	V	0.025
Channel Islands						
Fremont Point	41	47	51	44	H	20
St. Helier	59	65	55	62	V	0.034
Les Touillets	54	52	56	48	H	2
Alderney	61	68	58	64	V	0.1
St. Peter Port	24	31	21	27	V	0.0014
Torteval	46	66	50	40	V	0.02
Gorey	23	29	54	26	V	0.006
Lancashire						
Winter Hill	59	65	55	62	H	500
Darwen	49	42	39	45	V	0.5
Pendle Forest	25	32	22	28	V	0.5
Haslingden	23	29	33	26	V	8
Elton	24	31	21	27	V	0.063
Saddleworth	49	42	52	45	V	0.5
Storeton	25	32	22	28	V	2.8
Bacup	43	53	40	46	V	0.25
Ladder Hill	23	29	33	26	V	1
Bidston	30	47	51	44	V	0.066
Birch Vale	43	53	40	46	V	0.25
Whitworth	25	32	22	28	V	0.05
Glossop	25	32	22	28	V	1
Buxton	24	31	21	27	V	1
Trawden	60	67	57	63	V	0.2
Whalley	43	53	40	46	V	0.05
Lees	32	25	22	28	H	0.010
Littleborough	24	31	21	27	V	0.5
North Oldham	24	31	21	27	V	0.04
Macclesfield	25	32	22	28	V	0.037
Congleton	41	47	51	44	V	0.2
Oakenhead	41	47	51	44	V	0.1

(continued overleaf)

	ITV	Channel 4	BBC 1	BBC 2	Polarisation	Max vision erp(kW)
Whitewell	60	67	57	63	V	0.08
Delph	23	29	33	26	V	0.003
Lancaster	24	21	31	27	V	10
Blackburn	41	47	51	44	V	0.008
Millom Park	25	32	22	28	V	0.25
Ramsbottom	56	68	48	66	V	0.08
Dalton	43	53	40	46	V	0.025
Over Biddulph	30	48	34	67	V	0.022
Haughton Green	43	53	40	46	H	0.007
Parbold	41	47	51	44	V	0.036
Chinley	61	67	57	64	V	0.012
Dog Hill	43	53	40	46	V	0.085
Romiley	41	47	51	44	V	0.011
Bollington	24	31	21	27	V	0.021
Langley	24	31	21	27	V	0.0045
Ribblesdale	41	47	51	44	V	0.03
Backbarrow	60	59	57	63	V	0.003
Kendall Fell	43	50	40	46	H	0.016
West Kirby	24	31	34	27	V	0.013
Brook Bottom	61	68	58	64	V	0.006
Staveley-in-Cartmel	54	53	40	46	V	0.01
Penny Bridge	23	29	33	26	V	0.031
Cartmel	25	32	22	28	H	0.0022
Urswick	41	47	51	44	V	0.0058
Melling	60	53	57	63	V	0.025
Austwick	49	42	39	45	V	0.032
Chatburn	23	29	33	26	V	0.007
Woodnook	49	52	39	45	V	0.003
Middleton	30	48	67	34	V	0.006
Wardle	25	32	22	28	H	0.003
Norden	30	57	34	67	V	0.009
Brinscall	24	31	27	21	V	0.0008
Newchurch	24	31	27	21	H	0.004
North-east Scotland						
Durris	25	32	22	28	H	500
Peterhead	59	65	55	62	V	0.1
Gartly Moor	61	54	58	64	V	2.2
Rosehearty	41	47	51	44	V	2
Balgownie	43	50	40	46	V	0.04
Tullich	59	65	55	62	V	0.07
Braemar	42	49	39	45	V	0.015
Tomintoul	43	50	40	46	V	0.0065
Banff	42	49	39	45	V	0.028
Ellon	49	42	39	45	V	0.0027
Brechin	43	50	40	46	V	0.0065
Boddam	42	49	39	45	V	0.006
Angus	60	53	57	63	H	100
Perth	49	42	39	45	V	1
Crieff	23	29	33	26	V	0.1
Cupar	41	47	51	44	V	0.02
Pitlochry	25	32	22	28	V	0.15

	ITV	Channel 4	BBC 1	BBC 2	Polarisation	Max vision erp(kW)
Kenmore	23	29	33	26	V	0.12
Blair Atholl	43	50	40	46	V	0.05
Tay Bridge	41	47	51	44	V	0.5
Killin	49	42	39	45	V	0.13
Auchtermuchty	49	42	39	45	V	0.05
Camperdown	23	29	33	26	V	0.002
Strathallan	49	42	39	45	V	0.029
Methven	25	32	22	28	V	0.006
Dunkeld	41	47	51	44	V	0.1
Balmullo	49	42	39	45	V	0.05
Balnaguard	39	45	42	49	V	0.01
Grandtully	61	54	58	64	V	0.008
Keelylang Hill (Orkney)	43	50	40	46	H	100
Pierowall	23	29	33	26	V	0.0072
Bressay	25	32	22	28	V	10
Fitful Head	49	42	39	45	V	0.094
Scalloway	59	65	55	62	V	0.029
Swinister	59	65	55	62	V	0.21
Baltasound	42	49	39	45	V	0.018
Fetlar	43	50	40	46	V	0.013
Collarfirth Hill	41	47	51	44	V	0.41
Weisdale	61	54	58	64	V	0.06
Burgar Hill	24	31	21	27	V	0.0055
Rumster Forest	24	21	31	27	H	100
Ben Tongue	49	42	39	45	V	0.04
Thurso	60	53	57	63	V	0.0027
Melvich	41	47	51	44	V	0.055
Durness	53	60	57	63	V	0.007
Knock More	23	29	33	26	H	100
Grantown	41	47	51	44	V	0.35
Kingussie	43	50	40	46	V	0.091
Craigellachie	60	53	57	63	V	0.07
Balblair Wood	59	65	55	62	V	0.083
Lairg	41	47	51	44	V	0.013
Avoch	53	60	63	57	V	0.004
Eitshal (Lewis)	23	29	33	26	H	100
Borve	32	25	22	28	V	0.008
Scoval	59	65	55	62	V	0.16
Clettraval	41	47	51	44	V	2
Daliburgh (South Uist)	60	53	57	63	V	0.03
Skriaig	24	31	21	27	V	1
Penifiler	49	42	39	45	V	0.04
Duncraig	41	47	51	44	V	0.16
Attadale	25	32	22	28	V	0.0088
Badachro	43	50	40	46	V	0.035
Ness of Lewis	41	47	51	44	V	0.032
Ullapool	49	52	39	45	V	0.078
Kilbride (South Uist)	49	42	39	45	V	0.13

(continued overleaf)

	ITV	Channel 4	BBC 1	BBC 2	Polarisation	Max vision erp(kW)
Uig	43	50	53	46	V	0.0033
Ardintoul	49	42	39	45	V	0.047
Tarbert (Harris)	49	52	39	45	V	0.047
Bruernish	43	50	40	46	V	0.0069
Poolewe	47	41	51	44	V	0.02
Lochinver	43	50	40	46	V	0.008
Rosemarkie	49	42	39	45	H	100
Auchmore Wood	25	32	22	28	V	0.1
Clovenfords	24	31	21	27	V	0.005
Fort Augustus	23	29	33	26	V	0.011
Fodderty	60	53	57	63	V	0.12
Wester Erchite	24	31	21	27	V	0.016
Glen Urquhart	41	47	51	44	V	0.09
Tomatin	25	32	22	28	V	0.012
Inverness	65	59	55	62	V	0.05
Tomich	24	31	21	27	V	0.014
Wales						
Wenvoe	41	47	44	51	H	500
Aberbeeg	43	50	40	46	V	0.008
Alltwen	43	50	40	46	V	0.008
Kilvey Hill	23	29	33	26	V	10
Rhondda	23	29	33	26	V	2.5
Mynydd Machen	23	29	33	26	V	2
Maesteg	25	32	22	28	V	0.25
Pontypridd	25	32	22	28	V	0.5
Aberdare	24	31	21	27	V	0.5
Merthyr Tydfil	25	32	22	28	V	0.13
Bargoed	24	31	21	27	V	0.3
Rhymney	60	53	57	63	V	0.15
Clydach	23	29	33	26	V	0.0035
Abertillery	25	32	22	28	V	0.28
Ebbw Vale	59	65	55	62	V	0.5
Ebbw Vale South	24	31	27	21	V	0.002
Blaenllechau	24	31	21	27	H	0.004
Blaina	43	50	40	46	V	0.1
Pontypool	24	31	21	27	V	0.25
Cilfrew	49	52	39	45	V	0.015
Blaenavon	60	53	57	63	V	0.15
Abergavenny	49	42	39	45	V	1
Ferndale	60	53	57	63	V	0.08
Porth	43	50	40	46	V	0.08
Wattsville	60	53	63	57	V	0.0052
Llangeinor	59	65	55	62	V	0.19
Treharris	52	68	56	48	V	0.05
Cwmafon	24	31	21	27	V	0.07
Llanfach	60	53	57	63	H	0.02
Llyswen	24	31	21	27	V	0.030
Nant-y-moel	31	24	21	27	V	0.002
Penrhiwceiber	53	60	57	63	V	0.002
Llanhilleth	49	42	39	45	V	0.03
Gilfach Goch	24	31	21	27	V	0.05

	ITV	Channel 4	BBC 1	BBC 2	Polarisation	Max vision erp(kW)
Taff's Well	59	65	55	62	V	0.052
Ogmore Vale	60	53	57	63	V	0.1
Abetridwr	60	53	57	63	V	0.05
Ynys Owen	59	65	55	62	V	0.08
Tonypandy	59	65	55	62	V	0.02
Fernhill	59	65	62	55	V	0.0031
Mynydd Bach	61	54	58	64	V	0.25
Bedlinog	24	31	21	27	V	0.01
Machen Upper	62	68	55	65	V	0.009
Cwm Ffrwd-Oer	43	50	39	46	V	0.003
Blaenau-Gwent	60	53	57	63	V	0.0028
Pennar	43	50	40	46	V	0.1
Brecon	61	54	58	64	V	1
Sennybridge	43	50	40	46	V	0.064
Clyro	41	47	51	44	V	0.16
Crickhowell	24	31	21	27	V	0.15
Blackmill	25	32	22	28	V	0.01
Pennorth	23	29	33	26	V	0.05
Pontardawe	61	68	58	64	V	0.13
Deri	25	32	22	28	V	0.05
Cwmaman	49	42	39	45	V	0.0014
South Tredegar	49	39	52	45	V	0.013
Trebanog	24	34	21	27	V	0.004
Ton Pentre	61	54	58	64	V	0.08
Trecastle	25	32	22	28	V	0.006
Monmouth	59	65	55	62	V	0.05
Cwmfelinfach	48	42	52	45	V	0.006
Llanfoist	60	53	57	63	V	0.018
Abercynon	58	54	64	66	H	0.0062
Tynewydd	59	65	55	62	V	0.02
Craig-Cefn-Parc	43	50	46	40	V	0.0063
Briton Ferry	43	50	46	40	V	0.02
Dowlais	61	54	58	64	V	0.013
Rhondda Fach	25	32	22	28	V	0.0015
Trefechan (Merthyr)	42	49	39	45	V	0.005
Crucorney	24	31	21	27	V	0.011
Tonyrefail	59	65	55	62	V	0.02
Efail Fach	49	52	39	45	V	0.0084
Llanharan	24	31	21	27	V	0.0017
Burry Port	61	54	58	64	V	0.0031
Rhondda 'B'	49	68	66	39	H	0.005
Gelli-Fendigaid	59	65	55	62	H	0.012
South Maesteg	59	65	55	62	V	0.0059
Upper Killay	24	31	21	27	V	0.004
Llanddona	60	53	57	63	H	100
Betws-y-Coed	24	31	21	27	V	0.5
Penmaen Rhos	25	32	22	28	H	0.141
Conwy	43	50	40	46	V	2
Bethesda	60	53	57	63	V	0.025
Bethesda North	25	32	28	22	V	0.008
Caergybi	24	31	21	27	V	0.50
Gronant	29	23	26	33	V	0.002

(continued overleaf)

	ITV	Channel 4	BBC 1	BBC 2	Polarisation	Max vision erp(kW)
Maentwrog	43	50	40	46	V	0.016
Trefor	49	42	39	45	V	0.01
Deiniolen	25	32	22	28	V	0.05
Arfon	41	47	51	44	V	3.6
Llandecwyn	61	54	58	64	V	0.3
Ffestiniog	25	32	22	28	V	1.2
Waunfawr	25	32	22	28	V	0.026
Amlwch	25	32	22	28	V	0.035
Cemaes	43	50	40	46	V	0.012
Mochdre	23	29	33	26	V	0.0017
Dolwyddelan	41	47	51	44	V	0.011
Llanengan	61	54	58	64	H	0.003
Coed Derw	41	47	51	44	V	0.025
Carmel	60	53	57	63	H	100
Bronwydd Arms	31	24	21	27	V	0.008
Cilycwm	31	24	21	27	V	0.007
Cwm-Trwch	24	31	21	27	V	0.008
Llanelli	49	67	39	45	V	0.1
Ystalyfera	49	42	39	45	V	0.05
Llandrindod Wells	49	42	39	45	V	2.25
Rhayader	23	29	33	26	V	0.1
Llanwrtyd Wells	24	31	21	27	V	0.01
Builth Wells	25	32	22	28	V	0.026
Tenby	49	42	39	45	V	0.032
Cwmgors	24	31	21	27	V	0.026
Abercraf	25	32	22	28	V	0.13
Mynydd Emroch	43	50	40	46	V	0.09
Greenhill	24	31	21	27	V	0.074
Penderyn	49	42	39	45	V	0.012
Talley	49	42	39	45	V	0.0065
Llansawel	32	25	22	28	V	0.0065
Presely	43	50	46	40	H	100
Croeserw	61	54	58	64	V	0.120
Cynwyl Eifed	25	32	22	28	V	0.005
Haverford West	56	68	52	66	H	0.050
Mynydd Pencarreg	61	54	58	64	V	0.12
Tregaron	56	66	62	68	V	0.015
Llandyfriog	25	32	22	28	V	0.11
St. Dogmaels	23	29	33	26	V	0.015
Trefin	25	32	22	28	V	0.056
Abergwynfi	24	31	21	27	V	0.003
Glyncorrwg	49	42	39	45	V	0.0007
Llwyn Onn	25	32	22	28	V	0.05
Dolgellau	59	65	55	62	V	0.03
Croeserw	61	54	58	64	V	0.12
Pencader	23	29	33	26	V	0.006
Llandysul	60	53	57	63	V	0.05
Broad Haven	61	54	58	64	V	0.006
Rheola	59	65	55	62	V	0.1
Newport Bay	60	67	57	63	V	0.013
Ferryside	24	31	21	27	V	0.007

	ITV	Channel 4	BBC 1	BBC 2	Polarisation	Max vision erp(kW)
Llangybi	25	32	22	28	V	0.012
Duffryn	25	32	22	28	V	0.004
Blaen-Plwyf	24	21	31	27	H	100
Machynlleth	60	53	57	63	V	0.02
Aberystwyth	61	54	58	64	V	0.023
Fishguard	61	54	58	64	V	0.056
Long Mountain	61	54	58	64	V	1
Llandinam	41	47	44	51	V	0.25
Llanidloes	25	32	22	28	V	0.005
Llanfyllin	25	32	22	28	V	0.13
Moel-y-Sant	24	31	34	27	V	0.11
Kerry	24	31	21	27	V	0.017
Carno	24	31	21	27	V	0.011
Castle Caereinion	43	50	40	46	V	0.008
Dolybont	61	54	58	64	V	0.032
Llanbrynmair	25	32	22	28	V	0.02
Afon Dyfi	25	32	22	28	V	0.0063
Llangurig	23	29	33	26	V	0.008
Trefilan	60	53	57	63	V	0.086
Llanrhaedr-ym-						
Mochnant	49	42	39	45	V	0.077
Bow Street	41	47	51	44	V	0.02
Ynys-Pennal	41	47	51	44	V	0.02
Llangadfan	25	32	22	28	V	0.0063
Tregynon	25	32	22	28	V	0.035
Corris	49	42	39	45	V	0.006
Llangynog	65	59	55	62	V	0.006
Broneirion	29	23	33	26	V	0.007
Ystumtuen	53	60	57	63	V	0.013
Moel-y-Parc	49	42	52	45	H	100
Llangollen	60	53	57	63	V	0.015
Glyn Ceiriog	61	54	58	64	V	0.007
Bala	23	29	33	26	V	0.2
Corwen	25	32	22	28	V	0.3
Pontfadog	25	32	22	28	V	0.0064
Cerrigydrudion	23	29	33	26	V	0.032
Wrexham-Rhos	–	67	39	–	V	0.40
Llanuwchllyn	43	50	40	46	V	0.03
Cefn-Mawr	41	47	51	44	V	0.034
Llanarmon-yn-ial	24	31	21	27	V	0.006
Llangernyw	32	25	22	28	V	0.007
Betws-yn-Rhos	24	31	21	27	V	0.013
Glyndyfrdwy	59	65	55	62	V	0.0056
Llandderfel	65	59	55	62	V	0.0065
Llanddulas	23	29	33	26	H	0.012
Pwll-Glas	23	29	33	26	V	0.007
Pen-y-Banc	24	31	21	27	V	0.004
West of England						
Mendip	61	54	58	64	H	500
Chalford Vale	43	50	68	46	V	0.008
Chilfrome	45	52	39	49	H	0.080

(continued overleaf)

	ITV	Channel 4	BBC 1	BBC 2	Polarisation	Max vision erp(kW)
Chitterne	43	50	40	46	V	0.010
Crockerton	41	47	51	44	V	0.077
Bath	25	32	22	28	V	0.25
Westwood	43	50	40	46	V	0.1
Avening	41	47	51	44	V	0.0056
Calne	24	31	21	27	V	0.05
Redcliff Bay	34	67	30	56	H	0.011
Bristol KWH	42	52	45	48	V	1
Bristol IC	43	50	40	46	V	0.5
Washford	39	68	49	66	V	0.062
Easter Compton	34	67	30	56	V	0.01
West Lavington	24	31	21	27	V	0.0012
Seagry Court (Swindon)	41	47	44	51	V	0.0025
Coleford	45	39	42	52	V	0.01
Monksilver	52	42	45	48	V	0.015
Ogbourne St. George	43	50	40	46	V	0.013
Wooton Courtenay	25	32	22	28	V	0.056
Stroud	42	52	48	45	V	0.5
Cirencester	23	29	33	26	V	0.25
Nailsworth	23	29	33	26	V	0.031
Chalford	24	31	21	27	V	0.13
Roadwater	24	31	21	27	H	0.012
Marlborough	25	32	22	28	V	0.1
Upavon	23	29	33	26	V	0.07
Porlock	42	52	48	45	V	0.025
Countisbury	49	67	39	56	H	0.11
Cerne Abbas	25	32	22	28	V	0.11
Hutton	39	68	49	66	V	0.14
Bristol (Montpelier)	23	29	33	26	V	0.01
Box	43	50	40	46	V	0.0068
Dursley (Uley)	43	50	40	46	V	0.055
Slad	23	29	33	26	H	0.0028
Frome	24	31	21	27	V	0.0072
Bristol (Barton House)	24	31	21	27	H	0.011
Bruton	43	50	40	46	V	0.0015
Kewstoke	34	67	30	56	V	0.012
Burrington	59	65	55	62	H	0.103
Ubley	24	31	21	27	V	0.079
Portishead	49	68	66	39	V	0.007
Backwell	25	32	22	28	V	0.094
Tintern	24	31	21	27	V	0.006
Chiseldon	34	67	30	49	V	0.02
Chepstow	24	31	21	27	V	0.0031
Pillowell	43	50	40	46	H	0.013
Redbrook	42	52	39	45	V	0.002
Siston	34	21	31	24	V	0.008
Blakeney	24	31	21	27	V	0.007
Lydbrook	43	50	40	46	V	0.0075
Parkend	41	47	51	44	V	0.0017

	ITV	*Channel 4*	*BBC 1*	*BBC 2*	*Polarisation*	*Max vision erp(kW)*
Clearwell	68	56	66	48	V	0.01
Woodcombe	24	31	21	27	V	0.0063
Exford	41	47	51	44	V	0.008
Kilve	39	68	49	66	H	0.008
Crewkerne	43	50	40	46	V	0.0016
Carhampton	30	56	34	67	V	0.008
London						
Crystal Palace	23	30	26	33	H	1000
Guildford	43	50	40	46	V	10
Hertford	61	54	58	64	V	2
Reigate	60	53	57	63	V	10
Hemel Hempstead	41	47	51	44	V	10
Woolwich	60	67	57	63	V	0.63
Greenwich	52	48	56	63	V	0.015
Hampstead Heath	47	41	51	44	H	0.001
Kensal Town	52	67	56	49	H	0.025
Poplar	49	68	45	66	V	0.020
High Wycombe	59	65	55	62	V	0.5
Wooburn	56	68	49	52	V	0.1
Henley-on-Thames	67	54	48	64	V	0.1
Bishops Stortford	59	49	55	62	V	0.029
Chesham	43	50	40	46	V	0.1
Welwyn	43	50	40	46	V	0.15
Gt. Missenden	61	54	58	64	V	0.085
Mickleham	58	68	61	55	V	0.09
Kenley	43	50	40	46	V	0.09
Chepping Wycombe	41	47	51	44	V	0.02
Hughenden	43	50	40	46	V	0.06
Forest Row	62	66	48	54	V	0.12
Chingford	52	48	56	50	V	0.0075
Hemel Hempstead (Town)	61	24	58	64	V	0.013
Walthamstow North	49	68	45	66	V	0.0017
Marlow Bottom	61	54	58	64	V	0.0011
Cane Hill	58	68	61	54	V	0.018
New Addington	54	68	64	48	V	0.017
West Wycombe	43	67	40	46	V	0.028
Otford	60	53	57	63	V	0.031
Lea Bridge	39	59	55	62	V	0.006
Micklefield	57	67	54	64	V	0.0062
Alexandra Palace	61	54	58	64	H	0.065
Dorking	41	47	51	44	H	0.055
Caterham	59	65	55	62	V	0.035
East Grinstead	46	59	40	56	V	0.117
Biggin Hill	49	67	45	52	V	0.008
Croydon (Old Town)	52	67	49	56	V	0.033
Skirmett	41	47	51	44	V	0.13
St. Albans	57	67	49	63	V	0.022
Gravesend	59	49	55	62	V	0.011
Wonersh	52	67	48	65	V	0.025
New Barnet	59	48	55	62	V	0.007

(continued overleaf)

	ITV	Channel 4	BBC 1	BBC 2	Polarisation	Max vision erp(kW)
Hammersmith	59	65	48	62	V	0.01
World's End	46	68	43	50	V	0.025
Central Scotland						
Black Hill	43	50	40	46	H	500
Bridge of Allan	23	29	33	26	V	0.010
Cannongate	61	54	58	64	H	0.006
Clachan	43	50	40	46	V	0.004
Deanston	63	57	53	60	V	0.002
Easdale	49	42	39	45	V	0.007
Kilmacolm	24	31	21	27	V	0.032
South Knapdale	60	53	57	63	V	1.45
Biggar	25	32	22	28	V	0.5
Abington	60	53	57	63	H	0.0051
Glasgow WC	56	66	68	62	V	0.032
Killearn	59	55	65	62	V	0.5
Callander	25	32	22	28	V	0.1
Cathcart	60	53	57	63	V	0.002
Netherton Braes	25	32	22	28	V	0.005
Gigha Island	41	47	51	44	V	0.06
Tarbert (Loch Fyne)	24	31	21	27	V	0.0036
Haddington	61	54	58	64	V	0.02
Strachur	23	29	33	26	V	0.035
Dollar	61	54	58	64	V	0.01
Ravenscraig	24	31	21	27	V	0.02
Kirkfieldbank	60	53	57	63	V	0.0058
Tillicoultry	60	53	57	63	V	0.005
Fintry	24	31	34	27	V	0.019
Twechar	25	32	22	28	V	0.007
Strathblane	24	31	21	27	V	0.0071
Broughton	24	31	21	27	V	0.007
Leadhills	61	54	58	64	V	0.003
Glespin	61	54	58	64	V	0.006
Cumbernauld Village	61	54	58	64	V	0.008
Kelvindale	30	48	34	52	V	0.002
Uplawmoor	61	54	58	64	V	0.032
Craigkelly	24	21	31	27	H	100
Kinross	64	54	61	67	V	0.125
Penicuik	61	54	58	64	V	2
West Linton	23	29	33	26	V	0.025
Aberfoyle	61	54	58	64	V	0.087
Darvel	23	29	33	26	H	100
Blackwaterfoot	43	41	51	44	V	0.008
High Keil	47	41	51	44	V	0.020
Muirkirk	41	47	51	44	V	0.1
Kirkconnel	61	54	58	64	V	0.25
West Kilbride	41	47	51	44	V	0.35
Lethanhill	60	53	57	63	V	0.25
Girvan	59	65	55	62	V	0.25
Campbeltown	60	53	57	63	V	0.13
Port Ellen	25	32	22	28	V	0.09

	ITV	Channel 4	BBC 1	BBC 2	Polarisation	Max vision erp(kW)
Bowmore	49	42	39	45	V	0.08
Millburn Muir	42	49	39	52	V	0.25
Rosneath	61	54	58	64	V	10
Rosneath	61	54	58	64	H	0.05
Millport	61	54	58	64	H	0.0027
Troon	61	54	58	64	V	0.02
Rothesay	25	32	22	28	V	2
Tighnabruaich	49	42	39	45	V	0.092
Lochwinnoch	60	53	57	63	H	0.086
New Cumnock	43	50	40	46	V	0.012
Rothesay Town	59	65	55	62	V	0.0054
Claonaig	59	65	55	62	V	0.074
Carradale	41	47	51	44	V	0.029
Ardentinny	49	52	39	45	V	0.07
Arrochar	24	31	21	27	V	0.006
Ardnadam	41	47	51	44	V	0.017
Garelochhead	41	47	51	44	V	0.012
Wanlockhead	47	41	51	44	V	0.002
Kirkoswald	25	32	22	28	V	0.032
Kirkmichael	49	52	39	45	V	0.019
Dunure	43	50	40	46	V	0.012
Holmhead	41	47	51	44	V	0.012
Largs	42	49	39	45	H	0.012
Sorn	43	50	40	46	V	0.0065
Torosay	25	32	22	28	V	20
Arisaig	23	29	33	26	V	0.008
Ballachulish	23	29	33	26	V	0.018
Bellanoch	42	49	39	45	V	0.050
Castlebay	24	31	21	27	V	0.0066
Cow Hill	43	50	40	46	V	0.065
Dalmally	41	47	51	44	V	0.041
Fiunary	43	50	40	46	V	0.050
Glengorm	48	54	56	52	V	0.10
Kinlochleven	59	65	55	62	V	0.012
Mallaig	40	50	43	46	V	0.018
Oban	41	47	51	44	V	0.012
Onich	61	54	58	64	V	0.017
Spean Bridge	24	31	21	27	V	0.070
Strontian	39	45	42	49	V	0.008
Taynuilt	43	50	40	46	V	0.008
Tayvallich	43	50	40	46	V	0.008
South of England						
Rowridge	27	21	31	24	H	500
Salisbury	60	53	57	63	V	10
Till Valley	43	50	46	40	V	0.075
Ventnor	49	42	39	45	V	2
Poole	60	53	57	63	V	0.1
Brighton	60	53	57	63	V	10
Brighton (Central)	41	47	39	45	H	0.63
Coldean	44	68	65	42	V	0.040

(*continued overleaf*)

333

	ITV	Channel 4	BBC 1	BBC 2	Polarisation	Max vision erp(kW)
Shrewton	41	47	51	44	V	0.0045
Findon	41	47	51	44	V	0.05
Patcham	43	50	46	40	H	0.069
Winterborne						
Stickland	43	50	40	46	V	1
Corfe Castle	41	47	51	44	V	0.014
Portslade	41	47	51	44	V	0.019
Westbourne	41	47	51	44	V	0.038
Ovingdean	44	68	65	42	V	0.019
Saltdean	55	47	51	66	V	0.014
Singleton	41	47	51	44	V	0.016
Donhead	41	47	51	44	V	0.029
Millbrook	41	47	51	44	V	0.035
Brighstone	41	47	51	44	V	0.14
Hangleton	49	42	39	45	V	0.0068
Lulworth	59	65	55	62	V	0.011
Piddletrenthide	49	42	39	45	V	0.056
Winterbourne	45	66	39	49	V	0.012
Steepleton						
Cheselbourne	53	60	57	63	V	0.0065
Brading	41	47	51	44	V	0.004
Luscombe Valley	49	42	39	45	V	0.008
Midhurst	58	68	61	55	H	100
Haslemere	25	32	22	28	V	0.015
Hannington	42	66	39	45	H	250
Tidworth	32	25	22	28	V	0.01
Chisbury	59	52	55	62	V	0.025
Sutton Row	25	32	22	28	V	0.25
Alton	59	52	49	62	V	0.01
Hemdean						
(Caversham)	56	59	49	52	V	0.022
Aldbourne	24	31	21	27	V	0.007
Lambourn	59	52	55	62	V	0.007
Luccombe (IOW)	59	34	56	62	V	0.025
Dover	66	53	50	56	H	100
Dover Town	23	30	33	26	V	0.1
Hythe	24	31	21	27	V	0.051
Chartham	24	31	21	27	V	0.1
Faversham	25	32	22	28	V	0.013
Folkestone	23	30	33	26	V	0.20
Rye	41	47	58	44	V	0.012
Newnham	24	31	21	27	V	0.035
Lyminge	25	32	22	28	V	0.0069
Horn Street	41	47	58	44	V	0.003
Elham	23	30	33	26	V	0.0035
Heathfield	64	67	49	52	H	100
Tunbridge Wells	41	47	51	44	V	10

	ITV	Channel 4	BBC 1	BBC 2	Polarisation	Max vision erp(kW)
St. Marks	60	53	57	63	V	0.063
(additional horizontally polarized signal of 0.032 kW)						
Newhaven	43	41	39	45	V	2
Hastings	28	32	22	25	V	1
Hastings (Old town)	42	55	45	39	V	0.010
Lewes	25	32	22	28	V	0.032
(horizontally polarized signal of 0.032 kW to SW)						
Lydden	39	64	42	68	V	0.008
Margate	25	32	22	28	V	0.020
Ramsgate	23	30	33	26	V	0.008
Eastbourne	23	30	33	26	V	0.125
Haywards Heath	43	41	39	45	V	0.037
Wye (Ashford)	25	32	22	28	V	0.031
East Dean	54	42	62	44	V	0.008
Hamstreet	23	30	33	26	V	0.0007
Lamberhurst	62	58	54	60	V	0.003
Mountfield	24	31	21	27	V	0.0035
Sedelscombe	23	30	33	26	V	0.007
Steyning	62	56	45	59	V	0.14
Bluebell Hill	43	65	40	46	H	30
Chatham Town	61	54	58	68	V	0.011
North-east England						
Pontop Pike	61	54	58	64	H	500
Byrness	31	24	21	27	V	0.013
Newton	23	29	33	26	V	2
Fenham	24	31	21	27	V	2
Weardale	41	47	44	51	V	1
Alston	49	42	52	45	V	0.4
Catton Beacon	43	50	40	46	V	0.14
Morpeth	25	32	22	28	V	0.044
Bellingham	24	31	21	27	V	0.05
Humshaugh	49	42	39	45	V	0.059
Haydon Bridge	41	47	51	44	V	0.1
Shotley Field	25	32	22	28	V	0.2
Durham	43	50	40	46	V	0.015
Ireshopeburn	59	65	55	62	V	0.011
Kielder	33	26	23	29	V	0.027
Hedleyhope	43	50	40	46	H	0.018
Seaham	41	47	51	44	V	0.059
Sunderland	43	50	40	46	V	0.013
Staithes	41	47	51	44	V	0.0017
Esh	49	42	39	45	V	0.012
Falstone	41	47	51	44	V	0.0063
Wall	43	50	40	46	H	0.025
Whitaside	41	47	51	44	V	0.015
Bilsdale	29	23	33	26	H	500
Castleton	59	65	55	62	V	0.008
Whitby	59	65	55	62	V	0.25
Bainbridge	60	53	57	63	V	0.031
Grinton Lodge	43	50	40	46	V	0.025
Guisborough	60	53	57	63	V	0.05

(*continued overleaf*)

	ITV	Channel 4	BBC 1	BBC 2	Polarisation	Max vision erp(kW)
Ravenscar	61	54	58	64	V	0.02
Limber Hill	43	50	40	46	V	0.05
Skinningrove	43	50	40	46	V	0.014
Romaldkirk	41	47	51	44	V	0.058
Rookhope	43	50	40	46	V	0.009
West Burton	43	50	40	46	V	0.013
Aislaby	52	49	39	45	V	0.04
Rosedale Abbey	43	50	40	46	V	0.007
Peterlee (Horden)	49	39	45	52	V	0.002
Eston Nab	43	50	40	46	V	0.02
Chatton	49	42	39	45	H	100
Rothbury	65	59	55	62	V	0.05
Northern Ireland						
Divis	24	21	31	27	H	500
Larne	49	42	39	45	V	0.5
Carnmoney Hill	43	50	40	46	V	0.1
Kilkeel	49	42	39	45	V	0.5
Newcastle	59	65	55	62	V	1
Armagh	49	42	39	45	V	0.12
Black Mountain	49	42	39	45	V	0.025
Whitehead	52	67	48	56	V	0.012
Bellair	52	67	48	56	V	0.04
Draperstown	49	42	39	45	V	0.0118
Leitrim	60	53	57	63	V	0.063
Moneymore	49	42	39	45	V	0.0067
Newry North	41	47	51	44	V	0.01
Rostrevor Forest	46	50	48	40	V	0.058
Newry South	49	42	39	45	V	0.02
Benagh	25	32	22	28	V	0.056
Cushendun	32	25	22	28	V	0.026
Cushendall	43	50	40	46	V	0.013
Glynn	61	54	58	64	V	0.0014
Newtownards	61	54	58	64	V	0.011
Banbridge	46	50	44	48	V	0.0061
Glenariff	61	54	58	64	V	0.011
Killowen Mountain	24	21	31	27	V	0.015
Bangor	59	65	62	55	V	0.003
Dromore	61	54	58	64	V	0.004
Limavady	59	65	55	62	H	100
Londonderry	41	47	51	44	V	10
Ballycastle Forest	49	42	39	45	V	0.012
Bushmills	41	47	51	44	V	0.0065
Strabane	49	42	39	45	V	2
Claudy	60	53	57	63	V	0.029
Gortnalee	24	31	21	27	V	0.032
Castlederg	65	59	55	62	V	0.011
Plumbridge	56	68	52	66	V	0.0125
Glenelly Valley	23	29	33	26	V	0.012
Ballintoy	49	42	39	45	V	0.0017
Buckna	41	47	51	44	V	0.013
Gortnageeragh	42	49	39	45	V	0.019

	ITV	Channel 4	BBC 1	BBC 2	Polarisation	Max vision erp(kW)
Muldonagh	32	25	22	28	V	0.012
Brougher Mountain	25	32	22	28	H	100
Belcoo	41	47	51	44	V	0.087
Derrygonnelly	47	66	51	44	V	0.006
Lisbellaw	59	65	55	62	V	0.0065
Edemy	62	55	65	59	V	0.06
South-west England						
Caradon Hill	25	32	22	28	H	500
Fowey	61	54	58	64	V	0.010
St. Austell	59	65	55	62	V	0.1
Looe	43	50	40	46	V	0.005
Hartland	52	66	48	56	V	0.029
Gunnislake	43	50	40	46	V	0.04
Plympton (Plymouth)	61	54	58	64	V	2
Downderry	59	65	55	62	V	0.026
Tavistock	60	53	57	63	V	0.1
Woolacombe	42	49	39	45	V	0.006
Penaligon Downs	49	42	39	45	V	0.1
Newton Ferrers	59	65	55	62	V	0.0065
North Hessary Tor	59	65	55	62	V	0.0125
Ilfracombe	61	54	58	64	V	0.25
Combe Martin	49	42	39	45	V	0.1
Okehampton	49	42	39	45	V	0.1
Ivybridge	42	49	39	45	V	0.5
Kingsbridge	43	50	40	46	V	0.2
Penryn	59	65	55	62	V	0.022
Plymouth (North Road)	43	50	40	46	V	0.012
Slapton	55	68	48	66	V	0.125
Truro	61	54	58	64	V	0.022
Weston Mill	39	45	49	42	V	0.004
Croyde	41	47	51	44	V	0.0015
Chambercombe	24	31	21	27	V	0.007
Salcombe	44	30	51	41	V	0.017
Polperro	60	53	57	63	V	0.0028
Mevagissey	43	50	40	46	H	0.0066
Lostwithiel	43	50	40	46	V	0.0063
Aveton Gifford	66	47	51	44	V	0.0015
Berrynarbor	25	32	22	28	V	0.008
Port Isaac	65	59	55	62	V	0.002
St. Neot	49	45	39	42	V	0.004
Stockland Hill	23	29	33	26	H	250
Branscombe	47	41	51	44	V	0.004
St. Thomas (Exeter)	41	47	51	44	V	0.25
Beer	59	65	55	62	V	0.0029
Tiverton	43	50	40	46	V	0.1
Bampton	45	52	39	49	V	0.03
Charmouth	41	47	51	44	V	0.008
Culm Valley	49	42	39	45	V	0.058
Honiton	52	45	49	39	V	0.004

(continued overleaf)

	ITV	Channel 4	BBC 1	BBC 2	Polarisation	Max vision erp(kW)
Pennsylvania	61	54	58	64	V	0.008
Preston	61	54	58	64	V	0.005
Rampisham UHF	43	50	40	46	H	0.0125
Bridport	41	47	51	44	V	0.1
Beaminster	59	65	55	62	V	0.02
Weymouth	43	50	40	46	V	2
Dawlish	59	65	55	62	V	0.0066
Stokeinteignhead	41	47	51	44	V	0.0063
Dunsford	39	49	45	67	V	0.006
Crediton	43	50	40	46	V	0.04
Beacon Hill	60	53	57	63	H	100
Bovey Tracy	49	42	39	45	V	0.008
Brixham	43	50	40	46	V	0.018
Chudleigh	41	47	51	44	V	0.008
Clennon Valley	49	42	39	45	V	0.004
Dartmouth	41	47	51	44	V	0.01
Ashburton	24	31	21	27	V	0.003
Teignmouth	45	67	39	49	V	0.025
Coombe	24	31	21	27	V	0.0065
Newton Abbot	43	50	40	46	V	0.003
Buckfastleigh	41	47	51	44	V	0.0062
Totnes	24	31	21	27	V	0.0034
Harbertonford	49	42	39	45	H	0.0018
Sidmouth	45	67	39	49	V	0.012
South Brent	43	50	40	46	V	0.004
Occombe Valley	24	31	21	27	V	0.0008
Torquay Town	41	47	51	44	V	0.04
Hele	43	50	40	46	H	0.006
Edginswell	45	67	39	49	V	0.004
Huntshaw Cross	59	65	55	62	H	100
Barnstaple	43	30	40	46	V	0.040
Swimbridge	23	29	33	26	V	0.0066
Westward Ho	24	31	21	27	V	0.032
Chagford	24	31	21	27	V	0.012
Great Torrington	49	42	39	45	V	0.008
Muddiford	41	47	51	44	V	0.008
Brushford	24	31	21	27	V	0.02
North Bovey	43	50	40	46	V	0.034
Redruth	41	47	51	44	H	100
Tedbum St. Mary	52	31	42	48	V	0.020
Isles of Scilly	24	31	21	27	V	0.5
St. Just	61	54	58	64	V	0.25
Helston	61	54	58	64	V	0.01
Bossiney	61	54	58	64	V	0.014
Boscastle	23	29	33	26	V	0.0056
Perranporth	59	65	55	62	V	0.008
Porthtowan	24	31	21	27	V	0.008
Portreath	23	29	33	26	V	0.0016
Praa Sands	59	65	55	62	V	0.01
Porthleven	23	29	33	26	H	0.0016

	ITV	Channel 4	BBC 1	BBC 2	Polarisation	Max vision erp(kW)
St. Anthony-in-Roseland	23	29	33	26	V	0.0017
Gulval	23	29	33	26	V	0.026
Yorkshire						
Emley Moor	47	41	44	51	H	1000
Bradford West	49	67	57	63	V	0.0125
Cowling	43	50	40	46	V	0.016
Elland	61	54	58	64	H	0.004
Wharfedale	25	32	22	28	V	2
Sheffield	24	21	31	27	V	5
Skipton	49	42	39	45	V	10
Chesterfield	23	29	33	26	V	2
Halifax	24	31	21	27	V	0.5
Keighley	61	54	58	64	V	10
Keighley Town	23	29	33	26	V	0.006
Shatton Edge	48	54	52	58	V	1
Hebden Bridge	25	32	22	28	V	0.25
Ripponden	61	54	58	64	V	0.06
Cop Hill	25	32	22	28	V	1
Idle	24	31	21	27	V	0.25
Headingley	61	54	58	64	H	0.011
Beecroft Hill	59	65	55	62	V	1
Oxenhope	25	32	22	28	V	0.2
Calver Peak	49	42	39	45	V	0.25
Tideswell Moor	60	66	56	63	V	0.25
Hope	25	32	22	28	V	0.012
Addingham	43	50	40	46	V	0.025
Luddenden	60	67	57	63	V	0.059
Dronfield	59	65	55	62	H	0.003
Hasland	60	53	57	63	V	0.0065
Edale	60	53	57	63	V	0.004
Totley Rise	49	42	39	45	V	0.012
Cullingworth	49	68	66	39	H	0.013
Skipton Town	24	31	21	27	V	0.013
Batley	60	67	57	63	V	0.013
Heyshaw	60	53	57	63	V	0.5
Primrose Hill	60	67	57	63	V	0.028
Armitage Bridge	61	54	58	64	V	0.0065
Wincobank	59	65	55	62	V	0.0015
Holmfirth	56	68	49	66	V	0.026
Hagg Wood	59	65	55	62	V	0.033
Keighley Town	23	29	33	26	V	0.006
Sutton-in-Craven	23	29	33	26	V	0.012
Cragg Vale	61	54	58	64	V	0.025
Stocksbridge	61	54	58	64	V	0.012
Oughtibridge	59	65	55	62	V	0.02
Holmfield	59	65	55	62	V	0.022
Grassington	23	29	33	26	V	0.06
Cornholme	61	54	58	64	V	0.042
Walsden	60	67	57	63	V	0.05
Todmorden	49	42	39	45	V	0.5

(continued overleaf)

	ITV	Channel 4	BBC 1	BBC 2	Polarisation	Max vision erp(kW)
Walsden South	43	53	40	46	V	0.06
Copley	59	65	55	62	V	0.0014
Kettlewell	39	45	49	42	V	0.08
Conisbrough	60	53	57	63	V	0.006
Oliver's Mount	60	53	57	63	V	1
Hunmanby	43	50	40	46	V	0.06
Bradford West	49	67	57	63	V	0.0125
Brockwell	68	49	66	39	V	0.0063
Belmont	25	32	22	28	H	500
Weaverthorpe	59	65	55	62	V	0.045

24.7 Characteristics of UHF terrestrial television systems

24.7.1 World systems

System	Number of lines	Channel width (MHz)	Vision bandwidth (MHz)	Vision/sound separation (MHz)	Vestigial side-band (MHz)	Vision modulation	Sound modulation	Field frequency
A	625	8	5.5	+6	1.25	Neg.	FM	50
B	625	8	5.0	+5.5	1.25	Neg.	FM	50
C	625	8	5.0	+5.5	0.75	Neg.	FM	50
D	625	8	6.0	+6.5	1.25	Pos.	AM	50
E	625	8	6.0	+6.5	0.75	Neg.	FM	50
F	525	6	4.2	+4.5	1.25	Neg.	FM	60

A – UK and Eire
B – Eastern Europe
C – Most of Western Europe, Australia, New Zealand
D – France
E – Russia and Eastern Europe
F – USA, most of Central and South America, Japan

24.7.2 European systems

Country	System	Colour
Austria	C	PAL
Belgium	B	PAL
Bulgaria	No UHF system	
Cyprus	C	PAL
Czechoslovakia	E	SECAM
Denmark	No UHF system	PAL
Finland	C	PAL
France	D	SECAM
Germany	C	PAL
German DR	C	SECAM
Greece	B	PAL
Holland	C	PAL
Hungary	E	SECAM
Iceland	No UHF system	
Ireland	A	PAL
Italy	C	PAL
Luxembourg	D	SECAM
Malta	B	PAL
Monaco	D	SECAM
Norway	C	PAL
Poland	E	SECAM
Portugal	C	PAL
Romania	E	PAL
Spain	C	PAL
Sweden	C	PAL
Switzerland	C	PAL
Turkey	No UHF system	
Uk	A	PAL
USSR	E	SECAM
Yugoslavia	B	PAL

24.7.3 UK 625-line television system specification

Channel bandwidth	8 MHz
Upper sideband (vision signal)	5.5 MHz
Lower sideband (vision signal)	1.25 MHz
Vision modulation	AM negative
Sound modulation	FM
Sound deviation (max.)	±50 kHz
Sound pre-emphasis	50 µs
Sound carrier relative to vision carrier	+
Aspect ratio	4 : 3
Blanking and black level	76%
White level	20% peak
Sync. level	100% peak
Video bandwidth	5.5 MHz
Field frequency	50 Hz
Line frequency	15 625 Hz
Field sync. signal	5 equalising then 5 broad pulses, followed by 5 equalising pulses in 7.5 line periods
Field sync. and flyback intervals	2 × 25 line periods
Line period (approx.)	64 µs
Line syn. pulses (approx.)	4.7 µs
Line blanking (approx.)	12 µs
Field sync. pules (broad)	27.3 µs
Field sync. pulses (equalising)	2.3 µs
Colour subcarrier frequency	4.43361875 MHz
Burst duration	2.25 µs
Burst amplitude	equal to sync
Burst phase	180° ± 45°

24.7.4 UK 625-line television system field blanking details

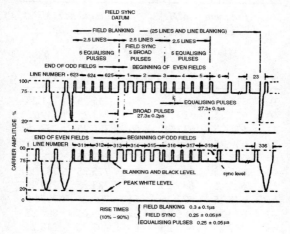

24.8 Terrestrial television channels

UK

BAND IV

Channel	Frequency (MHz)		Channel	Frequency (MHz)	
	Vision	Sound		Vision	Sound
21	471.25	477.25	28	527.25	533.25
22	479.25	485.25	29	535.25	541.25
23	487.25	493.25	30	543.25	549.25
24	495.25	501.25	31	551.25	557.25
25	503.25	509.25	32	559.25	565.25
26	511.25	517.25	33	567.25	573.25
27	519.25	525.25	34	575.25	581.25

BAND V

Channel	Frequency (MHz)		Channel	Frequency (MHz)	
	Vision	Sound		Vision	Sound
39	615.25	621.25	54	735.25	741.25
40	623.25	629.25	55	743.25	749.25
41	631.25	637.25	56	751.25	757.25
42	639.25	645.25	57	759.25	765.25
43	647.25	653.25	58	767.25	773.25
44	655.25	661.25	59	775.25	781.25
45	663.25	669.25	60	783.25	789.25
46	671.25	677.25	61	791.25	797.25
47	679.25	685.25	62	799.25	805.25
48	687.25	693.25	63	807.25	813.25
49	695.25	701.25	64	815.25	821.25
50	703.25	709.25	65	823.25	829.25
51	711.25	717.25	66	831.25	837.25
52	719.25	725.25	67	839.25	845.25
53	727.25	733.25	68	847.25	853.25

Republic of Ireland

Channel	Frequency (MHz)		Channel	Frequency (MHz)	
	Vision	Sound		Vision	Sound
IA	45.75	51.75	IF	191.25	197.25
IB	53.75	59.75	IG	199.25	205.25
IC	61.75	67.75	IH	207.25	213.25
ID	175.25	81.25	IJ	215.25	221.25
IE	183.25	189.25			

South Africa

Channel	Frequency (MHz) Vision	Sound	Channel	Frequency (MHz) Vision	Sound
4	175.25	181.25	9	215.25	221.25
5	183.25	189.25	10	223.25	229.25
6	191.25	197.25	11	231.25	237.25
7	199.25	205.25	13	247.43	253.43
8	207.25	213.25			

Australia

Channel	Frequency (MHz) Vision	Sound	Channel	Frequency (MHz) Vision	Sound
0	46.25	51.75	6	175.25	180.75
1	57.25	62.75	7	182.25	187.75
2	64.25	69.75	8	189.25	194.75
3	86.25	91.75	9	196.25	201.75
4	95.25	100.75	10	209.25	214.75
5	102.25	107.75	11	216.25	221.75
5A	138.25	143.75			

New Zealand

Channel	Frequency (MHz) Vision	Sound	Channel	Frequency (MHz) Vision	Sound
1	45.25	50.75	6	189.25	194.75
2	55.25	60.75	7	196.25	201.75
3	62.25	67.75	8	203.25	208.75
4	175.25	180.75	9	210.25	215.75
5	182.25	187.75			

USA

Channel	Frequency (MHz)		Channel	Frequency (MHz)	
	Vision	Sound		Vision	Sound
2	55.25	59.75	43	645.25	649.75
3	61.25	65.75	44	651.25	655.75
4	67.25	71.75	45	657.25	661.75
5	77.25	81.75	46	663.25	667.75
6	83.25	87.75	47	669.25	673.75
7	175.25	179.75	48	675.25	679.75
8	181.25	185.75	49	681.25	685.75
9	187.25	191.75	50	687.25	691.75
10	193.25	197.75	51	693.25	697.75
11	199.25	203.75	52	699.25	703.75
12	205.25	209.75	53	705.25	709.75
13	211.25	215.75	54	711.25	715.75
14	471.25	475.75	55	717.25	721.75
15	477.25	481.75	56	723.25	727.75
16	483.25	487.75	57	729.25	733.75
17	489.25	493.75	58	735.25	739.75
18	495.25	499.75	59	741.25	745.75
19	501.25	505.75	60	747.25	751.75
20	507.25	511.75	61	753.25	757.75
21	513.25	517.75	62	759.25	763.75
22	519.25	523.75	63	765.25	769.75
23	525.25	529.75	64	771.25	775.75
24	531.25	535.75	65	777.25	781.75
25	537.25	541.75	66	783.25	787.75
26	543.25	547.75	67	789.25	793.75
27	549.25	553.75	68	795.25	799.75
28	555.25	559.75	69	801.25	805.75
29	561.25	565.75	70	807.25	811.75
30	567.25	571.75	71	813.25	817.75
31	573.25	577.75	72	819.25	823.75
32	579.25	583.75	73	825.25	829.75
33	585.25	589.75	74	831.25	835.75
34	591.25	595.75	75	837.25	841.75
35	597.25	601.75	76	843.25	847.75
36	603.25	607.75	77	849.25	853.75
37	609.25	613.75	78	855.25	859.75
38	615.25	619.75	79	861.25	865.75
39	621.25	625.75	80	867.25	871.75
40	627.25	631.75	81	873.25	877.75
41	633.25	637.75	82	879.25	883.75
42	639.25	643.75	83	885.25	889.75

24.9 Terrestrial television aerial dimensions

Channel UHF Groups	Dimensions in cm										Channels covered in the uhf groups are:		
	A	B	C	D	E	F	G	H	a	b	c	Group letter	Colour Code	Channels
A	30.1	30	24.1	23	22.8	21.1	20.4	19.9	10.3	10.3	1.8	A	Red	21–34
B	26.5	21.7	18.9	18	17.8	16.5	16	15.5	8.9	8.9	1.8	B	Yellow	39–51
C	23.2	18.2	16	15.3	15	14	13.3	12.2	7.5	7.5	1.8	C	Green	50–66
D	26.1	23.5	18.4	16	15.5	14.8	13.8	13	7.6	7.6	1.8	D	Blue	49–68
E	27	26.5	21.1	18.6	17.9	17.6	16	15.8	15.8	15.8	1.8	E	Brown	39–68

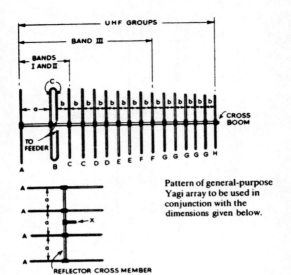

Pattern of general-purpose Yagi array to be used in conjunction with the dimensions given below.

24.10 AM broadcast station classes (USA)

The US AM broadcast band is 540 kHz to 1700 kHz, with 10 kHz channel spacings with centre frequencies divisible by ten (e.g. 780 kHz or 1540 kHz). Other countries in the western hemisphere operate on either 10 kHz channel spacings with frequencies ending in '5', or 9 kHz spacing. The Domestic Class is generally the class of station defined in 47 CFR Section 73.21. The Region 2 Class is generally the class of station as defined in the Region 2 [Western Hemisphere] AM Agreement. This class also corresponds to the class in the 1984 US–Canadian AM Agreement and the 1986 US–Mexican Agreement.

24.10.1 Class A station

A Class A station is an unlimited time station (that is, it can broadcast 24 hours per day) that operates on a clear channel. The operating power shall be not less than 10 kilowatts (kW) or more than 50 kW.

24.10.2 Class B station

A Class B station is an unlimited time station. Class B stations are authorized to operate with a minimum power of 250 watts and a maximum power of 50 kW. (If a Class B station operates with less than 250 W, the RMS must be equal to or greater than 141 mV/m at 1 km for the actual power.) If the station is authorized to operate in the expanded band (1610 to 1700 kHz), the maximum power is 10 kW.

24.10.3 Class C station

A Class C station is an unlimited time station that operates on a local channel. The power shall not be less than 250 W nor more than 1 kW. Class C stations that are licensed to operate with 100 W may continue to operate as licensed.

24.10.4 Class D station

A Class D station operates either daytime, limited time, or unlimited time with a night-time power less than 250 W and an equivalent RMS antenna field less than 141 mV/m at 1 km for the actual power. Class D stations shall operate with daytime powers not less than 0.250 kW nor more than 50 kW. *Note*: If a station is an existing daytime-only station, its class will be Class D.

24.11 FM broadcast frequencies and channel numbers (USA)

From US Code USC 47 CFR 73. The FM broadcast band consists of that portion of the radio frequency spectrum between 88 MHz and 108 MHz. It is divided into 100 channels of 200 kHz each. For convenience, the frequencies available for FM broadcasting (including those assigned to non-commercial educational broadcasting) are given numerical designations which are shown in the table below:

Frequency (MHz)	Channel No.
88.1	201
88.3	202
88.5	203
88.7	204
88.9	205
89.1	206
89.3	207
89.5	208
89.7	209
89.9	210
90.1	211
90.3	212
90.5	213
90.7	214
90.9	215
91.1	216
91.3	217
91.5	218
91.7	219
91.9	220
92.1	221
92.3	222
92.5	223
92.7	224
92.9	225
93.1	226
93.3	227
93.5	228
93.7	229
93.9	230
94.1	231
94.3	232
94.5	233
94.7	234
94.9	235
95.1	236
95.3	237
95.5	238
95.7	239
95.9	240
96.1	241
96.3	242
96.5	243
96.7	244
96.9	245
97.1	246
97.3	247
97.5	248
97.7	249
97.9	250

Frequency (MHz)	Channel No.
98.1	251
98.3	252
98.5	253
98.7	254
98.9	255
99.1	256
99.3	257
99.5	258
99.7	259
99.9	260
100.1	261
100.3	262
100.5	263
100.7	264
100.9	265
101.1	266
101.3	267
101.5	268
101.7	269
101.9	270
102.1	271
102.3	272
102.5	273
102.7	274
102.9	275
103.1	276
103.3	277
103.5	278
103.7	279
103.9	280
104.1	281
104.3	282
104.5	283
104.7	284
104.9	285
105.1	286
105.3	287
105.5	288
105.7	289
105.9	290
106.1	291
106.3	292
106.5	293
106.7	294
106.9	295
107.1	296
107.3	297
107.5	298
107.7	299
107.9	300

24.12 US television channel assignments

Channel No.	Frequency (MHz)
1	(No Ch. 1 assigned)
2	54−60
3	60−66
4	66−72
5	76−82
6	82−88
7	174−180
8	180−186
9	186−192
10	192−198
11	198−204
12	204−210
13	210−216
14	470−476
15	476−482
16	482−488
17	488−494
18	494−500
19	500−506
20	506−512
21	512−518
22	518−524
23	524−530
24	530−536
25	536−542
26	542−548
27	548−554
28	554−560
29	560−566
30	566−572
31	572−578
32	578−584
33	584−590
34	590−596
35	596−602
36	602−608
37	608−614
38	614−620
39	620−626
40	626−632
41	632−638
42	638−644
43	644−650
44	650−656
45	656−662
46	662−668
47	668−674

Channel No.	Frequency (MHz)
48	674–680
49	680–686
50	686–692
51	692–698
52	698–704
53	704–710
54	710–716
55	716–722
56	722–728
57	728–734
58	734–740
59	740–746
60	746–752
61	752–758
62	758–764
63	764–770
64	770–776
65	776–782
66	782–788
67	788–794
68	794–800
69	800–806

Notes:

1. In Alaska, television broadcast stations operating on Channel 5 (76–82 MHz) and on Channel 6 (82–88 MHz) shall not cause harmful interference to and must accept interference from non-Government fixed operations authorized prior to 1 January 1982.
2. Channel 37, 608–614 MHz is reserved exclusively for the radio astronomy service.
3. In Hawaii, the frequency band 488–494 MHz is allocated for non-broadcast use. This frequency band (Channel 17) will not be assigned in Hawaii for use by television broadcast stations.

24.13 Calculating radio antenna great circle bearings

Aiming radio antennas to target a particular area of the world requires calculation of the *great circle bearing* between your location and the other stations' location.

That bearing is calculated from some simple spherical trigonometry using a hand-held calculator or a computer program. Before talking about the maths, however, we need to establish a frame of reference that makes the system work.

24.13.1 Latitude and longitude

The need for navigation on the surface of the Earth caused the creation of a grid system uniquely to locate points on the surface of our globe. *Longitude* lines run from the north pole to the south pole, i.e. from north to south.

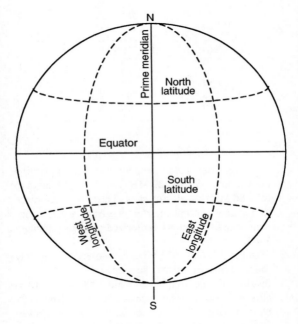

Fig. 24.1 *Lines of longitude and latitude*

The reference point (longitude zero), called the *prime meridian*, runs through Greenwich, England. The longitude of the prime meridian is 0 degrees. Longitudes west of the prime meridian are given a plus sign (+), while longitudes east of the prime are given a minus (−) sign. If you continue the prime meridian through the poles to the other side of the Earth it has a longitude of 180 degrees. Thus, the longitude values run from −180 degrees to +180 degrees, with ±180 degrees being the same line.

The observatory at Greenwich is also the point against which relative time is measured. Every 15 degree change of longitude is equivalent to a one hour difference with the Greenwich time. To the west, subtract one hour for each 15 degrees and to the east add one hour for each 15 degrees. Thus, the time on the East Coast of the United States is −5 hours relative to Greenwich time. At one time, we called time along the prime meridian *Greenwich mean time* (GMT), also called *Zulu time* to simplify matters for CW operators.

Latitude lines are measured against the equator, with distances north of the equator being taken as positive, and distances south of the equator being negative. The equator is 0 degrees latitude, while the north pole is +90 degrees latitude and the south pole is −90 degrees latitude.

Long ago navigators learned that the latitude can be measured by 'shooting' the stars and consulting a special atlas to compare the angle of certain stars with tables that translate to latitude numbers. The longitude measurement, however, is a bit different. For centuries sailors could measure latitude, but had to guess longitude (often with tragic results). In the early eighteenth century, the British government offered a large cash prize to anyone who could design a chronometer that could be taken to sea. By keeping the chronometer set accurately to Greenwich mean time, and comparing GMT against local time (i.e. at a time like high noon when the position of the sun is easy to judge), the longitude could be calculated. If you are interested in this subject, then most decent libraries have books on celestial navigation.

24.13.2 The great circle

On the surface of a globe, a curved line called a *great circle path* is the shortest distance between two points.

Consider two points on a globe: 'A' is your location, while 'B' is the other station's location. The distance 'D' is the great circle path between 'A' and 'B'.

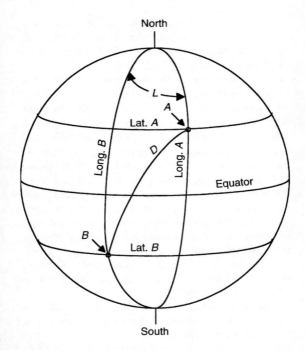

Fig. 24.2 *Great circle path*

The great circle path length can be expressed in either degrees or distance (e.g. miles, nautical miles or kilometres). To calculate the distance, it is necessary to find the difference in longitude (L) between your longitude (LA) and the other station's longitude (LB): $L = LA - LB$. Keep the signs straight. For example, if your longitude (LA) is 40 degrees, and the other station's longitude (LB) is -120 degrees, then $L = 40 - (-120) = 40 + 120 = 160$. The equation for distance (D) is:

$$\cos D = (\sin A \times \sin B) + (\cos A \times \cos B \times \cos L)$$

Where:

> D is the angular great circle distance
> A is your latitude
> B is the other station's latitude.

To find the actual angle, take the arccos of the above equation, i.e.

$$D = \arccos(\cos D)$$

In the next equation you will want to use D in angular measure, but later on will want to convert D to miles. To do this, multiply D in degrees by 69.4. Or, if you prefer metric measures, then $D \times 111.2$ yields kilometres. This is the approximate distance in statute miles between 'A' and 'B'.

To find the bearing from true north, work the equation below:

$$C = \arccos \left[\frac{\sin B - (\sin A \times \cos D)}{(\cos A \times \sin D)} \right]$$

However, this equation won't always give you the right answer unless you make some corrections.

The first problem is the 'same longitude error', i.e. when both stations are on the same longitude line. In this case, $L = LA - LB = 0$. If LAT $A >$ LAT B, then $C = 180$ degrees, but if LAT $A <$ LAT B, then $C = 0$ degrees. If LAT $A =$ LAT B, then what's the point of all these calculations?

The next problem is found when the condition $-180° \leq L \leq +180°$ is not met, i.e. when the absolute value of L is greater than $180°$, ABS(L) $> 180°$. In this case, either add or subtract 360 in order to make the value between $\pm 180°$:

$$\text{If } L > +180, \text{ then } L = L - 360$$
$$\text{If } L < -180, \text{ then } L = L + 360$$

One problem seen while calculating these values on a computer or hand calculator is the fact that the $\sin(X)$ and $\cos(X)$ cover different ranges. The $\sin(X)$ function returns values from $0°$ to $360°$, while the $\cos(X)$ function returns values only over $0°$ to $180°$. If L is positive, then the result bearing C is accurate, but if L is negative then the actual value of $C = 360 - C$.

357

The following test is necessary:

$$\text{If } L < 0 \text{ then}$$
$$L = 360 - L$$
$$\text{Else } L = L$$
$$\text{End if}$$

Another problem is seen whenever either station is in a high latitude near either pole ($\pm 90°$), or where both locations are very close together, or where the two locations are antipodal (i.e. on opposite points on the Earth's surface). The best way to handle these problems is to use a different equation that multiplies by the cosecant of D (i.e. $\operatorname{cosec}(D)$), rather than dividing by sine of D (i.e. $\sin(D)$).

25

ABBREVIATIONS AND SYMBOLS

25.1 Abbreviations

Many abbreviations are found as either capital or lower case letters, depending on publishers' styles. Symbols should generally be standard, as shown.

A	Ampere or anode
ABR	Auxiliary bass radiator
a.c.	Alternating current
A/D	Analogue to digital
ADC	Analogue to digital converter
Ae	Aerial
a.f.	Audio frequency
a.f.c.	Automatic frequency control
a.g.c.	Automatic gain control
a.m.	Amplitude modulation
AMPS	Advanced mobile phone system
ASA	Acoustical Society of America
ASCII	American Standard Code for Information Interchange
a.t.u.	Aerial tuning unit
AUX	Auxiliary
a.v.c.	Automatic volume control
b	Base of transistor
BAF	Bonded acetate fibre
B & S	Brown & Sharpe (U.S.) wire gauge
b.p.s.	Bits per second
BR	Bass reflex
BSI	British Standards Institution
C	Capacitor, cathode, centigrade, coulomb
c	Collector of transistor, speed of light
CB	Citizen's band
CCD	Charge coupled device
CCIR	International Radio Consultative Committee
CCITT	International Telegraph and Telephone Consultative Committee
CCTV	Closed circuit television
CDMA	Code division multiple access
chps	Characters per second
CPU	Central processor unit
CTD	Charge transfer device
CLK	Clock signal
CrO_2	Chromium dioxide
CMOS	Complementary metal oxide semiconductor
CTCSS	Continuous tone controlled signalled system
c.w.	Continuous wave
D	Diode
d	Drain of an f.e.t.

D/A	Digital to analogue
DAC	Digital to analogue converter
dB	Decibel
d.c.	Direct current
DCC	Double cotton covered
DCE	Data circuit-terminating equipment
DF	Direction finding
DIL	Dual-in-line
DIN	German standards institute
DMA	Direct memory access
DPDT	Double pole, double throw
DPSK	Differential phase shift keying
DPST	Double pole, single throw
dsb	or dsbam. Double sideband amplitude modulation
DSRR	Digital short range radio
DTE	Data terminal equipment
DTL	Diode-transistor logic
DTMF	Dual tone multi-frequency
DX	Long distance reception
e	Emitter of transistor
EAROM	Electrically alterable read only memory
ECL	Emitter coupled logic
e.h.t.	Extremely high tension (voltage)
e.m.f.	Electromotive force
en	Enamelled
EPROM	Erasable programmable read only memory
EQ	Equalisation
ERP	Effective radiated power
EROM	Erasable read only memory
ETACS	Extended total access communications system
F	Farad, fahrenheit or force
f	Frequency
FDM	Frequency division multiplex
FDMA	Frequency division multiple access
Fe	Ferrous
FeCr	Ferri-chrome
f.e.t.	Field effect transistor
FFSK	Fast frequency shift keying
f.m.	Frequency modulation
f.r.	Frequency response or range
f.s.d.	Full-scale deflection
FSK	Frequency shift keying
G	Giga (10^9)
g	Grid, gravitational constant
GMSK	Gaussian minimum shift keying
GSM	Global system mobile
H	Henry
h.f.	High frequency
Hz	Hertz (cycles per second)
I	Current
IB	Infinite baffle
i.c.	Integrated circuit
IF	Intermediate frequency
IHF	Institute of High Fidelity (U.S.)
$I^2L(HL)$	Integrated injection logic
i.m.d.	Intermodulation distortion
i/p	Input
i.p.s.	Inches per second

k	Kilo (10^3) or cathode
K	Kilo, in computing terms ($= 2^{10} = 1024$), or degrees Kelvin
L	Inductance or lumens
l.e.d.	Light emitting diode
l.f.	Low frequency
LIN	Linear
LOG	Logarithmic
LS	Loudspeaker
LSI	Large scale integration
l.w.	Long wave (approx. 1100–2000 m)
M	Mega (10^6)
m	Milli (10^{-3}) or metres
MHz	Megahertz
m.c.	Moving coil
mic	Microphone
MOS	Metal oxide semiconductor
MPU	Microprocessor unit
MPX	Multiplex
MSK	Minimum shift keying
m.w.	Medium wave (approx. 185–560 m)
n	Nano (10^9)
NAB	National Association of Broadcasters
Ni-Cad	Nickel-cadmium
n/c	Not connected; normally closed
n/o	Normally open
NMOS	Negative channel metal oxide semiconductor
o/c	Open channel; open circuit
o/p	Output
op-amp	Operational amplifier
p	Pico (10^{-12})
PA	Public address
PABX	Private automatic branch exchange
PAL	Phase alternation, line
p.a.m.	Pulse amplitude modulation
PCB	Printed circuit board
PCM	Pulse code modulation
PCN	Personal communications network
PLA	Programmable logic array
PLL	Phase locked loop
pm	Phase modulation
PMOS	Positive channel metal oxide semiconductor
P.P.M.	Peak programme meter
p.r.f.	Pulse repetition frequency
PROM	Programmable read only memory
PSK	Phase shift keying
PSS	Packet SwitchStream
PSTN	Public Switched Telephone Network
PSU	Power supply unit
PTFE	Polytetrafluoroethylene
PU	Pickup
PUJT	Programmable unijunction transistor
Q	Quality factor; efficiency of tuned circuit, charge
QAM	Quadruature (or quaternary) amplitude modulation
QPSK	Quadrature (or quaternary) phase shift keying
R	Resistance
RAM	Random access memory

RCF	Recommended crossover frequency
RIAA	Record Industry Association of America
r.f.	Radio frequency
r.f.c.	Radio frequency choke (coil)
r.m.s.	Root mean square
ROM	Read only memory
RTL	Resistor transistor logic
R/W	Read/write
RX	Receiver
S	Siemens
s	Source of an f.e.t.
s/c	Short circuit
SCR	Silicon-controlled rectifier
s.h.f.	Super high frequency
SI	International system of units
S/N	Signal-to-noise.
SPL	Sound pressure level
SPST	Single pole, single throw
SPDT	Single pole, double throw
ssb	Single sideband amplitude modulation
ssbdc	Single sideband diminished carrier
ssbsc	Single sideband suppressed carrier
SSI	Small scale integration
s.w.	Short wave (approx. 10–60 m)
s.w.g.	Standard wire gauge
s.w.r.	Standing wave ratio
T	Tesla
TACS	Total access communications system
TDM	Time division multiplex
TDMA	Time division multiple access
t.h.d.	Total harmonic distortion
t.i.d.	Transient intermodulation distortion
TR	Transformer
t.r.f.	Tuned radio frequency
TTL	Transistor transistor logic
TTY	Teletype unit
TVI	Television interface; television interference
TX	Transmitter
UART	Universal asynchronous receiver transmitter
u.h.f.	Ultra high frequency (approx. 470–854 MHz)
u.j.t.	Unijunction transistor
ULA	Uncommitted logic array
V	Volts
VA	Volt-amps
v.c.a.	Voltage controlled amplifier
v.c.o.	Voltage controlled oscillator
VCT	Voltage to current transactor
v.h.f.	Very high frequency (approx. 88–216 MHz)
v.l.f.	Very low frequency
VU	Volume unit
W	Watts
Wb	Weber
W/F	Wow and flutter
w.p.m.	Words per minute
X	Reactance
Xtal	Crystal
Z	Impedance
ZD	Zener diode

25.2 Letter symbols by unit name

Unit	Symbol	Notes
ampere	A	SI unit of electric current
ampere (turn)	At	SI unit of magnetomotive force
ampere-hour	Ah	
ampere per metre	Am^{-1}	SI unit of magnetic field strength
angstrom	Å	$1\text{Å} = 10^{-10}$ m
apostilb	asb	$1 \text{ asb} (1/\pi) \text{cd m}^{-2}$ A unit of luminance. The SI unit, candela per square metre, is preferred.
atmosphere:		
standard atmosphere	atm	$1 \text{ atm} = 101\,325\,\text{N m}^{-2}$
technical atmosphere	at	$1 \text{ at} = 1\,\text{kgf cm}^{-2}$
atomic mass unit (unified)	u	The (unified) atomic mass unit is defined as one-twelfth of the mass of an atom of the ^{12}C nuclide. Use of the old atomic mass unit (amu), defined by reference to oxygen, is deprecated.
bar	bar	$1 \text{ bar} = 100\,000\,\text{N m}^{-2}$
barn	b	$1 \text{ b} 10^{-28}\,\text{m}^2$
baud	Bd	Unit of signalling speed equal to one element per second.
becquerel	Bq	$1 \text{ Bq} = 1\,\text{s}^{-1}$ SI unit of radioactivity.
bel	B	
bit	b	
British thermal unit	Btu	
calorie (International Table calorie)	cal$_{IT}$	$1 \text{ cal}° - = 4.1868$ J The 9th Conférence Générale des Poids et Mesures adopted the joule as the unit of heat, avoiding the use of the calorie as far as possible.
calorie (thermochemical calorie)	cal	$1 \text{ cal} = 4.1840$ J (See note for International Table calorie.)
candela	cd	SI unit of luminous intensity.
candela per square inch	$cd \text{ in}^{-2}$	Use of the SI unit, candela per square metre, is preferred.
candela per square metre	$cd \text{ m}^{-2}$	SI unit of luminance. The name nit has been used.
candle		The unit of luminous intensity has been given the name *candela*; use of the word *candle* for this purpose is deprecated.
centimetre	cm	
circular mil	cmil	$1 \text{ cmil} = (\pi/4)10^{-6}\,\text{in}^2$
coulomb	C	SI unit of electrical charge.
cubic centimetre	cm^3	
cubic foot	ft^3	
cubic foot per minute	$ft^3 \text{ min}^{-1}$	
cubic foot per second	$ft^3 \text{s}^{-1}$	
cubic inch	in^3	
cubic metre	m^3	

363

cubic metre per second	$m^3 s^{-1}$	
cubic yard	yd^3	
curie	Ci	Unit of activity in the field of radiation dosimetry.
cycle	c	
cycle per second	cs^{-1}	Deprecated. Use hertz
decibel	dB	
degree (plane angle)	°	
degree (temperature):		Note that there is no space between the symbol ° and the letter. The use of the word *centigrade* of the
degree Celsius	°C	Celsius temperature scale was
degree Fahrenheit	°F	abandoned by the Conférence Générale des Poids et Mesures in 1948.
degree Kelvin		See Kelvin.
degree Rankine	°R	
dyne	dyn	
electronvolt	eV	
erg	erg	
erlang	E	Unit of telephone traffic.
farad	F	SI unit of capacitance.
foot	ft	
footcandle	fc	Use of the SI unit of illuminance, the lux (lumen per square metre), is preferred.
footlambert	fL	Use of the SI unit, the candela per square metre, is preferred.
foot per minute	ft min^{-1}	
foot per second	ft s^{-1}	
foot per second squared	ft s^{-2}	
foot pound-force	ft lb$_f$	
gal	Gal	$1 \, Gal = 1 \, cms^{-2}$
gallon	gal	The gallon, quart, and pint differ in the US and the UK, and their use is deprecated.
gauss	G	The gauss is the electromagnetic CGS (Centimetre Gram Second) unit of magnetic flux density. The SI unit, tesla, is preferred.
gigaelectronvolt	GeV	
gigahertz	GHz	
gilbert	Gb	The gilbert is the electromagnetic CGS (Centimetre Gram Second) unit of magnetomotive force. Use of the SI unit, the ampere (or ampere-turn), is preferred.
grain	gr	
gram	g	
gray	Gy	$1 \, Gy = 1 \, J \, kg^{-1}$ SI unit of absorbed dose.
henry	H	
hertz	Hz	SI unit of frequency.
horsepower	hp	Use of the SI unit, the watt, is preferred.
hour	h	Time may be designated as in the following example; $9^h46^m30^s$.
inch	in	
inch per second	in s^{-1}	
joule	J	SI unit of energy.

joule per Kelvin	JK^{-1}	SI unit of heat capacity and entropy.
Kelvin	K	SI unit of temperature (formerly called *degree Kelvin*). The symbol K is now used without the symbol°.
kiloelectronvolt	KeV	
kilogauss	kG	
kilogram	kg	SI unit of mass.
kilogram-force	kg$_f$	In some countries the name *kilopond* (kp) has been adopted for this unit.
kilohertz	kHz	
kilojoule	kJ	
kilohm	kΩ	
kilometre	km	
kilometre per hour	km h^{-1}	
kilopond	kp	See kilogram-force.
kilovar	kvar	
kilovolt	kV	
kilovoltampere	kVA	
kilowatt	kW	
kilowatthour	kWh	
knot	kn	1 kn = 1 nmi h^{-1}
lambert	L	The lambert is the CGS (Centimetre Gram Second) unit of luminance. The SI unit, candela per square metre, is preferred.
litre	l	
litre per second	ls^{-1}	
lumen	lm	SI unit of luminous flux.
lumen per square foot	lm ft^{-2}	Use of the SI unit, the lumen per square metre, is preferred.
lumen per square metre	lm m^{-2}	SI unit of luminous excitance
lumen per watt	lm W^{-1}	SI unit of luminous efficacy.
lumen second	lms	SI unit of quantity of light.
lux	lx	1 lx = 1 lm m^{-2} SI unit of illuminance.
maxwell	Mx	The maxwell is the electromagnetic CGS (Centimetre Gram Second) unit of magnetic flux. Use of the SI unit, the weber, is preferred.
megaelectronvolt	MeV	
megahertz	MHz	
megavolt	MV	
megawatt	MW	
megohm	MΩ	
metre	m	SI unit of length, 1 mho = 1Ω^{-1} = 1 S
mho	mho	
microampere	μA	
microbar	μbar	
microfarad	μF	
microgram	μg	
microhenry	μH	
micrometre	μm	
micron		The name *micrometre* (μm) is preferred.
microsecond	μS	
microwatt	μW	
mil	mil	1 mil = −0.001 in.
mile		
nautical	nmi	
statute	mi	
mile per hour	mi h^{-1}	
milliampere	mA	

millibar	mbar	mb may be used.
milligal	mGal	
milligram	mg	
millihenry	mH	
millilitre	ml	
millimetre	mm	
conventional millimetre of mercury	mm Hg	1 mm Hg = 133.322 N m^{-2}.
millimicron		The name *nanometre* (nm) is preferred.
millisecond	ms	
millivolt	mV	
milliwatt	mW	
minute (plane angle)	...$'$	
minute (time)	min	Time may be designated as in the following example: $9^h46^m30^s$.
mole	mol	SI unit of amount of substance.
nanoampere	nA	
nanofarad	nF	
nanometre	nm	
nanosecond	ns	
nanowatt	nW	
nautical mile	nmi	
neper	Np	
newton	N	SI unit of force.
newton metre	Nm	
newton per square metre	Nm^{-2}	See pascal.
nit	nt	1 nt = 1 cd m^{-2} See candela per square metre.
oersted	Oe	The oersted is the electromagnetic CGS (Centimetre Gram Second) unit of magnetic field strength. Use of the SI unit, the ampere per metre, is preferred.
ohm	Ω	SI unit of electrical resistance.
ounce (avoirdupois)	oz	
pascal	Pa	SI unit of pressure or stress. 1 Pa = 1 N m^{-2}
picoampere	pA	
picofarad	pF	
picosecond	ps	
picowatt	pW	
pint	pt	The gallon, quart, and pint differ in the US and the UK, and their use is deprecated.
pound	lb	
poundal	pdl	
pound-force	lb$_f$	
pound-force feet	lb$_f$ ft	
pound-force per square inch	lb$_f$in^{-2}	
pound per square inch		Although use of the abbreviation psi is common, it is not recommended See pound-force per square inch.
Quart	qt	The gallon, quart, and pint differ in the US and the UK, and their use is deprecated.

rad	rd	Unit of absorbed dose in the field of radiation dosimetry.
revolution per minute	r min^{-1}	Although use of the abbreviation rpm is common, it is not recommended.
revolution per second	rs^{-1}	
roentgen	R	Unit of exposure in the field of radiation dosimetry.
second (plane angle)	...$''$	
second (time)	s	SI unit of time. Time may be designated as in the following example: $9^h46^m30^s$.
siemens	S	SI unit of conductance. $1\,S = 1\,\Omega^{-1}$
square foot	ft^2	
square inch	in^2	
square metre	m^2	
square yard	yd^2	
steradian	sr	SI unit of solid angle.
stilb	sb	$1\,sb = 1\,cd\,cm^{-2}$ A CGS unit of luminance. Use of the SI unit, the candela per square metre, is preferred.
tesla	T	SI unit of magnetic flux density. $1\,T = 1\,Wb\,m^{-2}$.
tonne	t	$1\,t = 1000\,kg$
(unified) atomic mass unit	u	See atomic mass unit (unified).
var	var	Unit of reactive power.
volt	V	SI unit of electromotive force.
voltampere	VA	SI unit of apparent power.
watt	W	SI unit of power.
watthour	Wh	
watt per steradian	W sr^{-1}	SI unit of radiant intensity.
watt per steradian square metre	W (sr m^2)$^{-1}$	SI unit of radiance.
weber	Wb	SI unit of magnetic flux. $1\,Wb = 1\,Vs$.
yard	yd	

25.3 Electric quantities

Quantity	Symbol	Unit	Symbol
Admittance	Y	siemens	S
Angular frequency	ω	hertz	Hz
Apparent power	S	watt	W
Capacitance	C	farad	F
Charge	Q	coulomb	C
Charge density	ρ	coulomb per square metre	Cm^{-2}
Conductance	G	siemens	S
Conductivity	κ, γ, σ	siemens per metre	Sm^{-1}
Current	I	ampere	A
Current density	j, J	ampere per square metre	Am^{-2}
Displacement	D	coulomb per square metre	Cm^{-2}
Electromotive force	E	volt	V
Energy	E	joule	J
Faraday constant	F	coloumb per mole	$Cmol^{-1}$
Field strength	E	volt per metre	Vm^{-1}
Flux	ψ	coulomb	C
Frequency	v, f	hertz	Hz
Impedance	Z	ohm	Ω
Light, velocity of in a vacuum	c	metre per second	ms^{-1}
Period	T	second	s
Permeability	μ	henry per metre	Hm^{-1}
Permeability of space	μ_o	henry per metre	Hm^{-1}
Permeance	Λ	henry	H
Permittivity	ε	farad per metre	Fm^{-1}
Permittivity	ε_o	farad per metre	Fm^{-1}
Phase	ϕ	–	–
Potential	V, U	volt	V
Power	P	watt	W
Quality factor	Q	–	–
Reactance	X	ohm	Ω
Reactive power	Q	watt	W
Relative permeability	μ_r	–	–
Relative permittivity	ε_r	–	–
Relaxation time	τ	second	s
Reluctance	R	reciprocal henry	H^{-1}
Resistance	R	ohm	Ω
Resistivity	ρ	ohm metre	Ωm
Susceptance	B	siemens	S
Thermodynamic temperature	T	kelvin	K
Time constant	τ	second	s
Wavelength	λ	metre	m

25.4 Transistor letter symbols

Bipolar

C_{cb}, C_{ce}, C_{eb} International capacitance (collector-to-base, collector-to-emitter, emitter-to-base).

C_{Ibo}, C_{ieo} Open-circuit input capacitance (common-base, common-emitter).

C_{ibs}, C_{ieo} Short-circuit input capacitance (common-base, common-emitter).

C_{obo}, C_{oeo} Open-circuit output capacitance (common-base, common-emitter).

C_{obs}, C_{oes} Short-circuit output capacitance (common-base, common-emitter).

C_{rbs}, C_{res} Short-circuit output reverse transfer capacitance (common-base, common-emitter).

C_{tc}, C_{te} Depletion-layer capacitance (collector, emitter).

f_{hfb}, f_{fe} Small-signal short-circuit forward current transfer ratio cutoff frequency (common-base, common-emitter).

f_{max} Maximum frequency of oscillation.

f_T Transition frequency or frequency at which small-signal forward current transfer ratio (common-emitter) extrapolates to unity.

F_1 Frequency of unity current transfer ratio.

G_{PB}, G_{PE} Large-signal insertion power gain (common-base, common-emitter).

G_{pb}, G_{pe} Small-signal insertion power gain (common-base, common-emitter).

G_{TB}, G_{TE} Large-signal transducer power gain (common-base, common-emitter).

G_{tb}, G_{te} Small-signal transducer power gain (common-base, common-emitter).

h_{FB}, h_{FE} Static forward current transfer ratio (common-base, common-emitter).

h_{fb}, h_{fe} Small-signal short-circuit forward current transfer ratio (common-base, common-emitter).

h_{ib}, h_{ie} Small-signal short-circuit input impedance (common-base, common-emitter).

$h_{ie(imag)}$ or $Im(h_{ie})$ Imaginary part of the small-signal short-circuit input impedance (common-emitter).

$h_{ie(real)}$ or $Re(h_{ie})$ Real part of the small-signal short-circuit input impedance (common-emitter).

h_{ob}, h_{oe} Small-signal open-circuit output admittance (common-base, common-emitter).

$h_{oe(imag)}$ or $Im(h_{oe})$ Imaginary part of the small-signal open-circuit output admittance (common-emitter).

$h_{oe(real)}$ or $Re(h_{oe})$ Real part of the small-signal open-circuit output admittance (common-emitter).

h_{rb}, h_{re} Small-signal open-circuit reverse voltage transfer ratio (common-base, common-emitter).

I_B, I_C, I_E Current, d.c. (base-terminal, collector-terminal, emitter-terminal).

I_b, I_c, I_e Current, r.m.s. value of alternating component (base-terminal, collector-terminal, emitter-terminal).

i_B, i_C, i_E Current, instantaneous total value (base-terminal, collector-terminal, emitter-terminal).

I_{BEV} Base cutoff current, d.c.

I_{CBO} Collector cutoff current, d.c., emitter open.

$I_{EIE2(off)}$ Emitter cutoff current.

I_{EBO} Emitter cutoff current, d.c., collector open.

$I_{Ec(ofs)}$ Emitter-collector offset current.

I_{ECS} Emitter cutoff current, d.c., base-short-circuited to collector.

P_{IB}, P_{IE} Large-signal input power (common-base, common-emitter).

P_{ib}, P_{ie} Small-signal input power (common-base, common-emitter).

P_{OB}, P_{OE} Large-signal output power (common-base, common-emitter).

P_{ob}, P_{oe} Small-signal output power (common-base, common-emitter).

P_T Total nonreactive power input to all terminals.

$r_b{'}C_c$ Collector-base time constant.

$r_{CE(sat)}$ Saturation resistance, collector-to-emitter.

$\mathrm{Re}(y_{ie})$

$\mathrm{Re}(y_{oe})$

$r_{ele2(on)}$ Small-signal emitter-emitter on-state resistance.

R_θ Thermal resistance.

T_j Junction temperature.

t_d Delay time.

t_r Fall time.

t_{off} Turn-off time.

t_{on} Turn-on time.

t_p Pulse time.

t_r Rise time.

t_s Storage time.

t_w Pulse average time.

V_{BB}, V_{CC}, V_{EE} Supply voltage, d.c. (base, collector, emitter).

V_{BC}, V_{BE}, V_{CB}, V_{CE}, V_{EB}, V_{EC} Voltage, d.c. or average (base-to-collector, base-to-emitter, collector-to-base, collector-to-emitter, emitter-to-base, emitter-to-collector).

v_{bc}, v_{be}, v_{cb}, v_{ce}, v_{eb}, v_{ec} Voltage instantaneous value of alternating component (base-to-collector, base-to-emitter, collector-to-base, collector-to-emitter, emitter-to-base, emitter-to-collector).

$V_{(BR)CBO}$ **(formerly BV_{CBO})** Breakdown voltage, collector-to-base, emitter open.

V_{RT} Reach-through (punch-through) voltage.

y_{fb}, y_{fe} Small-signal short-circuit forward-transfer admittance (common-base, common-emitter).

y_{ib}, y_{ie} Small-signal short-circuit input admittance (common-base, common-emitter).

$y_{ie(imag)}$ or $Im(y_{ie})$ Imaginary part of the small-signal short-circuit input admittance (common-emitter).

$y_{ie(real)}$ or $Re(y_{ie})$ Real part of the small-signal short-circuit input admittance (common-emitter).

y_{ob}, y_{oe} Small-signal short-circuit output admittance (common-base, common-emitter).

$y_{oe(imag)}$ or $Im(y_{oe})$ Imaginary part of the small-signal short-circuit output admittance (common-emitter).

$y_{oe(real)}$ or $Re(y_{oe})$ Real part of the small-signal short-circuit output admittance (common-emitter).

y_{rb}, y_{re} Small-signal short-circuit reverse transfer admittance (common-base, common-emitter).

Unijunction

η Intrinsic standoff ratio.

$I_{B_2(mod)}$ Interbase modulated current.

I_{EB_2O} Emitter reverse current.

I_p Peak-point current.

I_v Valley-point current.

r_{BB} Interbase resistance.

T_j Junction temperature.

t_p Pulse time.

t_w Pulse average time.

$V_{B_2B_1}$ Interbase voltage.

$V_{EB_1(sat)}$ Emitter saturation voltage.

V_{OB_1} Base-1 peak voltage.

V_p Peak-point voltage.

V_v Valley-point voltage.

Field Effect

b_{fs}, b_{is}, b_{os}, b_{rs} Common-source small-signal (forward transfer, input, output, reverse transfer) susceptance.

C_{ds} Drain-source capacitance.

C_{du} Drain-substrate capacitance.

C_{iss} Short-circuit input capacitance, common-source.

C_{oss} Short-circuit output capacitance, common-source.

C_{ras} Short-circuit reverse transfer capacitance, common-source.

$\overline{F}$ or F Noise figure, average or spot.

g_{ts}, g_{is}, g_{os}, g_{rs} Signal (forward transfer, input, output, reverse transfer) conductance.

G_{pg}, G_{ps} Small-signal insertion power gain (common-gate, common-source).

G_{tg}, G_{ts} Small-signal transducer power gain (common-gate, common-source).

$I_{D(off)}$ Drain cutoff current.

$I_{D(on)}$ On-state drain current.

I_{DSS} Zero-gate-voltage drain current.

I_G Gate current, d.c.

I_{GF} Forward gate current.

I_{GR} Reverse gate current.

I_{GSS} Reverse gate current, drain short-circuited to source.

I_{GSSF} Forward gate current, drain short-circuited to source.

I_{GSSR} Reverse gate current, drain short-circuited to source.

I_n Noise current, equivalent input.

$\mathrm{Im}(y_{rs})$, $\mathrm{Im}(y_{is})$, $\mathrm{Im}(y_{os})$, $\mathrm{Im}(y_{rs})$.

I_s Source current, d.c.

$I_{S(off)}$ Source cutoff current.

I_{SDS} Zero-gate-voltage source current.

$r_{ds(on)}$ Small-signal drain-source on-state resistance.

$r_{DS(on)}$ Static drain-source on-state resistance.

$t_{d(on)}$ Turn-on delay time.

f_r Fall time.

t_{off} Turn-off time.

t_{on} Turn-on time.

t_p Pulse time.

t_r Rise time.

t_w Pulse average time.

$V_{(BR)GSS}$ Gate-source breakdown voltage.

$V_{(BR)GSSF}$ Forward gate-source breakdown voltage.

$V_{(BR)GSSR}$ Reverse gate-source breakdown voltage.

V_{DD}, V_{GG}, V_{SS} Supply voltage, d.c. (drain, gate, source).

V_{DG} Drain-gate voltage.

V_{DS} Drain-source voltage.

$V_{DS(on)}$ Drain-source on-state voltage.

V_{DU} Drain-substrate voltage.

V_{GS} Gate-source voltage.

V_{GSF} Forward gate-source voltage.

V_{GSR} Reverse gate-source voltage.

$V_{GS(off)}$ Gate-source cutoff voltage.

$V_{GS(th)}$ gate-source threshold voltage.

V_{GU} Gate-substrate voltage.

V_n Noise voltage equivalent input.

V_{SU} Source-substrate voltage.

y_{fs} Common-source small-signal short-circuit forward transfer admittance.

y_{is} Common-source small-signal short-circuit input admittance.

y_{os} Common-source small-signal short-circuit output admittance.

25.5 Component symbols

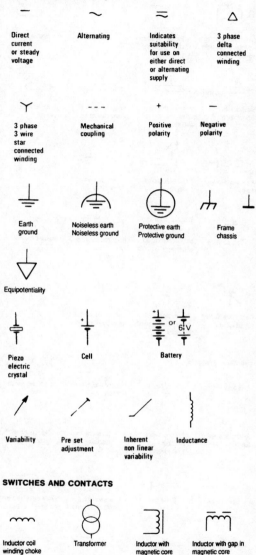

—	~	≂	△
Direct current or steady voltage	Alternating	Indicates suitability for use on either direct or alternating supply	3 phase delta connected winding

Y	- - -	+	—
3 phase 3 wire star connected winding	Mechanical coupling	Positive polarity	Negative polarity

Earth ground

Noiseless earth Noiseless ground

Protective earth Protective ground

Frame chassis

Equipotentiality

Piezo electric crystal

Cell

Battery

or 6 V

Variability

Pre set adjustment

Inherent non linear variability

Inductance

SWITCHES AND CONTACTS

Inductor coil winding choke

Transformer

Inductor with magnetic core

Inductor with gap in magnetic core

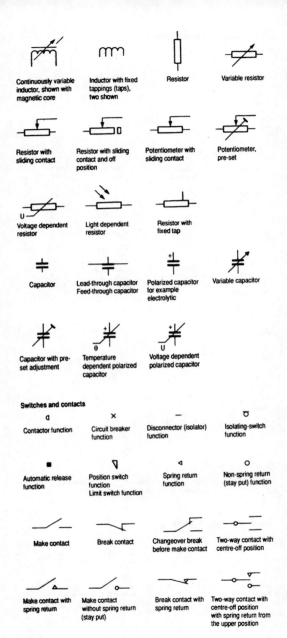

Continuously variable inductor, shown with magnetic core

Inductor with fixed tappings (taps), two shown

Resistor

Variable resistor

Resistor with sliding contact

Resistor with sliding contact and off position

Potentiometer with sliding contact

Potentiometer, pre-set

Voltage dependent resistor

Light dependent resistor

Resistor with fixed tap

Capacitor

Lead-through capacitor Feed-through capacitor

Polarized capacitor for example electrolytic

Variable capacitor

Capacitor with pre-set adjustment

Temperature dependent polarized capacitor

Voltage dependent polarized capacitor

Switches and contacts

Contactor function

Circuit breaker function

Disconnector (isolator) function

Isolating-switch function

Automatic release function

Position switch function
Limit switch function

Spring return function

Non-spring return (stay put) function

Make contact

Break contact

Changeover break before make contact

Two-way contact with centre-off position

Make contact with spring return

Make contact without spring return (stay put)

Break contact with spring return

Two-way contact with centre-off position with spring return from the upper position

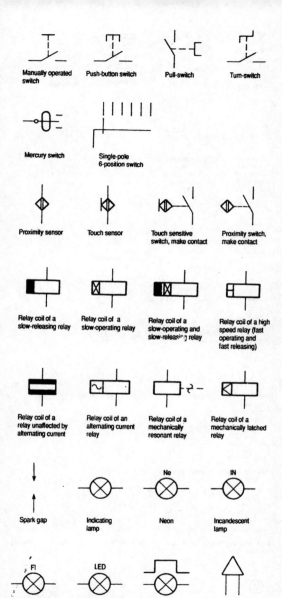

Manually operated switch

Push-button switch

Pull-switch

Turn-switch

Mercury switch

Single-pole 6-position switch

Proximity sensor

Touch sensor

Touch sensitive switch, make contact

Proximity switch, make contact

Relay coil of a slow-releasing relay

Relay coil of a slow-operating relay

Relay coil of a slow-operating and slow-releasing relay

Relay coil of a high speed relay (fast operating and fast releasing)

Relay coil of a relay unaffected by alternating current

Relay coil of an alternating current relay

Relay coil of a mechanically resonant relay

Relay coil of a mechanically latched relay

Spark gap

Indicating lamp

Neon

Incandescent lamp

Fluorescent lamp

Light emitting diode

Flashing lamp

Siren

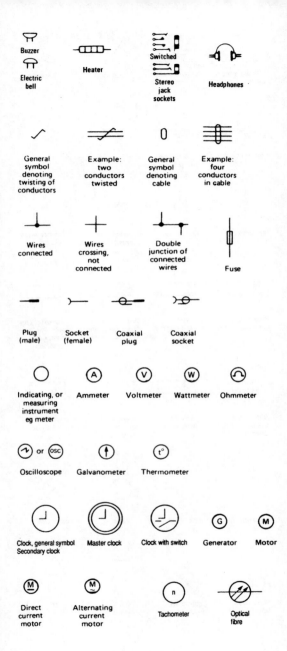

Buzzer

Electric bell

Heater

Switched

Stereo jack sockets

Headphones

General symbol denoting twisting of conductors

Example: two conductors twisted

General symbol denoting cable

Example: four conductors in cable

Wires connected

Wires crossing, not connected

Double junction of connected wires

Fuse

Plug (male)

Socket (female)

Coaxial plug

Coaxial socket

Indicating, or measuring instrument eg meter

Ammeter

Voltmeter

Wattmeter

Ohmmeter

Oscilloscope

Galvanometer

Thermometer

Clock, general symbol Secondary clock

Master clock

Clock with switch

Generator

Motor

Direct current motor

Alternating current motor

Tachometer

Optical fibre

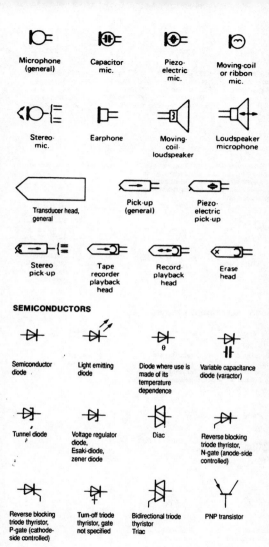

Microphone (general)

Capacitor mic.

Piezo-electric mic.

Moving-coil or ribbon mic.

Stereo-mic.

Earphone

Moving-coil loudspeaker

Loudspeaker microphone

Transducer head, general

Pick-up (general)

Piezo-electric pick-up

Stereo pick-up

Tape recorder playback head

Record playback head

Erase head

SEMICONDUCTORS

Semiconductor diode

Light emitting diode

Diode where use is made of its temperature dependence

Variable capacitance diode (varactor)

Tunnel diode

Voltage regulator diode, Esaki-diode, zener diode

Diac

Reverse blocking triode thyristor, N-gate (anode-side controlled)

Reverse blocking triode thyristor, P-gate (cathode-side controlled)

Turn-off triode thyristor, gate not specified

Bidirectional triode thyristor Triac

PNP transistor

NPN transistor with collector connected to the envelope

Unijunction transistor with P-type base

Unijunction transistor, N-type base

Junction field effect transistor with N-type channel

Junction field effect transistor with P-type channel

IGFET enhancement type, single gate, P-type channel without substrate connection

IGFET enhancement type, single gate, N-type channel without substrate connection

IGFET enhancement type, single gate, P-type channel with substrate connection brought out

IGFET enhancement type, single gate, N-type channel with substrate internally connected to source

IGFET, depletion type with two gates, N-type channel with substrate connection brought out

Photodiode

Photovoltaic cell

Phototransistor, PNP type shown

Hall generator with four ohmic connections

Magnetoresistor, linear type shown

Opto isolator shown with light emitting diode and photo-transistor

SOUND ELECTRONIC DEVICES

Recording or reproducing, arrow points in direction of energy transfer

Recording and reproducing, radiating and receiving

Magnetostriction type

Moving coil or ribbon type

Moving iron type

Stereo type

Low audio frequencies

High audio frequencies

Disk

Tape or film

Drum

25.6 Radiocommunications symbols

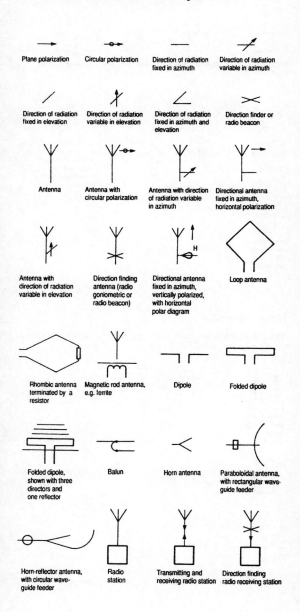

Plane polarization

Circular polarization

Direction of radiation fixed in azimuth

Direction of radiation variable in azimuth

Direction of radiation fixed in elevation

Direction of radiation variable in elevation

Direction of radiation fixed in azimuth and elevation

Direction finder or radio beacon

Antenna

Antenna with circular polarization

Antenna with direction of radiation variable in azimuth

Directional antenna fixed in azimuth, horizontal polarization

Antenna with direction of radiation variable in elevation

Direction finding antenna (radio goniometric or radio beacon)

Directional antenna fixed in azimuth, vertically polarized, with horizontal polar diagram

Loop antenna

Rhombic antenna terminated by a resistor

Magnetic rod antenna, e.g. ferrite

Dipole

Folded dipole

Folded dipole, shown with three directors and one reflector

Balun

Horn antenna

Paraboloidal antenna, with rectangular wave-guide feeder

Horn-reflector antenna, with circular wave-guide feeder

Radio station

Transmitting and receiving radio station

Direction finding radio receiving station

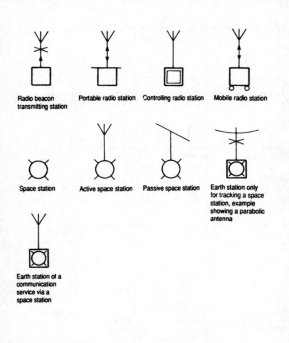

Radio beacon transmitting station

Portable radio station

Controlling radio station

Mobile radio station

Space station

Active space station

Passive space station

Earth station only for tracking a space station, example showing a parabolic antenna

Earth station of a communication service via a space station

Logic elements

Where two symbols are shown for a logic element, the second symbol is not recognised in B.S. 3939.

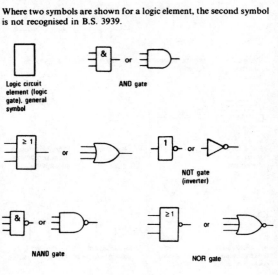

Logic circuit element (logic gate), general symbol

or **AND gate**

or

or **NOT gate (inverter)**

or **NAND gate**

or **NOR gate**

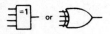

Exclusive OR gate

Logic identity gate; produces a logic 1 output if, and only if, all inputs are the same

Wired connection where a number of elements are wired together to achieve the effect of an AND or an OR operation without the use of an explicit element

Wire AND connection

Amplifier for logic diagrams

Symbol grouping to save space

Schmitt trigger

RS bistable element

Delay element, general symbol; this element produces a logic 1 output a set period of time after its input has changed from logic 0 to logic 1 and changes back to a logic 0 output a set period of time after its input has reverted to logic 0

Common control block; to make diagram clearer, inputs common to a number of related elements may be shown connected to a common control block

 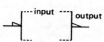

Direction of data flow should normally be from top to bottom. This symbol is used to indicate exceptions to the normal flow direction

Input/output polarity indicator indicating that the logic 1 state is the less positive level, ie negative logic is in force at this point

383

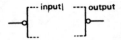

Logic negator input/
output, indicating the
state of the logic
variable is reversed at
the input

Inhibiting input; when
standing at its logic 1
state, prevents a logic 1
output (or a logic 0
output if the output is
negated) whatever the
state of the other
input variables

Negated inhibiting
input; when standing
at logic 0, prevents
a logic 1 output (or a
logic 0 output if the
output is negated)

Input or output
not carrying logic
information

Dynamic input

The (transitory)
internal 1-state
corresponds with the
transition from the
external 0-state to the
external 1-state. At all
other times, the internal
logic state is 0

Dynamic input with
logic negation

Bi-threshold input
Input with hysteresis
e.g. Schmitt trigger

Open-circuit output
(e.g. open-collector,
open-emitter, open-
drain, open-source)

3-state output

Monostable,
retriggerable (during
the output pulse)

Monostable, non
triggerable (during
the output pulse)

Astable

Synchronously
starting

25.7 Block diagram symbols

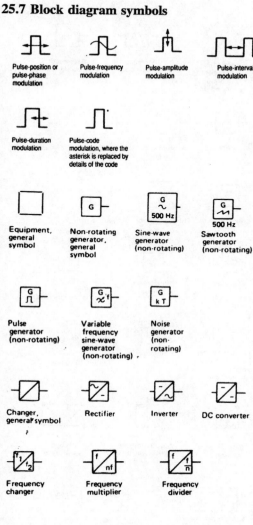

Pulse-position or pulse-phase modulation

Pulse-frequency modulation

Pulse-amplitude modulation

Pulse-interval modulation

Pulse-duration modulation

Pulse-code modulation, where the asterisk is replaced by details of the code

Equipment, general symbol

Non-rotating generator, general symbol

Sine-wave generator (non-rotating)

Sawtooth generator (non-rotating)

Pulse generator (non-rotating)

Variable frequency sine-wave generator (non-rotating)

Noise generator (non-rotating)

Changer, general symbol

Rectifier

Inverter

DC converter

Frequency changer

Frequency multiplier

Frequency divider

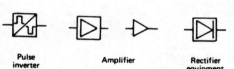

Pulse inverter

Amplifier

Rectifier equipment

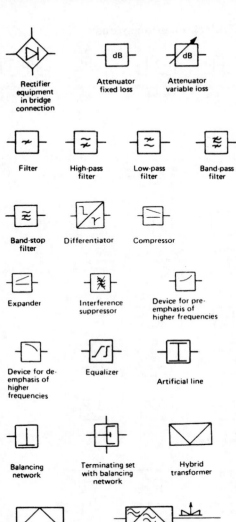

Rectifier equipment in bridge connection

Attenuator fixed loss

Attenuator variable loss

Filter

High-pass filter

Low-pass filter

Band-pass filter

Band-stop filter

Differentiator

Compressor

Expander

Interference suppressor

Device for pre-emphasis of higher frequencies

Device for de-emphasis of higher frequencies

Equalizer

Artificial line

Balancing network

Terminating set with balancing network

Hybrid transformer

Modulator, demodulator or discriminator

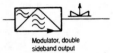

Modulator, double sideband output

Demodulator, single sideband with suppressed carrier to audio

25.8 Frequency spectrum symbols

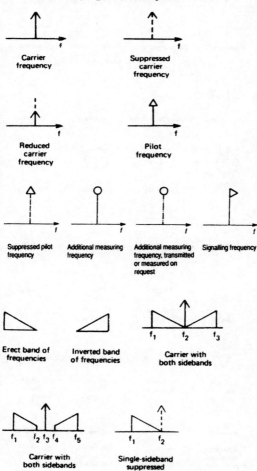

Carrier frequency

Suppressed carrier frequency

Reduced carrier frequency

Pilot frequency

Suppressed pilot frequency

Additional measuring frequency

Additional measuring frequency, transmitted or measured on request

Signalling frequency

Erect band of frequencies

Inverted band of frequencies

Carrier with both sidebands

Carrier with both sidebands

Single-sideband suppressed carrier

25.9 Equipment marking symbols (BS6217)

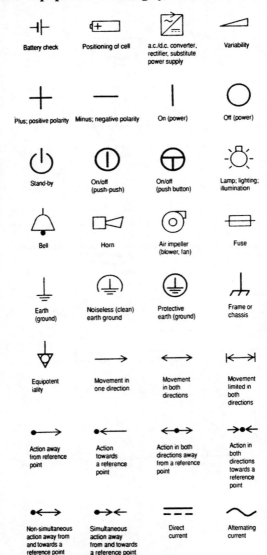

Battery check	Positioning of cell	a.c./d.c. converter, rectifier, substitute power supply	Variability
Plus; positive polarity	Minus; negative polarity	On (power)	Off (power)
Stand-by	On/off (push-push)	On/off (push button)	Lamp; lighting; illumination
Bell	Horn	Air impeller (blower, fan)	Fuse
Earth (ground)	Noiseless (clean) earth ground	Protective earth (ground)	Frame or chassis
Equipotentiality	Movement in one direction	Movement in both directions	Movement limited in both directions
Action away from reference point	Action towards a reference point	Action in both directions away from a reference point	Action in both directions towards a reference point
Non-simultaneous action away from and towards a reference point	Simultaneous action away from and towards a reference point	Direct current	Alternating current

Both direct
and alternating
current

Input

Output

Dangerous
voltage

Treble
control

Bass control

Aerial

Dipole

Frame aerial

Tuner

Signal
strength
attenuation

Tuning

Automatic
frequency
control

Muting

Colour
(qualifying
symbol)

TV, video

Colour TV

TV monitor

Colour TV
monitor

TV
receiver

Colour TV
receiver

Focus

Brightness

Contrast

Colour
saturation

Crispener

Hue

Horizontal
synchronization

Vertical
synchron-
ization

Horizontal
picture
shift

Vertical
picture shift

Horizontal
picture
amplitude

Vertical
picture
amplitude

Picture size
adjustment

Horizontal
linearity

Vertical
linearity

Monophonic	Stereophonic	Balance	Omni-directional microphone
Bidirectional microphone	Unidirectional microphone	Earphone	Headphones
Stereo headphones	Headset	Loudspeaker	Loudspeaker microphone
Microphone	Stereophonic microphone	Amplifier	Music
Pick-up for disc records	Stereophonic pick-up for disc records	Piezo-electric pick-up, crystal or ceramic	Dynamic pick-up
Telephone adapter	High-pass filter	Low-pass filter	Tape recorder
Magnetic tape stereo sound recorder	Recording on tape	Play-back or reading from tape	Erasing from tape
Monitoring at the input	Monitoring from tape after recording on tape	Monitoring during play-back	Recording lock on tape recorders

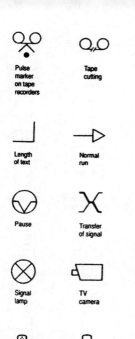

Pulse marker on tape recorders	Tape cutting	Start	instruction
Length of text	Normal run	Fast run	Stop
Pause	Transfer of signal	Rejection	Foot switch
Signal lamp	TV camera	Colour TV camera	Video tape recorder
Colour video tape recorders	Video recording	Colour video recording	Video play back
Colour video play back	Slow run	Recapitulate	Heading marker
Aerial rotation	Short pulse	Long pulse	Bearing marker
Ship's head-up presentation	North up presentation	Anti-sea clutter	Anti-rain clutter

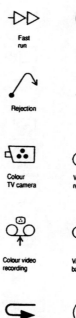

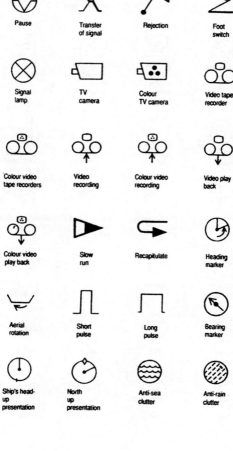

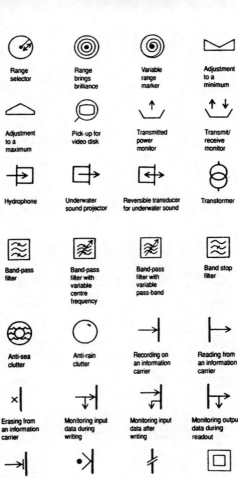

Range selector	Range brings brilliance	Variable range marker	Adjustment to a minimum
Adjustment to a maximum	Pick-up for video disk	Transmitted power monitor	Transmit/ receive monitor
Hydrophone	Underwater sound projector	Reversible transducer for underwater sound	Transformer
Band-pass filter	Band-pass filter with variable centre frequency	Band-pass filter with variable pass-band	Band stop filter
Anti-sea clutter	Anti-rain clutter	Recording on an information carrier	Reading from an information carrier
Erasing from an information carrier	Monitoring input data during writing	Monitoring input data after writing	Monitoring output data during readout
Recording lock	Marker	Cutting	Class II equipment

Fast start	Fast stop	Test voltage	Variability in steps

Sound	Clock	Rejection filter	Rectifier

392

Teacher

Student

Group of students

All students

Frame adjustment

Graphical recorder

Printer

d.c./a.c. converter

Variable band-stop filter

Gyro indicator

Gyro indicator setting

Gyro-compass true bearing

Relative bearing

Bearing ruler setting

Phase calibration

Angle calibration

Sense aerial switch

Speak

Listen

Morse key

Link unit

Travelling wave tube amplifier

Signalling sender

Signalling receiver

Demodulator

Modulator

Modem

Principal control panel

'On' for a part of equipment

'Off' for a part of equipment

Stand-by state for a part of equipment

In position of a bi-stable push control

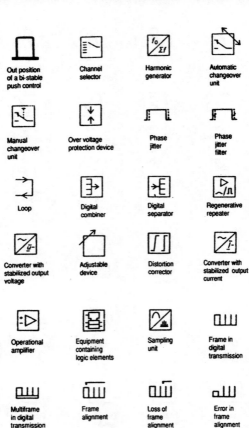

Out position of a bi-stable push control	Channel selector	Harmonic generator	Automatic changeover unit
Manual changeover unit	Over voltage protection device	Phase jitter	Phase jitter filter
Loop	Digital combiner	Digital separator	Regenerative repeater
Converter with stabilized output voltage	Adjustable device	Distortion corrector	Converter with stabilized output current
Operational amplifier	Equipment containing logic elements	Sampling unit	Frame in digital transmission
Multiframe in digital transmission	Frame alignment	Loss of frame alignment	Error in frame alignment
Two-level signal	Three-level signal	Binary coded signal	Indirect lighting
Low intensity lighting	Hand-held switch		

26

MISCELLANEOUS DATA

26.1 Fundamental constants

Constant	Symbol	Value
Boltzmann constant	k	1.38062×10^{-23} JK^{-1}
Electron charge, proton charge	e	$\pm 1.60219 \times 10^{-19}$ C
Electron charge-to-mass ratio	e/m	1.7588×10^{11} Ckg^{-1}
Electron mass	m_e	9.10956×10^{-31} kg
Electron radius	r_e	2.81794×10^{-15} m
Faraday constant	F	9.64867×10^{4} Cmol^{-1}
Neutron mass	m_n	1.67492×10^{-27} kg
Permeability of space	μ_o	$4\pi \times 10^{-7}$ Hm^{-1}
Permittivity of space	ε_o	8.85419×10^{-12} Fm^{-1}
Planck constant	h	6.6262×10^{-34} Js
Proton mass	m_p	1.67251×10^{-27} kg
Velocity of light	c	2.99793×10^{8} ms^{-1}

26.2 Electrical relationships

Amperes $\times$ ohms = **volts**
Volts $\div$ amperes = **ohms**
Volts $\div$ ohms = **amperes**
Amperes $\times$ volts = **watts**
(Amperes)^2 $\times$ ohms = **watts**
(Volts)^2 $\div$ ohms = **watts**
Joules per second = **watts**
Coulombs per second = **amperes**
Amperes $\times$ seconds = **coulombs**
Farads $\times$ volts = **coulombs**
Coulombs $\div$ volts = **farads**
Coulombs $\div$ farads = **volts**
Volts $\times$ coulombs = **joules**
Farads $\times$ (volts)^2 = **joules**

26.3 Dimensions of physical properties

Length: metre [L]. Mass: kilogram [M]. Time: second [T]. Quantity of electricity: coulomb [Q]. Area: square metre $[L^2]$. Volume: cubic metre $[L^3]$.

Velocity: metre per second	$[LT^{-1}]$
Acceleration: metre per second^2	$[LT^{-2}]$
Force: newton	$[MLT^{-2}]$
Work: joule	$[ML^2T^{-2}]$
Power: watt	$[ML^2T^{-3}]$
Electric current: ampere	$[QT^{-1}]$
Voltage: volt	$[ML^2T^{-2}Q^{-1}]$
Electric resistance: ohm	$[ML^2T^{-1}Q^{-2}]$
Electric conductance: siemens	$[M^{-1}L^{-2}TQ^2]$
Inductance: henry	$[ML^2Q^{-2}]$
Capacitance: farad	$[M^{-1}L^{-2}T^2Q^2]$
Current density: ampere per metre^2	$[L^{-2}T^{-1}Q]$
Electric field strength: volt per metre	$[MLT^{-2}Q^{-1}]$
Magnetic flux: weber	$[MLT^2T^{-1}Q^{-1}]$
Magnetic flux density: tesla	$[MT^{-1}Q^{-1}]$
Energy: joule	$[ML^2T^{-2}]$
Frequency: hertz	$[T^{-1}]$
Pressure: pascal	$[ML^{-1}T^{-2}]$

26.4 Fundamental units

Quantity	Unit	Symbol
Amount of a substance	mole	mol
Charge	coulomb	C
Length	metre	m
Luminous intensity	candela	cd
Mass	kilogram	kg
Plane angle	radian	rad
Solid angle	steradian	sr
Thermodynamic temperature	kelvin	K
Time	second	s

26.5 Greek alphabet

Capital letters	Small letters	Greek name	English equivalent	Capital letters	Small letters	Greek name	English equivalent
A	α	Alpha	a	N	ν	Nu	n
B	β	Beta	b	Ξ	ξ	Xi	x
Γ	γ	Gamma	g	O	o	Omicron	ŏ
Δ	δ	Delta	d	Π	π	Pi	p
E	ε	Epsilon	e	P	ρ	Rho	r
Z	ζ	Zeta	z	Σ	σ	Sigma	s
H	η	Eta	é	T	τ	Tau	t
Θ	θ	Theta	th	Y	υ	Upsilon	u
I	ι	Iota	i	Φ	φ	Phi	ph
K	κ	Kappa	k	X	χ	Chi	ch
Λ	λ	Lambda	l	Ψ	ψ	Psi	ps
M	μ	Mu	m	Ω	ω	Omega	ŏ

26.6 Standard units

Ampere Unit of electric current, the constant current which, if maintained in two straight parallel conductors of infinite length of negligible circular cross-section and placed one metre apart in a vacuum, will produce between them a force equal to 2×10^{-7} newton per metre length.

Ampere-hour Unit of quantity of electricity equal to 3,600 coulombs. One unit is represented by one ampere flowing for one hour.

Candela Unit of luminous intensity. It is the luminous intensity, in the perpendicular direction, of a surface of $1/600,000$ m^{-2} of a full radiator at the temperature of freezing platinum under a pressure of 101,325 newtons m^{-2}.

Coulomb Unit of electric charge, the quantity of electricity transported in one second by one ampere.

Decibel (dB) Unit of acoustical or electrical power ratio. Although the bel is officially the unit, this is usually regarded as being too large, so the decibel is preferred. The difference between two power levels is P_1 and P_2, is given as

$$10 \log_{10} \frac{P_1}{P_2} \text{ decibels}$$

Farad Unit of electric capacitance. The capacitance of a capacitor between the plates of which there appears a difference of potential of one volt when it is charged by one coulomb of electricity. Practical units are the microfarad (10^{-6} farad), the nanofared (10^{-9}) and the picofarad (10^{-12} farad).

Henry Unit of electrical inductance. The inductance of a closed circuit in which an electromotive force of one volt is produced when the electric current in the circuit varies uniformly at the rate of one ampere per second. Practical units are the microhenry (10^{-6} henry) and the millihenry (10^{-3} henry).

Hertz Unit of frequency. The number of repetitions of a regular occurrence in one second.

Joule Unit of energy, including work and quantity of heat. The work done when the point of application of a force of one newton is displaced through a distance of one metre in the direction of the force.

Kilovolt-ampere 1,000 volt-amperes.

Kilowatt 1,000 watts.

Lumen m^{-2}, lux Unit of illuminance of a surface.

Mho Unit of conductance, see Siemens.

Newton Unit of force. That force which, applied to a mass of one kilogram, gives it an acceleration of one metre per second per second.

Ohm Unit of electric resistance. The resistance between two points of a conductor when a constant difference of potential of one volt, applied between these two points, produces in the conductor a current of one ampere.

Pascal Unit of sound pressure. Pressure is usually quoted as the root mean square pressure for a pure sinusoidal wave.

Siemens Unit of conductance, the reciprocal of the ohm. A body having a resistance of 4 ohms would have a conductance of 0.25 siemens.

Tesla Unit of magnetic flux density, equal to one weber per square metre of circuit area.

Volt Unit of electric potential. The difference of electric potential between two points of a conducting wire carrying a constant current of one ampere, when the power dissipated between these points is equal to one watt.

Volt-ampere The product of the root-mean-square volts and root-mean-square amperes.

Watt Unit of power, equal to one joule per second. Volts times amperes equals watts.

Weber Unit of magnetic flux. The magnetic flux which, linking a circuit of one turn, produces in it an electromotive force of one volt as it is reduced to zero at a uniform rate in one second.

Light, velocity of Light waves travel at 300,000 kilometres per second (approximately). Also the velocity of radio waves.

Sound, velocity of Sound waves travel at 332 metres per second in air (approximately) at sea level.

26.7 Decimal multipliers

Prefix	Symbol	Multiplier	Prefix	Symbol	Multiplier
tera	T	10^{12}	centi	c	10^{-2}
giga	G	10^{9}	milli	m	10^{-3}
mega	M	10^{6}	micro	μ	10^{-6}
kilo	k	10^{3}	nano	n	10^{-9}
hecto	h	10^{2}	pico	p	10^{-12}
deka	da	10	femto	f	10^{-15}
deci	d	10^{-1}	atto	a	10^{-18}

26.8 Electronic multiple and sub-multiple conversion

	These multiply by the figures below													
To convert these to →	Pico-	Nano-	Micro-	Milli-	Centi-	Deci-	Units	Deka-	Hekto-	Kilo-	Myria-	Mega-	Giga-	Tera-
Pico-		0.001	10^{-6}	10^{-9}	10^{-10}	10^{-11}	10^{-12}	10^{-13}	10^{-14}	10^{-15}	10^{-16}	10^{-18}	10^{-21}	10^{-24}
Nano-	1,000		0.001	10^{-6}	10^{-7}	10^{-8}	10^{-9}	10^{-10}	10^{-11}	10^{-12}	10^{-13}	10^{-15}	10^{-18}	10^{-21}
Micro-	10^6	1,000		0.001	0.0001	10^{-5}	10^{-6}	10^{-7}	10^{-8}	10^{-9}	10^{-10}	10^{-12}	10^{-15}	10^{-16}
Milli-	10^9	10^6	1,000		0.1	0.01	0.001	0.0001	10^{-5}	10^{-6}	10^{-7}	10^{-9}	10^{-12}	10^{-15}
Centi-	10^{10}	10^7	10,000	10		0.1	0.01	0.001	0.0001	10^{-5}	10^{-6}	10^{-8}	10^{-11}	10^{-14}
Deci-	10^{11}	10^8	10^5	100	10		0.1	0.01	0.001	0.0001	10^{-5}	10^{-7}	10^{-10}	10^{-13}

Units	10^{12}	10^9	10^6	1,000	100	10		0.1	0.01	0.001	0.0001	10^{-6}	10^{-9}	10^{-12}
Deka-	10^{13}	10^{10}	10^7	10,000	1,000	100	10		0.1	0.01	0.001	10^{-5}	10^{-8}	10^{-11}
Hekto-	10^{14}	10^{11}	10^8	10^5	10,000	1,000	100	10		0.1	0.01	0.0001	10^{-7}	10^{-10}
Kilo-	10^{15}	10^{12}	10^9	10^6	10^5	10,000	1,000	100	10		0.1	0.001	10^{-6}	10^{-9}
Myria-	10^{16}	10^{13}	10^{10}	10^7	10^6	10^5	10,000	1,000	100	10		0.01	10^{-5}	10^{-8}
Mega-	10^{18}	10^{15}	10^{12}	10^9	10^8	10^7	10^6	10^5	10,000	1,000	100		0.001	10^{-6}
Giga-	10^{21}	10^{18}	10^{15}	10^{12}	10^{11}	10^{10}	10^9	10^8	10^7	10^6	10^5	1,000		0.001
Tera-	10^{24}	10^{21}	10^{18}	10^{15}	10^{14}	10^{13}	10^{12}	10^{11}	10^{10}	10^9	10^8	10^6	1,000	

26.9 Useful formulae

Boolean Algebra (laws of)

Absorption:	$A + (A.B)$	$= A$
	$A.(A + B)$	$= A$
Annulment:	$A + 1$	$= 1$
	$A.0$	$= 0$
Association:	$(A + B) + C$	$= A + (B + C)$
	$(A.B).C$	$= A.(B.C)$
Commutation:	$A + B$	$= B + A$
	$A.B$	$= B.A$
Complements:	$A + \overline{A}$	$= 1$
	$A.\overline{A}$	$= 0$
De Morgan's:	$\overline{(A + B)}$	$= \overline{A}.\overline{B}$
	$\overline{(A.B)}$	$= \overline{A} + \overline{B}$
Distributive:	$A.(B + C)$	$= (A.B) + (A.C)$
	$A + (B.C)$	$= (A + B).(A + C)$
Double negation:	$\overline{\overline{A}}$	$= A$
Identity:	$A + O$	$= A$
	$A.1$	$= A$
Tautology:	$A.A$	$= A$
	$A + A$	$= A$

Capacitance

The capacitance of a parallel plate capacitor can be found from

$$C = \frac{0.885 \, KA}{d}$$

C is in picofarads, K is the dielectric constant (air = 1). A is the area of the plate in square cm and d the thickness of the dielectric.

Calculation of overall capacitance with:
Parallel capacitors $-C = C_1 + C_2 + \cdots$
Series capacitors $-\dfrac{1}{C} = \dfrac{1}{C_1} + \dfrac{1}{C_2} + \cdots$

Characteristic impedance

$$\text{(open wire)} \, Z = 276 \log \frac{2D}{d} \, \text{ohms}$$

where

$$\left. \begin{array}{l} D = \text{wire spacing} \\ d = \text{wire diameter} \end{array} \right\} \text{in same units.}$$

$$\text{(coaxial)} \, Z = \frac{138}{\sqrt{(K)}} \log \frac{d_o}{d_i} \, \text{ohms}$$

where K = dielectric constant, d_o = outside diameter of inner conductor, d_i = inside diameter of outer conductor.

Dynamic resistance

In a parallel-tuned circuit at resonance the dynamic resistance is

$$R_d = \frac{L}{Cr} = Q\omega L = \frac{Q}{\omega C}\text{ohms}$$

where L = inductance (henries), C = capacitance (farads), r = effective series resistance (ohms). Q = Q-value of coil, and $\omega = 2\pi \times$ frequency (hertz).

Frequency – wavelength – velocity

(See also Resonance).

The velocity of propagation of a wave is

$$v = f\lambda \text{ metres per second}$$

where f = frequency (hertz) and λ = wavelength (metres).

For electromagnetic waves in free space the velocity of propagation v is approximately 3×10^8 m/sec, and if f is expressed in kilohertz and λ in metres

$$f = \frac{300,000}{\lambda}\text{kilohertz} \qquad f = \frac{300}{\lambda}\text{megahertz}$$

or

$$\lambda = \frac{300,000}{f}\text{metres} \qquad \lambda = \frac{300}{f}\text{metres}$$

f in kilohertz $\qquad\qquad$ f in megahertz

Radio horizon distance

The radio horizon at VHF/UHF and up is approximately 15% further than the optical horizon. Several equations are used in calculating the disance. If D is the distance to the radio horizon, and H is the antenna height, then:

$$D = k\sqrt{H}$$

1. When D is in statute miles (5280 feet) and H in feet, then $K = 1.42$.
2. When D is in nautical miles (6000 feet) and H in feet, then $K = 1.23$.
3. When D is in kilometres and H is in metres, then $K = 4.12$.

Impedance

The impedance of a circuit comprising inductance, capacitance and resistance in series is

$$Z = \sqrt{R^2 + \left(\omega L - \frac{1}{\omega C}\right)^2}$$

where R = resistance (ohms), $\omega = 2\pi \times$ frequency (hertz). L = inductance (henries), and C = capacitance (farads).

Inductance

Single layer coils

$$L(\text{in microhenries}) = \frac{a^2 N^2}{9a + 10l} \text{ approximately}$$

If the desired inductance is known, the number of turns required may be determined by the formula

$$N = \frac{5L}{na^2}\left[1 + \sqrt{\left(1 + \frac{0.36n^2a^3}{L}\right)}\right]$$

where N = number of turns, a = radius of coil in inches, n = number of turns per inch. L = inductance in microhenries (μH) and l = length of coil in inches.

Calculation of overall inductance with:
Series inductors $-L = L_1 + L_2 + \cdots$
Parallel inductors $-\dfrac{1}{L} = \dfrac{1}{L_1} + \dfrac{1}{L_2} + \cdots$

Meter conversions

Increasing range of ammeters or milliammeters

Current range of meter can be increased by connecting a shunt resistance across meter terminals. If R_m is the resistance of the meter; R_s the value of the shunt resistance and n the number of times it is wished to multiply the scale reading, then

$$R_s = \frac{R_m}{(n-1)}$$

Increasing range of voltmeters

Voltage range of meter can be increased by connecting resistance in series with it. If this series resistance is R_s and R_m and n as before, then $R_s = R_m \times (n-1)$.

Negaive feedback

Voltage feedback

$$\text{Gain with feedback} = \frac{A}{1 + Ab}$$

where A is the original gain of the amplifier section over which feedback is applied (including the output transformer if included) and b is the fraction of the output voltage fed back.

Distortion with feedback $= \dfrac{d}{1 + Ab}$ approximately

where d is the original distortion of the amplifier.

Ohm's Law

$$I = \frac{V}{R} \quad V = IR \quad R = \frac{V}{I}$$

where $I =$ current (amperes), $V =$ voltage (volts), and $R =$ resistance (ohms).

Power

In a d.c. circuit the power developed is given by

$$W = VI = \frac{V^2}{R} = I^2 R \text{ watts}$$

where $V =$ voltage (volts). $I =$ current (amperes), and $R =$ resistance (ohms).

Power ratio

$$P = 10 \log \frac{P_2}{P_2}$$

where $P =$ ratio in decibels, P_1 and P_2 are the two power levels.

Q

The Q value of an inductance is given by

$$Q = \frac{\omega L}{R}$$

Reactance

The reactance of an inductor and a capacitor respectively is given by

$$X_L = \omega L \text{ ohms} \qquad X_C = \frac{1}{\omega C} \text{ ohms}$$

where $\omega = 2\pi \times$ frequency (hertz), $L =$ inductance (henries), and $C =$ capacitance (farads).

The total resistance of an inductance and a capacitance in series is $X_L - X_C$.

Resistance

Calculation of overall resistance with:

Series resistors $- R = R_1 + R_2 + \cdots$

Parallel resistors $-\dfrac{1}{R} = \dfrac{1}{R_1} + \dfrac{1}{R_2} + \cdots$

Resonance

The resonant frequency of a tuned circuit is given by

$$f = \dfrac{1}{2\pi\sqrt{LC}}\text{hertz}$$

where L = inductance (henries), and C = capacitance (farads). If L is in microhenries (μH) and C is pico-farads, this becomes

$$f = \dfrac{10^6}{2\pi\sqrt{LC}}\text{kilohertz}$$

The basic formula can be rearranged

$$L = \dfrac{1}{4\pi^2 f^2 C}\text{henries} \qquad C = \dfrac{1}{4\pi^2 f L}\text{farads}.$$

Since $2\pi f$ is commonly represented by ω, these expressions can be written

$$L = \dfrac{1}{\omega^2 C}\text{henries} \qquad C = \dfrac{1}{\omega^2 L}\text{farads}.$$

Time constant

For a combination of inductance and resistance in series the time constant (i.e. the time required for the current to reach 63% of its final value) is given by

$$\tau = \dfrac{L}{R}\text{seconds}$$

where L = inductance (henries), and R = resistance (ohms).

For a combination of capacitance and resistance in series the time constant (i.e. the time required for the voltage across the capacitance to reach 63% of its final value) is given by

$$\tau = CR \text{ seconds}$$

where C = capacitance (farads), and R = resistance (ohms).

Transformer ratios

The ratio of a transformer refers to the ratio of the number of turns in one winding to the number of turns in the other winding. To avoid confusion it is always desirable to state in which sense the ratio is being expressed: e.g. the 'primary-to-secondary' ratio n_p/n_s. The turns ratio is related to the impedance ratio thus

$$\frac{n_p}{n_s} \sqrt{\frac{Z_p}{Z_s}}$$

where n_p = number of primary turns, n_s = number of secondary turns, Z_p = impedance of primary (ohms), and Z_s = impedance of seconndary (ohms).

Wattage rating

If resistance and current values are known,

$$W = I^2 R \text{ when } I \text{ is in amperes}$$

or

$$W = \frac{\text{Milliamps}^2}{1,000,000} \times R$$

If wattage rating and value of resistance are known, the safe current for the resistor can be calculated from

$$\text{milliampers} = 1.000 \times \sqrt{\frac{\text{Watts}}{\text{Ohms}}}$$

Wavelength of tuned circuit

Formula for the wavelength in metres of a tuned oscillatory circuit is: $1885 \sqrt{LC}$, where L = inductance in microhenries and C = capacitance in microfarads.

26.10 Colour codes

26.10.1 Resistor and capacitor colour coding

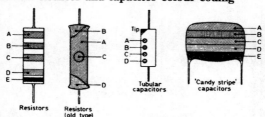

Resistors   Resistors (old type)   Tubular capacitors   'Candy stripe' capacitors

Tantalum capacitors

	1	2	3	4
Black	—	0	× 1	10 V
Brown	1	1	× 10	
Red	2	2	× 100	
Orange	3	3	—	
Yellow	4	4	—	6·3 V
Green	5	5	—	16 V
Blue	6	6	—	20 V
Violet	7	7	—	
Grey	8	8	× 0·01	25 V
White	9	9	× 0·1	3 V
				(Pink 35 V)

Resistor and capacitor colour coding

Colour	Brand A	Brand B	Band C (multiplier)		Band D (tolerance)			Band E	
			Resistors	Capacitors	Resistors	Capacitors Up to 10 pF	Over 10 pF	Resistors	Polyester capacitors
Black	–	0	1	1	–	2 pF	±20%	–	–
Brown	1	1	10	10	±1%	0.1 pF	±1%	–	–
Red	2	2	100	100	±2%	–	±2%	–	250 v.w.
Orange	3	3	1,000	1,000	–	–	±2.5%	–	–
Yellow	4	4	10,000	10,000	–	–	–	–	–
Green	5	5	100,000	–	–	0.5 pF	±5%	–	–
Blue	6	6	1,000,000	–	–	–	–	–	–
Violet	7	7	10,000,000	–	–	0.25 pF	–	–	–
Grey	8	8	10^8	0.01 µF	–	1 pF	–	–	–
White	9	9	10^9	0.1 µF	–	–	±10%	–	–
Silver	–	–	0.01	–	±10%	–	–	–	–
Gold	–	–	0.1	–	±5%	–	–	–	–
Pink	–	–	–	–	–	–	–	Hi-Stab.	–
None	–	–	–	–	±20%	–	–	–	–

Note that adjacent bands may be of the same colour unseparated.

Preferred values

E12 Series
1.0 1.2 1.5 1.8 2.2 2.7
3.3 3.9 4.7 5.6 6.8 8.2
and their decades

E24 Series
1.0 1.1 1.2 1.3 1.5 1.6
1.8 2.0 2.2 2.4 2.7 3.0
3.3 3.6 3.9 4.3 4.7 5.1
5.6 6.2 6.8 7.5 8.2 9.1
and their decades

26.10.2 Resistor and capacitor letter and digit code table (BS 1852)

Resistor values are indicated as follows:

0.47 Ω	marked	R47
1 Ω		1R0
4.7 Ω		4R7
47 Ω		47R
100 Ω		100R
1 kΩ		1K0
10 kΩ		10K
10 MΩ		10M

A letter following the value shows the tolerance.

$F = \pm1\%$; $G = \pm2\%$; $J = \pm5\%$; $K = \pm10\%$; $M = \pm20\%$; R33M $= 0.33\Omega \pm 20\%$; 6K8F $= 6.8k\Omega \pm 1\%$.

Capacitor values are indicated as:

0.68 pF	marked	p68
6.8 pf		6p8
1000 pF		1n0
6.8 nf		6n8
1000 nF		1μ0
6.8 μF		6μ8

Tolerance is indicated by letters as for resistors. Values up to 999 pF are marked in pF, from 1000 pf to 999 000 pF ($=$ 999 nF) as nF (1000 pF $=$ InF) and from 1000 nF ($=$ 1 μF) upwards as μF.

Some capacitors are marked with a code denoting the value in pF (first two figures) followed by a multiplier as a power of ten ($3 = 10^3$). Letters denote tolerance as for resistors but $C = \pm0.25$ pf. E.g. $123 J = 12 \, pF \times 10^3 \pm 5\% = 12\,000 \, pF$ (or 0.12μF).

26.10.3 Stereo pick-up lead colour codes

Number of leads	Right high	Right low	Left high	Left low	Ground
3	Red	–	White	–	Black
4	Red	Green	White	Blue	–
5	Red	Green	White	Blue	Black

26.11 RC time constants

Time (sec.)	(Capacitance (microfarads))									
	0.1	0.2	0.3	0.4	0.5	0.6	0.7	0.8	0.9	1.0
0.1	1.0 M	500 k	333 k	250 k	200 k	166 k	143 k	125 k	111 k	100 k
0.15	1.5 M	750 k	500 k	375 k	300 k	250 k	214 k	188 k	167 k	150 k
0.2	2.0 M	1.00 M	666 k	500 k	400 k	333 k	286 k	250 k	222 k	200 k
0.25	2.5 M	1.25 M	833 k	625 k	500 k	417 k	357 k	313 k	278 k	250 k
0.3	3.0 M	1.50 M	1.00 M	750 k	600 k	500 k	429 k	375 k	333 k	300 k
0.35	3.5 M	1.75 M	1.17 M	875 k	700 k	583 k	500 k	438 k	389 k	350 k
0.4	4.0 M	2.00 M	1.33 M	1.00 M	800 k	666 k	571 k	500 k	444 k	400 k
0.45	4.5 M	2.25 M	1.50 M	1.13 M	900 k	750 k	643 k	563 k	500 k	450 k
0.5	5.0 M	2.50 M	1.67 M	1.25 M	1.0 M	833 k	714 k	625 k	500 k	500 k
0.55	5.5 M	2.75 M	1.83 M	1.38 M	1.1 M	917 k	786 k	688 k	555 k	550 k
0.6	6.0 M	3.00 M	2.00 M	1.50 M	1.2 M	1.00 M	857 k	750 k	666 k	600 k
0.65	6.5 M	3.25 M	2.17 M	1.63 M	1.3 M	1.08 M	929 k	813 k	722 k	650 k
0.7	7.0 M	3.50 M	2.33 M	1.75 M	1.4 M	1.17 M	1.00 M	875 k	778 k	700 k
0.75	7.5 M	3.75 M	2.50 M	1.88 M	1.5 M	1.25 M	1.07 M	938 k	833 k	750 k

0.8	8.0 M	4.00 M	2.67 M	2.00 M	1.6 M	1.33 M	1.14 M	1.00 M	889 k	800 k
0.85	8.5 M	4.25 M	2.83 M	2.13 M	1.7 M	1.42 M	1.21 M	1.06 M	944 k	850 k
0.9	9.0 M	4.50 M	3.00 M	2.25 M	1.8 M	1.50 M	1.29 M	1.13 M	1.00 M	900 k
0.95	9.5 M	4.75 M	3.17 M	2.38 M	1.9 M	1.58 M	1.36 M	1.19 M	1.06 M	950 k
1.0	10.0 M	5.00 M	3.33 M	2.50 M	2.0 M	1.67 M	1.43 M	1.25 M	1.11 M	1.0 M
1.5	15.0 M	7.50 M	5.00 M	3.75 M	3.0 M	2.50 M	2.14 M	1.88 M	1.67 M	1.5 M
2.0	20.0 M	10.00 M	6.66 M	5.00 M	4.0 M	3.33 M	2.86 M	2.50 M	2.22 M	2.0 M
2.5	25.0 M	12.50 M	8.33 M	6.25 M	5.0 M	4.17 M	3.57 M	3.13 M	2.78 M	2.5 M
3.0	30.0 M	15.00 M	10.00 M	7.50 M	6.0 M	5.00 M	4.29 M	3.75 M	3.33 M	3.0 M
3.5	35.0 M	17.50 M	11.66 M	8.75 M	7.0 M	5.83 M	5.00 M	4.38 M	3.89 M	3.5 M
4.0	40.0 M	20.00 M	13.33 M	10.00 M	8.0 M	6.66 M	5.71 M	5.00 M	4.44 M	4.0 M
4.5	45.0 M	22.50 M	15.00 M	11.25 M	9.0 M	7.50 M	6.43 M	5.63 M	5.00 M	4.5 M
5.0	50.0 M	25.00 M	16.67 M	12.50 M	10.0 M	8.33 M	7.14 M	6.25 M	5.55 M	5.0 M
5.5	55.0 M	27.50 M	18.33 M	13.75 M	11.0 M	9.17 M	7.86 M	6.88 M	6.11 M	5.5 M
6.0	60.0 M	30.00 M	20.00 M	15.00 M	12.0 M	10.00 M	8.57 M	7.50 M	6.66 M	6.0 M
6.5	65.0 M	32.50 M	21.67 M	16.25 M	13.0 M	10.83 M	9.29 M	8.13 M	7.22 M	6.5 M
7.0	70.0 M	35.00 M	23.33 M	17.50 M	14.0 M	11.67 M	10.00 M	8.75 M	7.78 M	7.0 M
7.5	75.0 M	37.50 M	25.00 M	18.75 M	15.0 M	12.50 M	10.71 M	9.38 M	8.33 M	7.5 M
8.0	80.0 M	40.00 M	26.67 M	20.00 M	16.0 M	13.33 M	11.43 M	10.00 M	8.89 M	8.0 M
9.0	90.0 M	45.00 M	30.00 M	22.50 M	18.0 M	15.00 M	12.86 M	11.25 M	10.00 M	9.0 M
10.0	100.0 M	50.00 M	33.33 M	25.00 M	20.0 M	16.66 M	14.28 M	12.50 M	11.11 M	10.0 M

k = kilohms   M = megohms

26.12 RL time constants

Time (sec.)	Inductance (henrys)									
	10	20	30	40	50	60	70	80	90	100
0.1	100.0	200.0	300.0	400.0	500.0	600.0	700.0	800.0	900.0	1000.0
0.15	66.7	133.3	200.0	266.7	333.3	400.0	466.7	533.3	600.0	666.7
0.2	50.0	100.0	150.0	200.0	250.0	300.0	350.0	400.0	450.0	500.0
0.25	40.0	80.0	120.0	160.0	200.0	240.0	280.0	320.0	360.0	400.0
0.3	33.3	66.7	100.0	133.3	166.7	200.0	233.3	266.6	300.0	333.3
0.35	28.6	57.1	86.6	114.3	142.9	171.4	200.0	228.6	257.1	285.7
0.4	25.0	50.0	75.0	100.0	125.0	150.0	175.0	200.0	225.0	250.0
0.45	22.2	44.4	66.7	88.9	111.1	133.3	155.6	177.8	200.0	222.2
0.5	20.0	40.0	60.0	80.0	100.0	120.0	140.0	160.0	180.0	200.0
0.55	18.2	36.4	54.5	72.7	90.9	109.1	127.3	145.5	163.6	181.8
0.6	16.7	33.3	50.0	66.7	83.3	100.0	116.7	133.3	150.0	166.7
0.65	15.4	30.8	46.2	61.5	76.9	92.3	107.7	123.1	138.5	153.8
0.7	14.3	28.6	42.9	57.1	71.4	85.7	100.0	114.3	128.7	142.9
0.75	13.3	26.7	40.0	53.3	66.7	80.0	93.3	106.7	120.0	133.3
0.8	12.5	25.0	37.5	50.0	62.5	75.0	87.5	100.0	112.5	125.0
0.85	11.8	23.5	35.3	47.1	58.8	70.6	82.3	94.1	105.9	117.6

0.9	11.1	22.2	33.3	44.4	55.5	66.6	77.8	88.9	100.0	111.1
0.95	10.5	21.1	31.6	42.1	52.6	63.2	73.7	84.2	94.7	105.3
1.0	10.0	20.0	30.0	40.0	50.0	60.0	70.0	80.0	90.0	100.0
1.5	6.7	13.3	20.0	26.7	33.3	40.0	46.7	53.3	60.0	66.7
2.0	5.0	10.0	15.0	20.0	25.0	30.0	35.0	40.0	45.0	50.0
2.5	4.0	8.0	12.0	16.0	20.0	24.0	28.0	32.0	36.0	40.0
3.0	3.3	6.7	10.0	13.3	16.7	20.0	23.3	26.7	30.0	33.3
3.5	2.9	5.7	8.7	11.4	14.3	17.1	20.0	22.9	25.7	28.6
4.0	2.5	5.0	7.5	10.0	12.5	15.0	17.5	20.0	22.5	25.0
4.5	2.2	4.4	6.7	8.9	11.1	13.3	15.6	17.8	20.0	22.2
5.0	2.0	4.0	6.0	8.0	10.0	12.0	14.0	16.0	18.0	20.0
5.5	1.8	3.6	5.5	7.3	9.1	10.9	12.7	14.6	16.4	18.2
6.0	1.7	3.3	5.0	6.7	8.3	10.0	11.7	13.3	15.0	16.7
6.5	1.5	3.1	4.6	6.2	7.7	9.2	10.8	12.3	13.9	15.4
7.0	1.4	2.9	4.3	5.7	7.1	8.6	10.0	11.4	12.9	14.3
7.5	1.3	2.7	4.0	5.3	6.7	8.0	9.3	10.7	12.0	13.3
8.0	1.2	2.5	3.8	5.0	6.3	7.5	8.8	10.0	11.3	12.5
9.0	1.1	2.2	3.3	4.4	5.5	6.7	7.8	8.9	10.0	11.1
10.0	1.0	2.0	3.0	4.0	5.0	6.0	7.0	8.0	9.0	10.0

All resistance values in ohms

26.13 Reactance of capacitors at spot frequencies

	50 Hz	100 Hz	1 kHz	10 kHz	100 kHz	1 MHz	10 MHz	100 MHz
1 pF	–	–	–	–	1.6 M	160 k	16 k	1.6 k
10 pF	–	–	–	1.6 M	160 k	16 k	1.6 k	160
50 pF	–	–	3.2 M	320 k	32 k	3.2 k	320	32
250 pF	–	6.4 M	640 k	64 k	6.4 k	640	64	6.4
1,000 pF	3.2 M	1.6 M	160 k	16 k	1.6 k	160	16	1.6
2,000 pF	1.6 M	800 k	80 k	8 k	800	80	8	0.8
0.01 µF	320 k	160 k	16 k	1.6 k	160	16	1.6	0.16
0.05 µF	64 k	32 k	3.2 k	320	32	3.2	0.32	–
0.1 µF	32 k	16 k	1.6 k	160	16	1.6	0.16	–
1 µF	3.2 k	1.6 k	160	16	1.6	0.16	–	–
2.5 µF	1.3 k	640	64	6.4	0.64	–	–	–
5 µF	640	320	32	3.2	0.32	–	–	–
10 µF	320	160	16	1.6	0.16	–	–	–
30 µF	107	53	5.3	0.53	–	–	–	–
100 µF	32	16	1.6	0.16	–	–	–	–
1,000 µF	3.2	1.6	0.16	–	–	–	–	–

Values above 10 MΩ and below 0.1 Ω not shown. Values in ohms.

26.14 Reactance of inductors at spot frequencies

	50 Hz	100 Hz	1 kHz	10 kHz	100 kHz	1 MHz	10 MHz	100 MHz
1 µH	–	–	–	–	0.63	6.3	63	630
5 µH	–	–	–	0.31	3.1	31	310	3.1 k
10 µH	–	–	–	0.63	6.3	63	630	6.3 k
50 µH	–	–	0.31	3.1	31	310	3.1 k	31 k
100 µH	–	–	0.63	6.3	63	630	6.3 k	63 k
250 µH	–	0.16	1.6	16	160	1.6 k	16 k	160 k
1 mH	0.31	0.63	6.3	63	630	6.3 k	63 k	630 k
2.5 mH	0.8	1.6	16	160	1.6 k	16 k	160 k	1.6 M
10 mH	3.1	6.3	63	630	6.3 k	63 k	630 k	6.3 M
25 mH	8	16	160	1.6 k	16 k	160 k	1.6 M	–
100 mH	31	63	630	6.3 k	63 k	630 k	6.3 M	–
1 H	310	630	6.3 k	63 k	630 k	6.3 M	–	–
5 H	1.5 k	3.1 k	31 k	310 k	3.1 M	–	–	–
10 H	3.1 k	6.3 k	63 k	630 k	6.3 M	–	–	–
100 H	31 k	63 k	630 k	6.3 M	–	–	–	–

Values above 10 MΩ and below 0.1 Ω not shown. Values in ohms.

26.15 Boundaries of sea areas

As used in BBC and BT weather forecasts

Stations whose latest reports are broadcast in the shipping forecasts on Radio 4 (198 kHz) at 0048, 0535, 1200 and 1754 (daily).

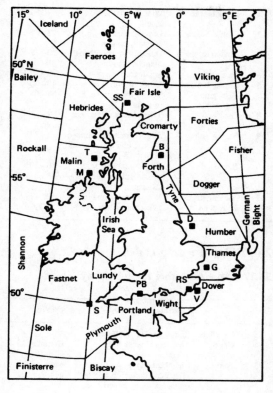

T	Tiree	RS	Royal Sovereign Light-tower
SS	Sule Skerry Lighthouse	PB	Portland Bill
B	Bell Rock Lighthouse	S	Scilly (St. Mary's)
D	Dowsing light-vessel	Va	Valentia
G	Galloper light-vessel	R	Ronaldsway
V	Varne light-vessel	M	Malin Head Lighthouse

26.16 The Beaufort scale

Force	Specification	Description	Speed (kmh^{-1})
0	Calm	Smoke rises vertically	Less than 1
1	Light air	Smoke drift shows wind direction	1–5
2	Light breeze	Wind can be felt on face	6–11
3	Gentle breeze	Leaves/twigs in constant motion	12–19
4	Moderate breeze	Dust blown about/small branches move	20–29
5	Fresh breeze	Small trees sway	30–39
6	Strong breeze	Large branches move	40–50
7	Near gale	Whole trees move, hard to walk	51–61
8	Gale	Twigs break, very hard to walk	62–74
9	Strong gale	Slight structural damage occurs, chimneys, slates blown off	75–87
10	Storm	Trees uprooted, considerable structural damage	88–101
11	Violent storm	Widespread damage	102–117
12	Hurricane	Catastrophic damage	>119

26.17 Signal rating codes

26.17.1 Overall rating for telephony

Symbol		Operating condition	Quality
5	Excellent	Signal quality unaffected	
4	Good	Signal quality slightly affected	Commercial
3	Fair	Signal quality seriously affected. Channel usable by operators or by experienced subscribers	Marginally commercial
2	Poor	Channel just usable by operators	Not commercial
1	Unusable	Channel unusable by operators	

26.17.2 The SINPFEMO code

Rating scale	S Signal strength	I Interference (QRM)	N Noise (QRN)	P Propagation disturbance	F Frequency of fading	E Quality	M Depth	O Overall rating
		Degrading effect of:				Modulation:		
5	Excellent	Nil	Nil	Nil	Nil	Excellent	Maximum	Excellent
4	Good	Slight	Slight	Slight	Slow	Good	Good	Good
3	Fair	Moderate	Moderate	Moderate	Moderate	Fair	Fair	Fair
2	Poor	Severe	Severe	Severe	Fast	Poor	Poor or nil	Poor
1	Barely audible	Extreme	Extreme	Extreme	Very fast	Very poor	Continuously overmodulated	Unusable

26.17.3 The SINPO code

Rating scale	S Signal Strength	I Interference (QRM)	N Noise (QRN)	P Propagation disturbance	O Overall readability (QRK)
5	Excellent	Nil	Nil	Nil	Excellent
4	Good	Slight	Slight	Slight	Good
3	Fair	Moderate	Moderate	Moderate	Fair
2	Poor	Severe	Severe	Severe	Poor
1	Barely audible	Extreme	Extreme	Extreme	Unusable

26.17.4 The SIO code

Rating scale	S Signal strength	I Interference	O Overall merit
4	Good	Nil or very slight	Good
3	Fair	Moderate	Fair
2	Poor	Heavy	Unusable

The SIO code is based on the SINPO code but in a simplified form.
Using the SIO code is perfectly acceptable, however.

26.18 World time

Difference between local time and coordinated universal time

The differences marked + indicate the number of hours
ahead of UTC. Differences marked − indicate the
number of hours behind UTC. Variations from summer
time during part of the year are decided annually and
may vary from year to year.

	Normal time	*Summer time*
Afghanistan	$+4\frac{1}{2}$	$+4\frac{1}{2}$
Alaska	-9	-8
	-10	-9
Albania	$+1$	$+2$
Algeria	UTC	$+1$
Andorra	$+1$	$+1$
Angola	$+1$	$+1$
Anguilla	-4	-4
Antigua	-4	-4
Argentina	-3	-3
Ascension Island	UTC	UTC
Australia		
Victoria &		
New South		
Wales	$+10$	$+11$
Queensland	$+10$	$+10$
Tasmania	$+10$	$+11$
N. Territory	$+9\frac{1}{2}$	$+9\frac{1}{2}$
S. Australia	$+9\frac{1}{2}$	$+10\frac{1}{2}$
W. Australia	$+8$	$+8$
Austria	$+1$	$+2$
Azores	-1	UTC
Bahamas	-5	-4
Bahrain	$+3$	$+3$
Bangladesh	$+6$	$+6$
Barbados	-4	-4
Belau	$+9$	$+9$
Belgium	$+1$	$+2$
Belize	-6	-6
Benin	$+1$	$+1$
Bermuda	-4	-3
Bhutan	$+6$	$+6$
Bolivia	-4	-4
Botswana	$+2$	$+2$
Brazil		
(a) Oceanic Isl.	-2	-2
(b) Ea & Coastal	-3	-3
(c) Manaos	-4	-4
(d) Acre	-5	-5

(*continued overleaf*)

	Normal time	Summer time
Brunei	+8	+8
Bulgaria	+2	+3
Burkina Faso	UTC	UTC
Burma	$+6\frac{1}{2}$	$+6\frac{1}{2}$
Burundi	+2	+2
Cameroon	+1	+1
Canada		
(a) Newfoundland	$-3\frac{1}{2}$	$-2\frac{1}{2}$
(b) Atlantic (Labrador, Nova Scotia)	−4	−3
(c) Ea (Ontario, Quebec)	−5	−4
(d) Ce (Manitoba)	−6	−5
(e) Mountain (Alberta) NWT (Mountain)	−7	−6
(f) Pacific (Br. Columbia	−8	−7
Yukon	−8	−7
Canary Isl.	UTC	+1
Cape Verde Isl.	−1	−1
Cayman Isl.	−5	−4
Central African Republic	+1	+1
Chad	+1	+1
Chile	−4	−3
China People's Rep.	+8	+8
Christmas Isl.	+7	+7
Cocos Isl.	$+6\frac{1}{2}$	$+6\frac{1}{2}$
Colombia	−5	−5
Comoro Rep.	+3	+3
Congo	+1	+1
Cook Isl.	−10	$-9\frac{1}{2}$
Costa Rica	−6	−6
Cuba	−5	−4
Cyprus	+2	+3
Czechoslovakia	+1	+2

	Normal time	Summer time
Denmark	+1	+2
Diego Garcia	+5	+5
Djibouti	+3	+3
Dominica	−4	−4
Dom. Rep.	−4	−4
Easter Isl.	−6	−5
Ecuador	−5	−5
Egypt	+2	+3
El Salvador	−6	−6
Equatorial Guinea	+1	+1
Ethiopia	+3	+3
Falkland Isl.	−4	−4
(Port Stanley)	−4	−3
Faroe Isl.	UTC	+1
Fiji	+12	+12
Finland	+2	+3
France	+1	+2
Gabon	+1	+1
Gambia	UTC	UTC
Germany	+1	+2
Ghana	UTC	UTC
Gibraltar	+1	+2
Greece	+2	+3
Greenland		
Scoresbysund	−1	UTC
Thule area	−3	−3
Other areas	−3	−2
Grenada	−4	−4
Guadeloupe	−4	−4
Guam	+10	+10
Guatemala	−6	−6
Guiana (French)	−3	−3
Guinea (Rep.)	UTC	UTC
Guinea Bissau	UTC	UTC
Guyana (Rep.)	−3	−3
Haiti	−5	−4
Hawaii	−10	−10
Honduras (Rep.)	−6	−6
Hong Kong	+8	+8

(continued overleaf)

421

	Normal time	Summer time
Hungary	+1	+2
Iceland	UTC	UTC
India	+5$\frac{1}{2}$	+5$\frac{1}{2}$
Indonesia		
(a) Java, Bali.		
Sumatra	+7	+7
(b) Kalimantan,		
Sulawesi,		
Timor	+8	+8
(c) Moluccas,		
We, Irian	+9	+9
Iran	+3$\frac{1}{2}$	+3$\frac{1}{2}$
Iraq	+3	+4
Ireland	UTC	+1
Israel	+2	+3
Italy	+1	+2
Ivory Coast	UTC	UTC
Jamaica	−5	−4
Japan	+9	+9
Johnston Isl.	−10	−10
Jordan	+2	+3
Kampuchea	+7	+7
Kenya	+3	+3
Kiribati	+12	+12
Korea	+9	+9
Kuwait	+3	+3
Laos	+7	+7
Lebanon	+2	+3
Lesotho	+2	+2
Liberia	UTC	UTC
Libya	+1	+2
Lord Howe Isl.	+10$\frac{1}{2}$	+11$\frac{1}{2}$
Luxembourg	+1	+2
Macau	+8	+8
Madagascar	+3	+3
Madeira	UTC	+1
Malawi	+2	+2
Malaysia	+8	+8
Maldive Isl.	+5	+5

	Normal time	Summer time
Mali	UTC	UTC
Malta	+1	+2
Marshall Isl.	+12	+12
Maritinique	−4	−4
Mauritania	UTC	UTC
Mauritius	+4	+4
Mayotte	+3	+3
Mexico		
(a) Campeche, Quintana Roo, Yucatan	−6	−5
(b) Sonora, Sinaloa, Nayarit, Baja California Sur	−7	−7
(c) Baja California Norte	−8	−7
(d) other states	−6	−6
Micronesia		
Truk, Yap	+10	+10
Ponape	+11	+11
Midway Isl.	−11	−11
Monaco	+1	+2
Mongolia	+8	+9
Monserrat	−4	−4
Morocco	UTC	UTC
Mozambique	+2	+2
Nauru	$+11\frac{1}{2}$	$+11\frac{1}{2}$
Nepal	+5.45	+5.45
Netherlands	+1	+2
Neth. Antilles	−4	−4
New Caledonia	+11	+11
New Zealand	+12	+13
Nicaragua	−6	−6
Niger	+1	+1
Nigeria	+1	+1
Niue	−11	−11
Norfolk Isl.	$+11\frac{1}{2}$	$+11\frac{1}{2}$

(continued overleaf)

26.18 (*continued*)

	Normal time	Summer time
N. Marianas	+10	+10
Norway	+1	+2
Oman	+4	+4
Pakistan	+5	+5
Panama	−5	−5
Papua N. Guinea	+10	+10
Paraguay	−4	−3
Peru	−5	−5
Philippines	+8	+8
Poland	+1	+2
Polynesia (Fr.)	−10	−10
Portugal	UTC	+1
Puerto Rico	−4	−4
Qatar	+3	+3
Reunion	+4	+4
Romania	+2	+3
Rwanda	+2	+2
Samoa Isl.	−11	−11
S. Tomé	UTC	UTC
Saudi Arabia	+3	+3
Senegal	UTC	UTC
Seychelles	+4	+4
Sierra Leone	UTC	UTC
Singapore	+8	+8
Solomon Isl.	+11	+11
Somalia	+3	+3
So. Africa	+2	+2
Spain	+1	+2
Sri Lanka	$+5\frac{1}{2}$	$+5\frac{1}{2}$
St. Helena	UTC	UTC
St. Kitts-Nevis	−4	−4
St. Lucia	−4	−4
St. Pierre	−3	−3
St. Vincent	−4	−4
Sudan	+2	+2
Surinam	$−3\frac{1}{2}$	$−3\frac{1}{2}$
Swaziland	+2	+2
Sweden	+1	+2

	Normal time	Summer time
Switzerland	+1	+2
Syria	+2	+3
Taiwan	+8	+8
Tanzania	+3	+3
Thailand	+7	+7
Togo	UTC	UTC
Tonga	+13	+13
Transkei	+2	+2
Trinidad & Tobago	−4	−4
Tristan da Cunha	UTC	UTC
Tunisia	+1	+1
Turks & Caicos	−4	−4
Turkey	+2	+3
Tuvalu	+12	+12
Uganda	+3	+3
United Arab Em.	+4	+4
United Kingdom	UTC	+1
Uruguay	−3	−3
USA		
(a) Eastern Zone*	−5	−4
(*) Indiana	−5	−5
(b) Central Zone	−6	−5
(c) Mountain Zone*	−7	−6
(*) Arizona	−7	−7
(d) Pacific Zone	−8	−7
USSR		
Moscow & Leningrad	+3	+4
Baku. Tbilisi	+4	+5
Sverdlovsk	+5	+6
Tashkent	+6	+7
Novobirsk	+7	+8
Irkutsk	+8	+9
Yakutsk	+9	+10
Khabarovsk	+10	+11

(continued overleaf)

	Normal time	Summer time
Magadan	+11	+12
Petropaviovsk	+12	+13
Anadyr	+13	+14
Vanuatu	+11	+12
Vatican	+1	+2
Venezuela	−4	−4
Vietnam	+7	+7
Virgin Isl.	−4	−4
Wake Isl.	+12	+12
Wallis & Futuna	+11	+11
Yemen	+3	+3
Yugoslavia	+1	+2
Zaire		
Kinshasa	+1	+1
Lubumbashi	+2	+2
Zambia	+2	+2
Zimbabwe	+2	+2

26.19 International allocation of call signs

The first character or the first two characters of a call sign indicate the nationality of the station using it.

AAA-ALZ	USA
AMA-AOZ	Spain
APA-ASZ	Pakistan
ATA-AWZ	India
AXA-AXZ	Australia
AYA-AZZ	Argentina
A2A-A2Z	Botswana
A3A-A3Z	Tonga
A5A-A5Z	Bhutan
BAA-BZZ	China
CAA-CEZ	Chile
CFA-CKZ	Canada
CLA-CMZ	Cuba
CNA-CNZ	Morocco
COA-COZ	Cuba
CPA-CPZ	Bolivia
CQA-CRZ	Portuguese Territories

CSZ-CUZ	Portugal
CVA-CXZ	Uruguay
CYA-CZZ	Canada
C2A-C2Z	Nauru
C3A-C3Z	Andorra
DAA-DTZ	Germany
DUA-DZZ	Philippines
EAA-EHZ	Spain
EIA-EJZ	Ireland
EKA-EKZ	USSR
ELA-ELZ	Liberia
EMA-EOZ	USSR
EPA-EQZ	Iran
ERA-ERZ	USSR
ESA-ESZ	Estonia (USSR)
ETA-ETZ	Ethiopia
EUA-EWZ	Belorussia (USSR)
EXA-EZZ	USSR
FAA-FZZ	France and Territories
GAA-GZZ	United Kingdom
HAA-HAZ	Hungary
HBA-HBZ	Switzerland
HCA-HDZ	Ecuador
HEA-HEZ	Switzerland
HFA-HFZ	Poland
HGA-HGZ	Hungary
HHA-HHZ	Haiti
HIA-HIZ	Dominican Republic
HJA-HKZ	Colombia
HLA-HMZ	Korea
HNA-HNZ	Iraq
HOA-HPZ	Panama
HQA-HRZ	Honduras
HSA-HSZ	Thailand
HTA-HTZ	Nicaragua
HUA-HUZ	El Salvador
HVA-HVZ	Vatican State
HWA-HYZ	France and Territories
HZA-HZZ	Saudi Arabia
IAA-IZZ	Italy and Territories
JAA-JSZ	Japan
JTA-JVZ	Mongolia
JWA-JXZ	Norway
JYA-JYZ	Jordan
JZA-JZZ	West Irian
KAA-KZZ	United States
LAA-LNZ	Norway

LOA-LWZ	Argentina
LXA-LXZ	Luxembourg
LYA-LYZ	Lithuania (USSR)
LZA-LZZ	Bulgaria
L2A-L9Z	Argentina
MAA-MZZ	United Kingdom
NAA-NZZ	United States
OAA-OCZ	Peru
ODA-ODZ	Lebanon
OEA-OEZ	Austria
OFA-OJZ	Finland
OKA-OMZ	Czechoslovakia
ONA-OTZ	Belgium
OUA-OZZ	Denmark
PAA-PIZ	Netherlands
PJA-PJZ	Netherlands West Indies
PKA-POZ	Indonesia
PPA-PYZ	Brazil
PZA-PZZ	Surinam
QAA-QZZ	(Service abbreviations)
RAA-RZZ	(USSR)
SAA-SMZ	Sweden
SNA-SRZ	Poland
SSA-SSM	Egypt
SSN-STZ	Sudan
SUA-SUZ	Egypt
SVA-SZZ	Greece
TAA-TCZ	Turkey
TDA-TDZ	Guatemala
TEA-TEZ	Costa Rica
TFA-TFZ	Iceland
TGA-TGZ	Guatemala
THA-THZ	France and Territories
TIA-TIZ	Costa Rica
TJA-TJZ	Cameroon
TKA-TKZ	France and Territories
TLA-TLZ	Central African Republic
TMA-TMZ	France and Territories
TNA-TNZ	Congo
TOA-TQZ	France and Territories
TRA-TRZ	Gabon
TSA-TSZ	Tunisia
TTA-TTZ	Chad
TUA-TUZ	Ivory Coast
TVA-TXZ	France and Territories

TYA-TYZ	Dahomey
TZA-TZZ	Mali
UAA-UQZ	USSR
URA-UTZ	Ukraine (CIS)
UUA-UZZ	USSR
VAA-VGZ	Canada
VHA-VNZ	Australia
VOA-VOZ	Canada
VPA-VSZ	British Territories
VTA-VWZ	India
VXA-VYZ	Canada
VZA-VZZ	Australia
WAA-WZZ	United States
XAA-XIZ	Mexico
XJA-XOZ	Canada
XPA-XPZ	Denmark
XQA-XRZ	Chile
XSA-XSZ	China
XTA-XTZ	Upper Volta
XUA-XUZ	Khmer Republic
XVA-XVZ	Vietnam
XWA-XWZ	Laos
XXA-XXZ	Portuguese Territories
XYZ-XZZ	Burma
YAA-YAZ	Afghanistan
YBA-YHZ	Indonesia
YIA-YIZ	Iraq
YJA-YJZ	New Hebrides
YKA-YKZ	Syria
YLA-YLZ	Latvia (USSR)
YMA-YMZ	Turkey
YNA-YNZ	Nicaragua
YOA-YRZ	Romania
YSA-YSZ	El Salvador
YTA-YUZ	Yugoslavia
YVA-YYZ	Venezuela
YZA-YZZ	Yugoslavia
ZAA-ZAZ	Albania
ZBA-ZJZ	British Territories
ZKA-ZMZ	New Zealand
ZNA-ZOZ	British Territories
ZPA-ZPZ	Paraguay
ZQA-ZQZ	British Territories
ZRA-ZUZ	South Africa
ZVA-ZZZ	Brazil
2AA-2ZZ	United Kingdom
3AA-3AZ	Monaco
3BA-3BZ	Mauritius
3CA-3CZ	Equatorial Guinea
3DA-3DM	Swaziland
3DN-3DZ	Fiji

3EA-3FZ	Panama
3GA-3GZ	Chile
3HA-3UZ	China
3VA-3VZ	Tunisia
3WA-3WZ	Vietnam
3XA-3XZ	Guinea
3YA-3YZ	Norway
3ZA-3ZZ	Poland
4AA-4CZ	Mexico
4DA-4IZ	Philippines
4JA-4LZ	USSR
4MA-4MZ	Venezuela
4NA-4OZ	Yugoslavia
4PA-4SZ	Sri Lanka
4TA-4TZ	Peru
4UA-4UZ	United Nations
4VA-4VZ	Haiti
4WA-4WZ	Yemen (YAR)
4XA-4XZ	Israel
4YA-4YZ	International Civil Aviation Organization
4ZA-4ZZ	Israel
5AA-5AZ	Libya
5BA-5BZ	Cyprus
5CA-5GZ	Morocco
5HA-5IZ	Tanzania
5JA-5KZ	Colombia
5LA-5MZ	Liberia
5NA-5OZ	Nigeria
5PA-5QZ	Denmark
5RA-5SZ	Malagasy Republic
5TA-5TZ	Mauretania
5UA-5UZ	Niger
5VA-5VZ	Togo
5WA-5WZ	Western Samoa
5XA-5XZ	Uganda
5YA-5ZZ	Kenya
6AA-6BZ	Egypt
6CA-6CZ	Syria
6DA-6JZ	Mexico
6KA-6NZ	Korea
6OA-6OZ	Somali Republic
6PA-6SZ	Pakistan
6TA-6UZ	Sudan
6VA-6WZ	Senegal
6XA-6XZ	Malagasy Republic
6YA-6YZ	Jamaica
6ZA-6ZZ	Liberia
7AA-7IZ	Indonesia
7JA-7NZ	Japan

7OA-7OZ	Yemen (PDRY)
7PA-7PZ	Lesotho
7QA-7QZ	Malawi
7RA-7RZ	Algeria
7SA-7SZ	Sweden
7TA-7YZ	Algeria
7ZA-7ZZ	Saudi Arabia
8AA-8IZ	Indonesia
8JA-8NZ	Japan
8OA-8OZ	Botswana
8PA-8PZ	Barbados
8QA-8QZ	Maldive Islands
8RA-8RZ	Guyana
8SA-8SZ	Sweden
8TA-8YZ	India
8ZA-8ZZ	Saudi Arabia
9AA-9AZ	San Marino
9BA-9DZ	Iran
9EA-9FZ	Ethiopia
9GA-9GZ	Ghana
9HA-9HZ	Malta
9IA-9JZ	Zambia
9KA-9KZ	Kuwait
9LA-9LZ	Sierra Leone
9MA-9MZ	Malaysia
9NA-9NZ	Nepal
9OA-9TZ	Zaire
9UA-9UZ	Burundi
9VA-9VZ	Singapore
9WA-9WZ	Malaysia
9XA-9XZ	Rwanda
9YA-9ZZ	Trindidad and Tobago

26.20 Amateur radio

26.20.1 Amateur bands in the UK

The Schedule of frequency bands, powers, etc, which, for the sake of convenience, appear in an identical format in both the Class A and Class B licences

Frequency bands in MHz	Status of allocations in the UK to: The Amateur Service	The Amateur Satellite Service	Maximum power Carrier PEP		Permitted type of transmission
1.810–1.850	Available to amateurs on a basis of non interference to other services		9dBW	15dBW	Morse Telephony RTTY Data Facsimile SSTV
1.850–2.000		No allocation			Morse Telephony Data Facsimile SSTV
3.500–3.800	Primary. Shared with other services	No allocation	20dBW	26dBW	Morse Telephony RTTY

Frequency (MHz)					Modes
7.000–7.100	Primary	Primary			Data Facsimile SSTV
10.100–10.150	Secondary	No allocation			
14.000–14.250	Primary	Primary			
14.250–14.350		No allocation			
18.068–16.168	Available to amateurs on a basis of non interference to other services. Antennas limited to horizontal polarisation, maximum gain 0dB with respect to a half-wave dipole	No allocation	10dBW	–	Morse, A1A only
21.000–21.450	Primary	Primary	20dBW	26dBW	Morse Telephony RTTY Data Facsimile SSTV

(continued overleaf)

26.20 (continued)

Frequency bands in MHz	Status of allocations in the UK to: The Amateur Service	The Amateur Satellite Service	Maximum power Carrier PEP		Permitted type of transmission
24.890–24.990	Available to amateurs on basis of non interference to other services. Antennas limited to horizontal polarisation, maximum gain 0dB with respect to a half-wave dipole	No allocation	10dBW	–	Morse, A1A only
28.000–29.700	Primary	Primary	20dBW	26dBW	Morse Telephony
50.000–50.500	Primary	No allocation	14dBW	20dBW	Morse Telephony

70.025–70.500	Secondary basis until further notice. Subject to not causing interference to other services. Use of any frequency shall cease immediately on demand of a government official	No allocation	16dBW	22dBW	RTTY Data Facsimile SSTV
144.0–146.0*	Primary	Primary	20dBW	26dBW	
430.0–431.0	Secondary. This band is not available for use within the area bounded by: 53 N 02 E, 55 N 02 E, 53 N 03 W, and 55 N 03 W	No allocation	10dBW e.r.p.	16dBW e.r.p.	Morse Telephony RTTY Data Facsimile SSTV Television

(continued overleaf)

26.20 (continued)

Frequency bands in MHz	Status of allocations in the UK to: The Amateur Service	The Amateur Satellite Service	Maximum power Carrier PEP		Permitted type of transmission
431.0–432.0	Secondary. This band is not available for use: a) Within the area bounded by: 53 N 02 E, 55 N 02 E, 55 N 03 W, 55 N 03 W b) Within a 100km radius of Charing Cross. 51 30'30"N 00.07'24"W	No allocation	10dBW e.r.p.	16dBW e.r.p.	Morse Telephony RTTY Data Facsimile SSTV Television
432.0–435.0	Secondary	No allocation	20dBW	26dBW	
435.0–438.0	Secondary	Secondary			
438.0–440.0	Secondary	No allocation			

Band		
1240–1260	Secondary	No allocation
1260–1270		Secondary Earth to Space only
1270–1325		No allocation
2310–2400		
2400–2450	Secondary. Users must accept interference from the ISM allocations in this band	Secondary Users must accept interference from the ISM allocations in this band
3400–3475	Secondary	No allocation
5650–5670		Secondary Earth to Space only
5670–5680		

(continued overleaf)

26.20 (continued)

Frequency bands in MHz	Status of allocations in the UK to: The Amateur Service	The Amateur Satellite Service	Maximum power Carrier PEP	Permitted type of transmission
5755–5765	Secondary. Users must accept interference from the ISM allocations in this band	No allocation		
5820–5830				
5830–5850		Secondary. Users must accept interference from the ISM allocations in this band. Space to Earth only		
10000–10450	Secondary	No allocation		
10450–10500		Secondary		
24000–24050	Primary	Primary		

Frequency (kHz)	Users must accept interference from the ISM allocations in this band	Users must accept interference from the ISM allocations in this band
24050 – 24250	Secondary. This band may only be used with the written consent of the Secretary of State. Users must accept interference from the ISM allocations in this band	No allocation
47000 – 47200	Primary	
75500 – 76000	Primary	
142000 – 144000		
248000 – 250000	Primary	

*Except in accordance with clause 1(2)(c)(ii) holders of the Amateur Radio Licence (B) are not permitted to use frequencies below 144 MHz, nor may they use the type of transmission known as morse (whether sent manually or automatically).

26.20.2 Dipole lengths for the amateur bands

Amateur band (metres)	Dipole length (metres)
80	39
40	20.2
20	10.1
15	6.7
10	5.0

26.20.3 Amateur radio emission designations

The first symbol specifies the modulation of the main carrier, the second symbol the nature of the signal(s) modulating the main carrier, and the third symbol the type of information to be transmitted.

Amplitude modulation

A1A Telegraphy by on-off keying without the use of a modulating audio frequency
A1B Automatic telegraphy by on-off keying, without the use of a modulating audio frequency
A2A Telegraphy by on-off keying of an amplitude modulating audio frequency or frequencies, or by on-off keying of the modulated emission
A2B Automatic telegraphy by on-off keying of an amplitude modulating audio frequency or modulated emission
A3E Telephony, double sideband
A3C Facsimile transmission
H3E Telephony using single sideband full carrier, amplitude modulation
R3E Telephony, single sideband, reduced carrier
J3E Telephony, single sideband, suppressed carrier
A3F/
C3F Slow scan and high definition television

Frequency modulation

F1A Telegraphy by frequency shift keying without the use of a modulating frequency: one of two frequencies being emitted at any instant
F1B Automatic telegraphy by frequency shift keying without the use of a modulating frequency
F2A Telegraphy by on-off keying of a frequency modulating audio frequency or on-off keying of an f.m. emission
F2B Automatic telegraphy by on-off keying of a frequency modulating audio frequency or of an f.m. emission
F3E Telephony
F3C Facsimile transmission
F3F Slow scan and high definition television

26.21 Microwave band designation system

All frequencies in gigahertz (GHz)

Old MOD		IEE (UK) Rec.		NATO		Old USA DoD		New USA Dos		IEEE (USA)	
P	0.08–0.39	A	0–0.25	L	1–2	P	0.225–0.39	A	0.1–25	HF	0.003–0.03
L2	0.39–1.0	B	0.25–0.50	S	2–4	L	0.39–1.55	B	0.25–0.50	VHF	0.03–0.30
L1	1.0–2.5	C	0.50–1.0	C	4–8	S	1.55–5.2	C	0.50–1.0	UHF	0.30–1.0
S	2.5–4.1	D	1–2	X	5.2–10.9	X	5.2–10.9	D	1–2	L	1–2
C	4.1–7.0	E	2–3	J	12–18	K	10.9–36	E	2–3	S	2–4
X	7.0–11.5	F	3–4	K	18–26	Q	36–46	F	3–4	C	4–8
J	11.5–18.0	G	4–6	Q	26–40	V	46–56	G	4–6	X	8–12
K	18–33	H	6–8	V	40–60	W	56–100	H	6–8	Ku	12–18
Q	33–40	I	8–10	O	60–90			I	8–10	K	18–27
O	40–60	J	10–20					J	10–20	Ka	27–40
V	60–90	K	20–40					K	20–40	MM	40–300
		L	40–60					L	40–60	Sub-MM	>300
		M	60–100					M	60–100		

26.22 International 'Q' code

Abbrev.	Question	Answer for advice
QRA	What is the name of your station?	The name of my station is ...
QRB	How far approximately are you from my station?	The approximate distance is ... miles
QRD	Where are you bound and where are you from?	I am bound for ... from ...
QRG	Will you tell me my exact frequency in kHz?	Your exact frequency is ... kHz.
QRH	Does my frequency vary?	Your frequency varies.
QRI	Is my note good?	Your note varies.
QRJ	Do you receive me badly? Are my signals weak?	I cannot receive you. Your signals are too weak.
QRK	Do you receive me well? Are my signals good?	I receive you well. Your signals are good.
QRL	Are you busy?	I am busy. Please do not interfere.
QRM	Are you being interfered with?	I am being interfered with.
QRN	Are you troubled by atmospherics?	I am troubled by atmospherics.
QRO	Shall I increase power?	Increase power.
QRP	Shall I decrease power?	Decrease power.
QRQ	Shall I send faster?	Send faster (... words per minute).
QRS	Shall I send more slowly?	Send more slowly (... words per minute).
QRT	Shall I stop sending?	Stop sending.
QRU	Have you anything for me?	I have nothing for you.
QRV	Are you ready?	I am ready.
QRX	Shall I wait? When will you call me again?	Wait (or wait until I have finished communicating with ...). I will call you at ... GMT.
QRZ	Who is calling me?	You are being called by ...
QSA	What is the strength of my signals? (1 to 5).	The strength of your signals is (1 to 5).
QSB	Does the strength of my signals vary?	The strength of your signals varies.
QSD	Is my keying correct? Are my signals distinct?	Your keying is indistinct. Your signals are bad.

Abbrev.	Question	Answer for advice
QSL	Can you give me acknowledgement of receipt?	I give you acknowledgement of receipt.
QSM	Shall I repeat the last telegram (message) I sent you?	Repeat the last telegram (message) you have sent me.
QSO	Can you communicate with ... direct (or through the medium of...)?	I can communicate with ... direct (or through the medium of ...).
QSP	Will you relay to ...?	I will relay to ...
QSV	Shall I send a series of V's?	Send a series of V's.
QSX	Will you listen for ... (call sign) on ... kHz?	I am listening for ... (call sign) on ... kHz.
QSZ	Shall I send each word or group twice?	Send each word or group twice.
QTH	What is your position in latitude and longitude?	My position is ... latitude ... longitude.
QTR	What is the exact time?	The exact time is ...

26.22.1 QSA Code (signal strength)

QSA1. Hardly perceptible; unreadable.
QSA2. Weak, readable now and then.
QSA3. Fairly good; readable, but with difficulty.
QSA4. Good; readable.
QSA5. Very good; perfectly readable.

26.22.2 QRK Code (audibility)

R1. Unreadable.
R2. Weak signals; barely readable.
R3. Weak signals; but can be copied.
R4. Fair signals; easily readable.
R5. Moderately strong signals.
R6. Good signals.
R7. Good strong signals.
R8. Very strong signals.
R9. Extremely strong signals.

26.23 RST Code

(readability)	(Signal strength)

(readability)

1. Unreadable.
2. Barely readable, occasional words distinguishable.
3. Readable with considerable difficulty.
4. Readable with practically no difficulty.
5. Perfectly rendable.

(Signal strength)

1. Faint, signals barely perceptible
2. Very weak signals.
3. Weak signals.
4. Fair signals.
5. Fairly good signals.
6. Good signals.
7. Moderately strong signals.
8. Strong signals.
9. Extremely strong signals.

(Tone)

1. Extremely rough hissing note.
2. Very rough AC note, no trace of musicality.
3. Rough, low-pitched AC note, slightly musical.
4. Rather rough AC note, moderately musical.
5. Musically modulated note.
6. Modulated note, slight trace of whistle.
7. Near DC note, smooth ripple.
8. Good DC note, just a trace of ripple.
9. Purest DC note.

(If the note appears to be crystal-controlled add an X after the appropriate number.)

26.24 International Morse Code

A	dit dah	.-	N	dah dit	-.
B	dah dit dit dit	-...	O	dah dah dah	- - -
C	dah dit dah dit	-.-.	P	dit dah dah dit	.- -.
D	dah dit dit	-..	Q	dah dah dit dah	- -.-
E	dit	.	R	dit dah dit	.-.
F	dit dit dah dit	..-.	S	dit dit dit	...
G	dah dah dit	- -.	T	dah	-
H	dit dit dit dit	....	U	dit dit dah	..-
I	dit dit	..	V	dit dit dit dah	...-
J	dit dah dah dah	.- - -	W	dit dah dah	.- -
K	dah dit dah	-.-	X	dah dit dit dah	-..-
L	dit dah dit dit	.-..	Y	dah dit dah dah	-.- -
M	dah dah	- -	Z	dah dah dit dit	- -..

Number code

1	dit dah dah dah dah	.- - - -	6	dah dit dit dit dit	-....
2	dit dit dah dah dah	..- - -	7	dah dah dit dit dit	- -...
3	dit dit dit dah dah	...- -	8	dah dah dah dit dit	- - -..
4	dit dit dit dit dah	....-	9	dah dah dah dah dit	- - - -.
5	dit dit dit dit dit	.....	0	dah dah dah dah dah	- - - - -

Note of interrogation	dit dit dah dah dit dit	..- -..
Note of exclamation	dah dah dit dit dah dah	- -..- -
Apostrophe	dit dah dah dah dah dit	.- - - -.
Hyphen	dah dit dit dit dit dah	-.... .-
Fractional bar	dah dit dit dah dit	-..-.
Brackets	dah dit dah dah dit dah	-.- -.-

Inverted commas	dit dah dit dit dah dit	.-..-.
Underline	dit dit dah dah dit dah	..--.-
Prelim. call	dah dit dah dit dah	-.-.-
Break sign	dah dit dit dit dah	-...-
End message	dit dah dit dah dit	.-.-.
Error	dit dit dit dit dit dit	......

Timing

The basic timing measurement is the dot pulse (dit), all other morse code timings are a function of this unit length:

Dot length (dit)	on unit
Dash length (dah)	three units
Pause between elements of one character	one unit
Pause between characters	three units
Pause between words	seven units

26.25 Phonetic alphabet

To avoid the possibility of the letters of a call-sign being misunderstood, it is usual to use the words given below in place of the letters. For example, G6PY would be given as G6 Papa Yankee.

Letter	Code word	Pronunciation	Letter	Code word	Pronunciation
A	Alfa	*AL*FAH	N	November	NO*VEM*BER
B	Bravo	*BRAH*VOH	O	Oscar	*OSS*CAH
C	Charlie	*CHAR*LEE	P	Papa	PAH*PAH*
D	Delta	*DELL*TAH	Q	Quebec	KEH*BECK*
E	Echo	*ECK*OH	R	Romeo	*ROW*MEOH
F	Foxtrot	*FOKS*TROT	S	Sierra	SEE*AIR*RAH
G	Golf	GOLF	T	Tango	*TANG*GO
H	Hotel	HOH*TELL*	U	Uniform	*YOU*NEEFORM
I	India	*IN*DEEAH	V	Victor	*VIK*TAH
J	Juliett	*JEW*LEE*ETT*	W	Whiskey	*WISS*KEY
K	Kilo	*KEY*LOH	X	X-ray	*ECKS*RAY
L	Lima	*LEE*MAH	Y	Yankee	*YANG*KEY
M	Mike	MIKE	Z	Zulu	*ZOO*LOO

Syllables in italic carry the accent.

445

26.26 Miscellaneous international abbreviations

C	Yes	GA	Resume sending
N	No	MN	Minute/minutes
W	Word	NW	I resume transmission
AA	All after ...	OK	Agreed
AB	All before ...	UA	Are we agreed?
AL	All that has just been sent	WA	Word after ...
		WB	Word before ...
BN	All between	XS	Atmospherics
CL	I am closing my station		

26.26.1 Amateur abbreviations

ABT	About	NIL	Nothing
AGN	Again	NM	No more
ANI	Any	NR	Number
BA	Buffer amplifier	NW	Now
BCL	Broadcast listener	OB	Old boy
BD	Bad	OM	Old man
BI	By	OT	Old timer
BK	Break in	PA	Power amplifier
BN	Been	PSE	Please
CK	Check	R	Received all sent
CKT	Circuit	RAC	Rectified AC
CLD	Called	RCD	Received
CO	Crystal oscillator	RX	Receiver
CUD	Could	SA	Say
CUL	See you later	SED	Said
DX	Long distance	SIGS	Signals
ECO	Electron-coupled oscillator	SIGN	Signature
		SSS	Single signal super-heterodyne
ES	And	SKD	Schedule
FB	Fine business (good work)	TKS	Thanks
		TMN	Tomorrow
FD	Frequency doubler	TNX	Thanks
FM	From	TPTG	Tuned plate tuned grid
GA	Go ahead, or Good afternoon	TX	Transmitter
		U	You
GB	Good-bye	UR	You are
GE	Good evening	VY	Very
GM	Good morning	WDS	Words
GN	Good night	WKG	Working
HAM	Radio amateur	WL	Will
HI	Laughter	WUD	Would
HR	Hear or here	WX	Weather
HRD	Heard	YF	Wife
HV	Have	YL	Young lady
LTR	Later	YR	Your
MO	Meter oscillator	73	Kind regards
ND	Nothing doing	88	Love and kisses

26.27 Post WARC-79 radio astronomy frequency allocations

13.360–13.410 MHz
25.55–25.67 MHz
37.5–38.25 MHz
73–74.6 MHz
150.05–153 MHz
322–328.6 MHz
406.1–410 MHz
1330–1400 MHz
1400–1427 MHz
1610.6–1613.8 MHz
1660–1668.4 MHz
1718.8–1722.2 MHz
2655–2700 MHz
4800–5000 MHz
10 600–10 700 MHz
14 470–14 500 MHz
15 350–15 400 MHz

26.27.1 Conversion factors

To convert from column one to column two *multiply* by the conversion factor.

To convert	Into	Multiply by
acres	square feet	4.356×10^4
acres	square metres	4047
acres	square yards	4.84×10^3
acres	hectares	0.4047
ampere-hours	coulombs	3600
amperes per sq cm	amperes per sq inch	6.452
ampere-turns	gilberts	1.257
ampere-turns per cm	ampere-turns per inch	2.540
angstroms	nanometres	10^{-1}
ares	square metres	10^2
atmospheres	bars	1.0133
atmospheres	mm or mercury at 0°C	760
atmospheres	feet of water at 4°C	33.90
atmospheres	inches of mercury at 0°C	29.92
atmospheres	kg per sq metre	1.033×10^4
atmospheres	newtons per sq metre	1.0133×10^5
atmospheres	pounds per sq inch	14.70
barns	square metres	10^{-28}

(*continued overleaf*)

447

To convert	Into	Multiply by
bars	newtons per sq metre	10^5
bars	hectopiezes	1
bars	baryes (dyne per sq cm)	10^6
bars	pascals (newtons per sq metre)	10^5
baryes	newtons per sq metre	10^{-1}
Btu	foot-pounds	778.3
Btu	joules	1054.8
Btu	kilogram-calories	0.2520
Btu	horsepower-hours	3.929×10^{-4}
bushels	cubic feet	1.2445
calories (I.T.)	joules	4.1868
calories (thermochem)	joules	4.184
carats (metric)	grams	0.2
Celsius (centigrade)	Fahrenheit (see pages 158–159)	
chains (surveyor's)	feet	66
circular mils	square centimetres	5.067×10^{-6}
circular mils	square mils	0.7854
cords	cubic metres	3.625
cubic feet	cords	7.8125×10^{-3}
cubic feet	litres	28.32
cubic inches	cubic centimetres	16.39
cubic inches	cubic feet	5.787×10^{-4}
cubic inches	cubic metres	1.639×10^{-5}
cubic metres	cubic feet	35.31
cubic metres	cubic yards	1.308
degrees (angle)	radians	1.745×10^{-2}
dynes	pounds	2.248×10^{-6}
dynes	newtons	10^{-5}
electron volts	joules	1.602×10^{-19}
ergs	foot-pounds	7.376×10^{-8}
ergs	joules	10^{-7}
fathoms	feet	6
fathoms	metres	1.8288
feet	centimetres	30.48
feet	varas	0.3594
feet of water at 4°C	inches of mercury at 0°C	0.8826
feet of water at 4°C	kg per sq metre	304.8
feet of water at 4°C	pounds per sq foot	62.43
fermis	metres	10^{-15}
footcandles	lumens per sq metre	10.764
footlamberts	candelas per sq metre	3.4263
foot-pounds	horsepower-hours	5.050×10^{-7}
foot-pounds	kilogram-metres	0.1383
foot-pounds	kilowatt-hours	3.766×10^{-7}
gallons (liq US)	gallons (liq Imp)	0.8327
gammas	teslas	10^{-9}
gausses	lines per sq inch	6.452
gausses	teslas	10^{-4}

To convert	Into	Multiply by
gilberts	amperes	7.9577×10^{-1}
grain (for humidity calculations)	pounds (avoirdupois)	1.429×10^{-4}
grams	dynes	980.7
grams	grains	15.43
grams	ounces (avoirdupois)	3.527×10^{-2}
grams	poundals	7.093×10^{-2}
grams per cm	pounds per inch	5.600×10^{-3}
grams per cu cm	pounds per cu inch	3.613×10^{-2}
grams per sq cm	pounds per sq foot	2.0481
hectares	square metres	10^4
hectares	acres	2.471
horsepower (boiler)	Btu per hour	3.347×10^4
horsepower (metric) (542.5 ft-lb per second)	Btu per minute	41.83
horsepower (metric) (542.5 ft-lb per second)	foot-lb per minute	3.255×10^4
horsepower (metric) (542.5 ft-lb per second)	kg-calories per minute	10.54
horsepower (550 ft-lb per second)	Btu per minute	42.41
horsepower (550 ft-lb per second)	foot-lb per minute	3.3×10^4
horsepower (550 ft-lb per second)	kilowatts	0.745
horsepower (metric) (542.5 ft-lb per second)	horsepower (550 ft-lb per second)	0.9863
horsepower (550 ft-lb per second)	kg-calories per minute	10.69
inches	centimetres	2.540
inches	feet	8.333×10^{-2}
inches	miles	1.578×10^{-5}
inches	mils	1000
inches	yards	2.778×10^{-2}
inches of mercury at 0°C	lbs per sq inch	0.4912
inches of water at 4°C	kg per sq metre	25.40
inches of water at 4°C	ounces per sq inch	0.5782
inches of water at 4°C	pounds per sq foot	5.202
inches of water at 4°C	in of mercury	7.355×10^{-2}
inches per ounce	metres per newton (compliance)	9.136×10^{-2}
joules	foot-pounds	0.7376
joules	ergs	10^7
kilogram-calories	kilogram-metres	426.9
kilogram-calories	kilojoules	4.186
kilogram-metres	joules	0.102
kilogram force	newtons	0.102
kilograms	tons, long (avdp 2240 lb)	9.842×10^{-4}
kilograms	tons, short (avdp 2000 lb)	1.102×10^{-3}

(*continued overleaf*)

To convert	Into	Multiply by
kilograms	pounds (avoirdupois)	2.205
kilograms per kilometre	pounds (avdp) per mile (stat)	3.548
kg per sq metre	pounds per sq foot	0.2048
kilometres	feet	3281
kilopond force	newtons	9.81
kilowatt-hours	Btu	3413
kilowatt-hours	foot-pounds	2.655×10^6
kilowatt-hours	joules	3.6×10^6
kilowatt-hours	kilogram-calories	860
kilowatt-hours	kilogram-metres	3.671×10^5
kilowatt-hours	pounds carbon oxidized	0.235
kilowatt-hours	pounds water evaporated from and at 212°F	3.53
kilowatt-hours	pounds water raised from 62° to 212°F	22.75
kips	newtons	4.448×10^3
knots* (naut mi per hour)	feet per second	1.688
knots	metres per minute	30.87
knots	miles (stat) per hour	1.1508
lamberts	candelas per sq cm	0.3183
lamberts	candelas per sq inch	2.054
lamberts	candelas per sq metre	3.183×10^3
leagues	miles (approximately)	3
links (surveyor's)	chains	0.01
links	inches	7.92
litres	bushels (dry US)	2.838×10^{-2}
litres	cubic centimetres	1000
litres	cubic metres	0.001
litres	cubic inches	61.02
litres	gallons (liq Imp)	0.2642
litres	pints (liq Imp)	1.816
$\log_e$ or ln	$\log_{10}$	0.4343
lumens per sq foot	foot-candles	1
lux	lumens per sq foot	0.0929
maxwells	webers	10^{-8}
metres	yards	1.094
metres	varas	1.179
metres per min	feet per minute	3.281
metres per min	kilometres per hour	0.06
microhms per cu cm	microhms per inch cube	0.3937
microhms per cu cm	ohms per mil foot	6.015
microns	metres	10^{-6}
miles (nautical)*	feet	6076.1
miles (nautical)	metres	1852
miles (nautical)	miles (statute)	1.1508
miles (statute)	feet	5280
miles (statute)	kilometres	1.609
miles per hour	kilometres per minute	2.682×10^{-2}
miles per hour	feet per minute	88

To convert	Into	Multiply by
miles per hour	kilometres per hour	1.609
millibars	inches of mercury (0°C)	0.02953
millibars	pounds per sq foot	2.089
(10^3 dynes per sq cm)		
mils	metres	2.54×10^{-5}
nepers	decibels	8.686
newtons	dynes	10^5
newtons	kilograms	0.1020
newtons	poundals	7.233
newtons	pounds (avoirdupois)	0.2248
oersteds	amperes per metre	7.9577×10
ounce-inches	newton-metres	7.062×10^{-3}
ounces (fluid)	quarts	3.125×10^{-2}
ounces (avoirdupois)	pounds	6.25×10^{-2}
pascals	newtons per sq metre	1
pascals	pounds per sq inch	1.45×10^{-4}
piezes	newtons per sq metre	10^3
piezes	sthenes per sq metre	1
pints	quarts	0.50
poises	newton-seconds per sq metre	10^{-1}
pounds of water (dist)	cubic feet	1.603×10^{-2}
pounds per inch	kg per metre	17.86
pounds per foot	kg per metre	1.488
pounds per mile (statute)	kg per kilometre	0.2818
pounds per cu foot	kg per cu metre	16.02
pounds per cu inch	ponds per cu foot	1728
pounds per sq foot	pounds per sq inch	6.944×10^{-3}
pounds per sq foot	kg per sq metre	4.882
pounds per sq inch	kg per sq metre	703.1
poundals	dynes	1.383×10^4
poundals	pounds (avoirdupois)	3.108×10^{-2}
quarts	gallons	0.25
rods	feet	16.5
slugs (mass)	pounds (avoirdupois)	32.174
sq inches	circular mils	1.273×10^6
sq inches	sq centimetres	6.452
sq feet	sq metres	9.290×10^{-2}
sq miles	sq yards	3.098×10^6
sq miles	acres	640
sq miles	sq kilometres	2.590
sq millimetres	circular mils	1973
steres	cubic metres	1
stokes	sq metres per second	10^{-4}
(temp rise, °C)× (US gal water)/minute	watts	264
tonnes	kilograms	10^3
tons, short (avoir 2000 lb)	tonnes (1000 kg)	0.9072
tons, long (avoir 2240 lb)	tonnes (1000 kg)	1.016
tons, long (avoir 2240 lb)	tons, short (avoir 2000 lb)	1.120

(*continued overleaf*)

26.27 (continued)

To convert	Into	Multiply by
tons (US shipping)	cubic feet	40
torrs	newtons per sq metre	133.32
watts	Btu per minute	5.689×10^{-2}
watts	ergs per second	10^7
watts	foot-lb per minute	44.26
watts	horsepower (550 ft-lb per second)	1.341×10^{-3}
watts	horsepower (metric) (542.5 ft-lb per second)	1.360×10^{-3}
watts	kg-calories per minute	1.433×10^{-2}
watt-seconds (joules)	gram-calories (mean)	0.2389
webers per sq metre	gausses	10^4
yards	feet	3

26.27.2 Fractions of an inch with metric equivalents

Fractions of an inch		Decimals of an inch	mm
	1/64	0.0156	0.397
1/32		0.0312	0.794
	3/64	0.0468	1.191
1/16		0.0625	1.588
	5/64	0.0781	1.985
3/32		0.0938	2.381
	7/64	0.1094	2.778
1/8		0.1250	3.175
	9/64	0.1406	3.572
5/32		0.1563	3.969
	11/64	0.1719	4.366
3/16		0.1875	4.762
	13/64	0.2031	5.159
7/32		0.2187	5.556
	15/64	0.2344	5.953
1/4		0.2500	6.350
	17/64	0.2656	6.747
9/32		0.2813	7.144
	19/64	0.2969	7.541
5/16		0.3125	7.937
	21/64	0.3281	8.334
11/32		0.3438	8.731
	23/64	0.3593	9.128
3/8		0.3750	9.525
	25/64	0.3906	9.922
13/32		0.4063	10.319
	27/64	0.4219	10.716
7/16		0.4375	11.112

Fractions of an inch		Decimals of an inch	mm
	29/64	0.4531	11.509
15/32		0.4687	11.906
	31/64	0.4844	12.303
1/2		0.5000	12.700
	33/64	0.5156	13.097
17/32		0.5313	13.494
	35/64	0.5469	13.891
9/16		0.5625	14.287
	37/64	0.5781	14.684
19/32		0.5938	15.081
	39/64	0.6094	15.478
5/8		0.6250	15.875
	41/64	0.6406	16.272
21/32		0.6563	16.668
	43/64	0.6719	17.065
11/16		0.6875	17.462
	45/64	0.7031	17.859
23/32		0.7188	18.256
	47/64	0.7344	18.653
3/4		0.7500	19.050
	49/64	0.7656	19.447
25/32		0.7813	19.843
	51/64	0.7969	20.240
13/16		0.8125	20.637
	53/64	0.8281	21.034
27/32		0.8438	21.431
	55/64	0.8594	21.828
7/8		0.8750	22.225
	57/64	0.8906	22.622
29/32		0.9062	23.019
	59/64	0.9219	23.416
15/16		0.9375	23.812
31/32		0.9688	24.606
	61/64	0.9531	24.209
	63/64	0.9844	25.003
		1.000	25.400

26.27.3 Code conversion tables

Dec	Octal	Hex	Binary bit pattern							ASCII character
			7	6	5	4	3	2	1	
0	0	0	0	0	0	0	0	0	0	NUL
1	1	1	0	0	0	0	0	0	1	SOH
2	2	2	0	0	0	0	0	1	0	STX

(*continued overleaf*)

Dec	Octal	Hex	Binary bit pattern							ASCII character
3	3	3	0	0	0	0	0	1	1	ETX
4	4	4	0	0	0	0	1	0	0	EOT
5	5	5	0	0	0	0	1	0	1	ENQ
6	6	6	0	0	0	0	1	1	0	ACK
7	7	7	0	0	0	0	1	1	1	BEL
8	10	8	0	0	0	1	0	0	0	BS
9	11	9	0	0	0	1	0	0	1	HT
10	12	A	0	0	0	1	0	1	0	LF
11	13	B	0	0	0	1	0	1	1	VT
12	14	C	0	0	0	1	1	0	0	FF
13	15	D	0	0	0	1	1	0	1	CR
14	16	E	0	0	0	1	1	1	0	SO
15	17	F	0	0	0	1	1	1	1	SI
16	20	10	0	0	1	0	0	0	0	DLE
17	21	11	0	0	1	0	0	0	1	DC1
18	22	12	0	0	1	0	0	1	0	DC2
19	23	13	0	0	1	0	0	1	1	DC3
20	24	14	0	0	1	0	1	0	0	DC4
21	25	15	0	0	1	0	1	0	1	NAK
22	26	16	0	0	1	0	1	1	0	SYN
23	27	17	0	0	1	0	1	1	1	ETB
24	30	18	0	0	1	1	0	0	0	CAN
25	31	19	0	0	1	1	0	0	1	EM
26	32	1A	0	0	1	1	0	1	0	SUB
27	33	1B	0	0	1	1	0	1	1	ESC
28	34	1C	0	0	1	1	1	0	0	FS
29	35	1D	0	0	1	1	1	0	1	GS
30	36	1E	0	0	1	1	1	1	0	RS
31	37	1F	0	0	1	1	1	1	1	US
32	40	20	0	1	0	0	0	0	0	SPACE
33	41	21	0	1	0	0	0	0	1	!
34	42	22	0	1	0	0	0	1	0	"
35	43	23	0	1	0	0	0	1	1	#
36	44	24	0	1	0	0	1	0	0	$
37	45	25	0	1	0	0	1	0	1	%
38	46	26	0	1	0	0	1	1	0	&
39	47	27	0	1	0	0	1	1	1	'
40	50	28	0	1	0	1	0	0	0	(
41	51	29	0	1	0	1	0	0	1	)
42	52	2A	0	1	0	1	0	1	0	*
43	53	2B	0	1	0	1	0	1	1	+
44	54	2C	0	1	0	1	1	0	0	,
45	55	2D	0	1	0	1	1	0	1	-
46	56	2E	0	1	0	1	1	1	0	.
47	57	2F	0	1	0	1	1	1	1	/
48	60	30	0	1	1	0	0	0	0	Ø
49	61	31	0	1	1	0	0	0	1	1
50	62	32	0	1	1	0	0	1	0	2
51	63	33	0	1	1	0	0	1	1	3
52	64	34	0	1	1	0	1	0	0	4
53	65	35	0	1	1	0	1	0	1	5
54	66	36	0	1	1	0	1	1	0	6

Dec	Octal	Hex	Binary bit pattern								ASCII character
55	67	37	0	1	1	0	1	1	1		7
56	70	38	0	1	1	1	0	0	0		8
57	71	39	0	1	1	1	0	0	1		9
58	72	3A	0	1	1	1	0	1	0		:
59	73	3B	0	1	1	1	0	1	1		;
60	74	3C	0	1	1	1	1	0	0		<
61	75	3D	0	1	1	1	1	0	1		=
62	76	3E	0	1	1	1	1	1	0		>
63	77	3F	0	1	1	1	1	1	1		?
64	100	40	1	0	0	0	0	0	0		@
65	101	41	1	0	0	0	0	0	1		A
66	102	42	1	0	0	0	0	1	0		B
67	103	43	1	0	0	0	0	1	1		C
68	104	44	1	0	0	0	1	0	0		D
69	105	45	1	0	0	0	1	0	1		E
70	106	46	1	0	0	0	1	1	0		F
71	107	47	1	0	0	0	1	1	1		G
72	110	48	1	0	0	1	0	0	0		H
73	111	49	1	0	0	1	0	0	1		I
74	112	4A	1	0	0	1	0	1	0		J
75	113	4B	1	0	0	1	0	1	1		K
76	114	4C	1	0	0	1	1	0	0		L
77	115	4D	1	0	0	1	1	0	1		M
78	116	4E	1	0	0	1	1	1	0		N
79	117	4F	1	0	0	1	1	1	1		O
80	120	50	1	0	1	0	0	0	0		P
81	121	51	1	0	1	0	0	0	1		Q
82	122	52	1	0	1	0	0	1	0		R
83	123	53	1	0	1	0	0	1	1		S
84	124	54	1	0	1	0	1	0	0		T
85	125	55	1	0	1	0	1	0	1		U
86	126	56	1	0	1	0	1	1	0		V
87	127	57	1	0	1	0	1	1	1		W
88	130	58	1	0	1	1	0	0	0		X
89	131	59	1	0	1	1	0	0	1		Y
90	132	5A	1	0	1	1	0	1	0		Z
91	133	5B	1	0	1	1	0	1	1		[
92	134	5C	1	0	1	1	1	0	0		\
93	135	5D	1	0	1	1	1	0	1		]
94	136	5E	1	0	1	1	1	1	0		↑
95	137	5F	1	0	1	1	1	1	1		←
96	140	60	1	1	0	0	0	0	0		–
97	141	61	1	1	0	0	0	0	1		a
98	142	62	1	1	0	0	0	1	0		b
99	143	63	1	1	0	0	0	1	1		c
100	144	64	1	1	0	0	1	0	0		d
101	145	65	1	1	0	0	1	0	1		e
102	146	66	1	1	0	0	1	1	0		f
103	147	67	1	1	0	0	1	1	1		g
104	150	68	1	1	0	1	0	0	0		h
105	151	69	1	1	0	1	0	0	1		i
106	152	6A	1	1	0	1	0	1	0		j

(*continued overleaf*)

Dec	Octal	Hex	Binary bit pattern								ASCII character
107	153	6B	1	1	0	1	0	1	1		k
108	154	6C	1	1	0	1	1	0	0		l
109	155	6D	1	1	0	1	1	0	1		m
110	156	6E	1	1	0	1	1	1	0		n
111	157	6F	1	1	0	1	1	1	1		o
112	160	70	1	1	1	0	0	0	0		p
113	161	71	1	1	1	0	0	0	1		q
114	162	72	1	1	1	0	0	1	0		r
115	163	73	1	1	1	0	0	1	1		s
116	164	74	1	1	1	0	1	0	0		t
117	165	75	1	1	1	0	1	0	1		u
118	166	76	1	1	1	0	1	1	0		v
119	167	77	1	1	1	0	1	1	1		w
120	170	78	1	1	1	1	0	0	0		x
121	171	79	1	1	1	1	0	0	1		y
122	172	7A	1	1	1	1	0	1	0		z
123	173	7B	0	1	1	1	1	0	1	1	
124	174	7C	0	1	1	1	1	1	0	0	
125	175	7D	0	1	1	1	1	1	0	1	
126	176	7E	0	1	1	1	1	1	1	0	
127	177	7F	0	1	1	1	1	1	1	1	DEL
128	200	80	1	0	0	0	0	0	0	0	
129	201	81	1	0	0	0	0	0	0	1	
130	202	82	1	0	0	0	0	0	1	0	
131	203	83	1	0	0	0	0	0	1	1	
132	204	84	1	0	0	0	0	1	0	0	
133	205	85	1	0	0	0	0	1	0	1	
134	206	86	1	0	0	0	0	1	1	0	
135	207	87	1	0	0	0	0	1	1	1	
136	210	88	1	0	0	0	1	0	0	0	
137	211	89	1	0	0	0	1	0	0	1	
138	212	8A	1	0	0	0	1	0	1	0	
139	213	8B	1	0	0	0	1	0	1	1	
140	214	8C	1	0	0	0	1	1	0	0	
141	215	8D	1	0	0	0	1	1	0	1	
142	216	8E	1	0	0	0	1	1	1	0	
143	217	8F	1	0	0	0	1	1	1	1	
144	220	90	1	0	0	1	0	0	0	0	
145	221	91	1	0	0	1	0	0	0	1	
146	222	92	1	0	0	1	0	0	1	0	
147	223	93	1	0	0	1	0	0	1	1	
148	224	94	1	0	0	1	0	1	0	0	
149	225	95	1	0	0	1	0	1	0	1	
150	226	96	1	0	0	1	0	1	1	0	
151	227	97	1	0	0	1	0	1	1	1	
152	230	98	1	0	0	1	1	0	0	0	
153	231	99	1	0	0	1	1	0	0	1	
154	232	9A	1	0	0	1	1	0	1	0	
155	233	9B	1	0	0	1	1	0	1	1	
156	234	9C	1	0	0	1	1	1	0	0	
157	235	9D	1	0	0	1	1	1	0	1	
158	236	9E	1	0	0	1	1	1	1	0	

Dec	Octal	Hex	Binary bit pattern								ASCII character
159	237	9F	1	0	0	1	1	1	1	1	
160	240	A0	1	0	1	0	0	0	0	0	
161	241	A1	1	0	1	0	0	0	0	1	
162	242	A2	1	0	1	0	0	0	1	0	
163	243	A3	1	0	1	0	0	0	1	1	
164	244	A4	1	0	1	0	0	1	0	0	
165	245	A5	1	0	1	0	0	1	0	1	
166	246	A6	1	0	1	0	0	1	1	0	
167	247	A7	1	0	1	0	0	1	1	1	
168	250	A8	1	0	1	0	1	0	0	0	
169	251	A9	1	0	1	0	1	0	0	1	
170	252	AA	1	0	1	0	1	0	1	0	
171	253	AB	1	0	1	0	1	0	1	1	
172	254	AC	1	0	1	0	1	1	0	0	
173	255	AD	1	0	1	0	1	1	0	1	
174	256	AE	1	0	1	0	1	1	1	0	
175	257	AF	1	0	1	0	1	1	1	1	
176	260	B0	1	0	1	1	0	0	0	0	
177	261	B1	1	0	1	1	0	0	0	1	
178	262	B2	1	0	1	1	0	0	1	0	
179	263	B3	1	0	1	1	0	0	1	1	
180	264	B4	1	0	1	1	0	1	0	0	
181	265	B5	1	0	1	1	0	1	0	1	
182	266	B6	1	0	1	1	0	1	1	0	
183	267	B7	1	0	1	1	0	1	1	1	
184	270	B8	1	0	1	1	1	0	0	0	
185	271	B9	1	0	1	1	1	0	0	1	
186	272	BA	1	0	1	1	1	0	1	0	
187	273	BB	1	0	1	1	1	0	1	1	
188	274	BC	1	0	1	1	1	1	0	0	
189	275	BD	1	0	1	1	1	1	0	1	
190	276	BE	1	0	1	1	1	1	1	0	
191	277	BF	1	0	1	1	1	1	1	1	
192	300	C0	1	1	0	0	0	0	0	0	
193	301	C1	1	1	0	0	0	0	0	1	
194	302	C2	1	1	0	0	0	0	1	0	
195	303	C3	1	1	0	0	0	0	1	1	
196	304	C4	1	1	0	0	0	1	0	0	
197	305	C5	1	1	0	0	0	1	0	1	
198	306	C6	1	1	0	0	0	1	1	0	
199	307	C7	1	1	0	0	0	1	1	1	
200	310	C8	1	1	0	0	1	0	0	0	
201	311	C9	1	1	0	0	1	0	0	1	
202	312	CA	1	1	0	0	1	0	1	0	
203	313	CB	1	1	0	0	1	0	1	1	
204	314	CC	1	1	0	0	1	1	0	0	
205	315	CD	1	1	0	0	1	1	0	1	
206	316	CE	1	1	0	0	1	1	1	0	
207	317	CF	1	1	0	0	1	1	1	1	
208	320	D0	1	1	0	1	0	0	0	0	
209	321	D1	1	1	0	1	0	0	0	1	
210	322	D2	1	1	0	1	0	0	1	0	

(*continued overleaf*)

Dec	Octal	Hex	Binary bit pattern								ASCII character
211	323	D3	1	1	0	1	0	0	1	1	
212	324	D4	1	1	0	1	0	1	0	0	
213	325	D5	1	1	0	1	0	1	0	1	
214	326	D6	1	1	0	1	0	1	1	0	
215	327	D7	1	1	0	1	0	1	1	1	
216	330	D8	1	1	0	1	1	0	0	0	
217	331	D9	1	1	0	1	1	0	0	1	
218	332	DA	1	1	0	1	1	0	1	0	
219	333	DB	1	1	0	1	1	0	1	1	
220	334	DC	1	1	0	1	1	1	0	0	
221	335	DD	1	1	0	1	1	1	0	1	
222	336	DE	1	1	0	1	1	1	1	0	
223	337	DF	1	1	0	1	1	1	1	1	
224	340	E0	1	1	1	0	0	0	0	0	
225	341	E1	1	1	1	0	0	0	0	1	
226	342	E2	1	1	1	0	0	0	1	0	
227	343	E3	1	1	1	0	0	0	1	1	
228	344	E4	1	1	1	0	0	1	0	0	
229	345	E5	1	1	1	0	0	1	0	1	
230	346	E6	1	1	1	0	0	1	1	0	
231	347	E7	1	1	1	0	0	1	1	1	
232	350	E8	1	1	1	0	1	0	0	0	
233	351	E9	1	1	1	0	1	0	0	1	
234	352	EA	1	1	1	0	1	0	1	0	
235	353	EB	1	1	1	0	1	0	1	1	
236	354	EC	1	1	1	0	1	1	0	0	
237	355	ED	1	1	1	0	1	1	0	1	
238	356	EE	1	1	1	0	1	1	1	0	
239	357	EF	1	1	1	0	1	1	1	1	
240	360	F0	1	1	1	1	0	0	0	0	
241	361	F1	1	1	1	1	0	0	0	1	
242	362	F2	1	1	1	1	0	0	1	0	
243	363	F3	1	1	1	1	0	0	1	1	
244	364	F4	1	1	1	1	0	1	0	0	
245	365	F5	1	1	1	1	0	1	0	1	
246	366	F6	1	1	1	1	0	1	1	0	
247	367	F7	1	1	1	1	0	1	1	1	
248	370	F8	1	1	1	1	1	0	0	0	
249	371	F9	1	1	1	1	1	0	0	1	
250	372	FA	1	1	1	1	1	0	1	0	
251	373	FB	1	1	1	1	1	0	1	1	
252	374	FC	1	1	1	1	1	1	0	0	
253	375	FD	1	1	1	1	1	1	0	1	
254	376	FE	1	1	1	1	1	1	1	0	
255	377	FF	1	1	1	1	1	1	1	1	

26.27.4 Wavelength-frequency conversion table

Metres to kilohertz

Metres	kHz	Metres	kHz	Metres	kHz
5	60,000	270	1,111	490	612.2
6	50,000	275	1,091	500	600
7	42,857	280	1,071	510	588.2
8	37,500	290	1,034	520	576.9
9	33,333	295	1,017	530	566
10	30,000	300	1,000	540	555.6
25	12,000	310	967.7	550	545.4
50	6,000	320	937.5	560	535.7
100	3,000	330	909.1	570	526.3
150	2,000	340	882.3	580	517.2
200	1,500	350	857.1	590	508.5
205	1,463	360	833.3	600	500
210	1,429	370	810.8	650	461.5
215	1,395	380	789.5	700	428.6
220	1,364	390	769.2	750	400
225	1,333	400	750	800	375
230	1,304	410	731.7	850	352.9
235	1,277	420	714.3	900	333.3
240	1,250	430	697.7	950	315.9
245	1,225	440	681.8	1,000	300
250	1,200	450	666.7	1,250	240
255	1,177	460	652.2	1,500	200
260	1,154	470	638.3	1,750	171.4
265	1,132	480	625	2,000	150

Note:- To convert kilohertz to wavelengths in metres, divide 300,000 by kilohertz.

To convert wavelengths in metres to kilohertz, divide 300,000 by the number of metres. One megahertz $= 1,000,000$ hertz or $= 1,000$ kilohertz. Thus, 30,000 kilohertz = 30 megahertz.

26.27.5 ASCII control characters

Decimal	Hexadecimal	ASCII character	Meaning	Keyboard entry
0	00	NUL	Null	CTRL-@
1	01	SOH	Start of heading	CTRL-A
2	02	STX	Start of text	CTRL-B
3	03	ETX	End of text	CTRL-C

(continued overleaf)

459

Decimal	Hexadecimal	ASCII character	Meaning	Keyboard entry
4	04	EOT	End of transmission	CTRL-D
5	05	ENQ	Enquiry	CTRL-E
6	06	ACK	Acknowledge	CTRL-F
7	07	BEL	Bell	CTRL-G
8	08	BS	Backspace	CTRL-H
9	09	HT	Horizontal tabulation	CTRL-I
10	0A	LF	Line feed	CTRL-J
11	0B	VT	Vetical tabulation	CTRL-K
12	0C	FF	Form feed	CTRL-L
13	0D	CR	Carriage return	CTRL-M
14	0E	SO	Shift out	CTRL-N
15	0F	SI	Shift in	CTRL-O
16	10	DLE	Data link escape	CTRL-P
17	11	DC1	Device control one	CTRL-Q
18	12	DC2	Device control two	CTRL-R
19	13	DC3	Device control three	CTRL-S
20	14	DC4	Device control four	CTRL-T
21	15	NAK	Negative acknowledge	CTRL-U
22	16	SYN	Synchronous idle	CTRL-V
23	17	ETB	End of transmission	CTRL-W
24	18	CAN	Cancel	CTRL-X
25	19	EM	End of medium	CTRL-Y
26	1A	SUB	Substitute	CTRL-Z
27	1B	ESC	Escape	CTRL-[
28	1C	FS	File separator	CTRL-\
29	1D	GS	Group separator	CTRL-]
30	1E	RS	Record separator	CTRL-^
31	1F	US	Unit separator	CTRL- –

26.28 Laws

Ampere's Rule Refers to the deflection direction of a magnetic pointer that is influenced by a current; an analogy being that if a person is assumed to be swimming with the current and facing the indicator, the north-seeking pole is deflected towards the left hand, the south pole being deflected in an opposite direction.

Ampere's Theorem The magnetic field from current flowing in a circuit is equivalent to that due to a simple magnetic shell, the outer edge coinciding with the electrical conductor with such strength that it equals that current strength.

Baur's Constant That voltage necessary to cause a discharge through a determined insulating material 1 mm thick. The law of dielectric strength is that breakdown voltage necessary to cause a discharge through a substance proportional to a 2/3 power of its thickness.

Coulomb's Law Implies that the mechanical force between two charged bodies is directly proportionate to the charges and inversely so to the squares of the distance separating them.

Faraday's Laws That of induction is that the e.m.f. induced in a circuit is proportional to the rate of change in the lines of force linking it. That of electrolysis is (1). That the quantity of a substance deposited in defined time is proportional to the current. (2) That different substances and quantities deposited by a single current in a similar time are proportional to the electro-chemical equivalents. The Faraday Effect states that when a light beam passes through a strong magnetic field the plane of polarisation is rotated.

Fleming's Rules By placing the thumb and first two fingers at right-angles respectively, the forefinger can represent the direction of magnetic field; the second finger, current direction; the thumb, motion direction. Use of the right hand in this way represents the relation in a dynamo; use of the left hand represents the relation in a motor.

Hall Effect If an electric current flows across the lines of flux of a magnetic field, an e.m.f. is observed at right-angles to the primary current and to the magnetic field. When a steady current flows in a magnetic field, e.m.f. tendencies develop at right-angles to the magnetic force and to the current, proportionately to the product of the current strength, the magnetic force and the sine of the angle between the direction of quantities.

Joule's Law As a formula this is I^2Rt joules. It refers to that heat developed by the current (I) which is proportional to the square of I multiplied by R and

461

t, letting R = resistance and t = time. If the formula is seen as $JH = RI^2t$ it equals EIt, letting J = joules equivalent of heat, and H = the number of heat units.

Kerr Effect Illustrates that an angle of rotation is proportional to a magnetisation intensity and applies to the rotation of polarisation plane of plain polarised light as reflected from the pole of a magnet. The number (a constant) varies for different wavelengths and materials, making necessary the multiplication of magnetisation intensity in order to find the angle of rotation forming the effect.

Lambert's Cosine Law For a surface receiving light obliquely, the illumination is proportional to the cosine of the angle which the light makes with the normal to the surface.

Lenz's Law That induced currents have such a direction that the reaction forces generated have a tendency to oppose the motion or action producing them.

Maxwell's Law (*a*) Any two circuits carrying current tend so to dispose themselves as to include the largest possible number of lines of force common to the two. (*b*) Every electro-magnetic system tends to change its configuration so that the exciting circuit embraces the largest number of lines of force in a positive direction.

Maxwell's Rule Maxwell's *unit tubes* of electric or magnetic induction are such that a *unit pole* delivers 4π unit tubes of force.

Miller Circuit A form of circuit in which the time-constant of a resistance-capacitance combination is multiplied by means of the Miller effect on the capacitance. Named after John M. Miller.

Miller Effect Implies that the grid input impedance of a valve with a load in the anode circuit is different from its input impedance with a zero anode load. Should the load in the anode be resistance, the input impedance is purely capacitive. If the load impedance has a reactive component, the input impedance will have a resistive component. In pre-detector amplification, with a.v.c. to signal grids, the capacity across the tuned grid circuits tends to vary with the signal strength, evidencing detuning, the effect causing a charge (electrostatic) to be induced by the anode on the grid.

Planck's Constant Quanta of energy radiated when atomic electrons transfer from one state to another, assuming both to be *energy states* with electromagnetic radiation. The constant (h) is given the value of 6.626×10^{-34} joule second. h is usually coupled to the symbol (v) to represent the frequency of the radiated energy in hertz. That is, the frequency of the radiated energy is determinable by the relation $W_1 - W_2$, this equalling hv. W_1 and W_2 equal the values of the internal energy of the atom in initial and final stages. This constant is also known as the *Quantum Theory*.

Sabine's Relation For an auditorium whose boundaries comprise areas $A_1, A_2, A_3 \ldots$ etc, of absorption coefficient $\alpha_1, \alpha_2, \alpha_3 \ldots$ etc, the reverberation time, t, is given by

$$t = \frac{0.16 V}{\sum \alpha A}$$

where V is the auditorium's volume, and $\sum \alpha A = z_1 A_1 + \alpha_1 A_1 + \cdots$ etc.

Snell's Law For light incident on the boundary between two media, the ratio of the sine of the angle of incidence to the sine of the angle of refraction is a constant; equal to the inverse ratio of the refractive indices of the two media.

Thévenin's Theorem The current through a resistance R connected across any two points A and B of an active network (i.e. a network containing one or more sources of e.m.f.) is obtained by dividing the p.d. between A and B, with R disconnected, by $(R + r)$, where r is the resistance of the network measured between points A and B with R disconnected and the sources of e.m.f. replaced by their internal resistances.

26.29 CCITT recommendations

Series	Description
A	CCITT organisation
B	Means of expression
C	Telecommunications statistics
D	Tariff principles for leased circuits
E	Telephones, quality of service and tariffs
F	Telegraph, quality of service and tariffs
G	Line transmission
H	Non-telephone signal transmission
J	Television and sound programme transmission
K	Protection against interference
L	Protection of cable sheaths and poles
M	Telephone, telegraph and data transmission maintenance
N	Television and sound programme transmission maintenance
O	Measuring equipment specification
P	Telephone transmission quality
Q	Telephone signalling and transmission
R	Telegraph transmission
S	Alphabetic telegraph and data terminal equipment
T	Facsimile transmission
U	Telegraph switching
V	Data transmission via public switched telephone networks
X	Data transmission via public data networks

26.30 Powers of numbers

26.30.1 Powers of 2

n	2^n	$-n$	2^{-n}
1	2	−1	.5
2	4	−2	.25
3	8	−3	.125
4	16	−4	.062 5
5	32	−5	.031 25
6	64	−6	.015 625
7	128	−7	.007 812 5
8	256	−8	.003 906 25
9	512	−9	.001 953 125
10	1 024	−10	.000 976 562 5
11	2 048	−11	.000 488 281 25
12	4 096	−12	.000 244 140 625
13	8 192	−13	.000 122 070 312 5
14	16 384	−14	.000 061 035 156 25
15	32 768	−15	.000 030 517 578 125

(continued overleaf)

26.30 (*continued*)

n	2^n	2^{-n}	$-n$
16	65 536	.000 015 258 789 062 5	−16
17	131 072	.000 007 629 394 531 25	−17
18	262 144	.000 003 814 679 265 625	−18
19	524 288	.000 001 907 348 632 812 5	−19
20	1 048 576	.000 000 953 674 316 406 25	−20
21	2 097 152	.000 000 476 837 158 203 125	−21
22	4 194 304	.000 000 238 418 579 101 562 5	−22
23	8 388 608	.000 000 119 209 289 550 781 25	−23
24	16 777 216	.000 000 059 604 644 775 390 625	−24
25	33 554 432	.000 000 029 802 322 387 695 312 5	−25
26	67 108 864	.000 000 014 901 161 193 847 656 25	−26
27	134 217 728	.000 000 007 450 580 596 923 828 125	−27
28	268 435 456	.000 000 003 725 290 298 461 914 062 5	−28
29	536 870 912	.000 000 001 862 645 149 230 957 031 25	−29
30	1 073 741 824	.000 000 000 931 322 574 615 478 515 625	−30
31	2 147 483 648	.000 000 000 465 661 287 307 739 257 812 5	−31
32	4 294 967 296	.000 000 000 232 830 643 653 869 628 906 25	−32

26.30.2 Powers of 10_{16}

	10^n	n	10^{-n}				
	1	0	1.0000	0000	0000	0000	
	A	1	0.1999	9999	9999	999A	
	64	2	0.28F5	C28F	5C28	F5C3	$\times$ 16^{-1}
	3E8	3	0.4189	374B	C6A7	EF9E	$\times$ 16^{-2}
	2710	4	0.68DB	8BAC	710C	B296	$\times$ 16^{-3}
1	86A0	5	0.A7C5	AC47	1B47	8423	$\times$ 16^{-4}
F	4240	6	0.10C6	F7A0	B5ED	8D37	$\times$ 16^{-4}
98	9680	7	0.1AD7	F29A	BCAF	4858	$\times$ 16^{-5}
5F5	E100	8	0.2AF3	1DC4	6118	73BF	$\times$ 16^{-6}
3B9A	CA00	9	0.44B8	2FA0	9B5A	52CC	$\times$ 16^{-7}

(continued overleaf)

26.30 (continued)

		10^n		n			10^{-n}			
	2	540B	E400	10	0.6DF3	7F67	5EF6	EADF	$\times$	16^{-8}
	17	4876	E800	11	0.AFEB	FF0B	CB24	AAFF	$\times$	16^{-9}
	E8	D4A5	1000	12	0.1197	9981	2DEA	1119	$\times$	16^{-9}
	918	4E72	A000	13	0.1C25	C268	4976	81C2	$\times$	16^{-10}
	5AF3	107A	4000	14	0.2D09	370D	4257	3604	$\times$	16^{-11}
3	8D7E	A4C6	8000	15	0.480E	BE7B	9D58	566D	$\times$	16^{-12}
23	86F2	6FC1	0000	16	0.734A	CA5F	6226	F0AE	$\times$	$16 \times^{-13}$
163	4578	5D8A	0000	17	0.B877	AA32	36A4	B449	$\times$	16^{-14}
DE0	B6B3	A764	0000	18	0.1272	5DD1	D243	ABA1	$\times$	16^{-14}
8AC7	2304	89E8	0000	19	0.1D83	C94F	B6D2	AC35	$\times$	16^{-15}

26.30.3 Powers of 16_{10}

16^n			n	16^{-n}			
		1	0	0.10000	00000	00000	× 10
		16	1	0.62500	00000	00000	× 10^{-1}
		256	2	0.39062	50000	00000	× 10^{-2}
		4 096	3	0.24414	06250	00000	× 10^{-3}
		65 536	4	0.15258	78906	25000	× 10^{-4}
	1	048 576	5	0.95367	43164	06250	× 10^{-6}
	16	777 216	6	0.59604	64477	53906	× 10^{-7}
	268	435 456	7	0.37252	90298	46191	× 10^{-8}

(continued overleaf)

26.30 (continued)

16^n	n	16^{-n}
4 294 967 296	8	$0.23283\ 06436\ 53869\ 62891 \times 10^{-9}$
68 719 476 736	9	$0.14551\ 91522\ 83668\ 51807 \times 10^{-10}$
1 099 511 627 776	10	$0.90949\ 47017\ 72928\ 23792 \times 10^{-12}$
17 592 186 044 416	11	$0.56843\ 41886\ 08080\ 14870 \times 10^{-13}$
281 474 976 710 656	12	$0.35527\ 13678\ 80050\ 09294 \times 10^{-14}$
4 503 599 627 370 496	13	$0.22204\ 46049\ 25031\ 30808 \times 10^{-15}$
72 057 594 037 927 936	14	$0.13877\ 78780\ 78144\ 56755 \times 10^{-16}$
1 152 921 504 606 846 976	15	$0.86736\ 17379\ 88403\ 54721 \times 10^{-18}$

26.31 Sound

26.31.1 Sounds and sound levels

Sound Pressure (mPa)	Pressure ratio		Intensity ratio	Sound level (dB)	Source or description of typical sound
0.2 (datum)	1	$(= 10^0)$	1	0	Sound-proof room (threshold of hearing)
0.063	3.16	$(= 10^{0.5})$	10^1	10	Rustle of leaves in a breeze
0.2	10	$(= 10^1)$	10^2	20	Whisper
0.63	31.6	$(= 10^{1.5})$	10^3	30	Quiet conversation
2	100	$(= 10^2)$	10^4	40	Suburban home
6.3	316	$(= 10^{2.5})$	10^5	50	Typical conversation
20	1 000	$(= 10^3)$	10^6	60	Large shop
63	3 160	$(= 10^{3.5})$	10^7	70	City street
200	10 000	$(= 10^4)$	10^8	80	Noisy office with typing
630	31 600	$(= 10^{4.5})$	10^9	90	Underground railway
2 000	100 000	$(= 10^5)$	10^{10}	100	Pneumatic drill at 3 m
6 300	316 000	$(= 10^{5.5})$	10^{11}	110	Prop aircraft taking off
20 000	1 000 000	$(= 10^6)$	10^{12}	120	Jet aircraft taking off (threshold of pain)

26.31.2 Velocity of sound

In air at various temperatures

Temperature (°C)	Speed (ms^{-1})
0	331.32
10	337.42
15	340.47
20	343.51
30	349.61

In liquids and solids

Material	Speed (ms^{-1})
Alcohol	1440
Aluminium	6220
Brass	4430
Copper	4620
Glass	5400
Lead	2430
Magnesium	5330
Mercury	1460
Nickel	5600
Polystyrene	2670
Quartz	5750
Steel	6110
Water	1450

26.31.3 Audible frequency range

Musical instruments

Instrument	Low (Hz)	High (Hz)
Bass clarinet	82.41	493.88
Bass tuba	43.65	349.23
Bass viola	41.20	246.94
Bassoon	61.74	493.88
Cello	65.41	987.77
Clarinet	164.81	1,567.00
Flute	261.63	3,349.30
French horn	110.00	880.00
Guitar	82.41	880.00
Oboe	261.63	1,568.00
Piano	27.50	4,186.00
Trombone	82.41	493.88
Trumpet	164.81	987.77
Viola	130.81	1,174.00
Violin	196.00	3,136.00

Human voices

Voice	Low (Hz)	High (Hz)
Alto	130.81	698.46
Baritone	98.00	392.00
Bass	87.31	349.23
Soprano	246.94	1,174.70
Tenor	130.81	493.88

26.31.4 Audible intensity

Musical instruments

Instrument	Range (dB)
Bass drum	35 to 115
Cymbal	40 to 110
Organ	35 to 110
Piano	60 to 100
Trumpet	55 to 95
Tympani	30 to 110
Violin	42 to 95

26.31.5 Musical notes frequencies

The range of notes on a piano keyboard is from 27.5 Hz
to 4186 Hz. Middle C (the centre note on a standard
keyboard) has a frequency of 261.6 Hz. Standard pitch
is A above middle C at a frequency of 440 Hz. Note
that raising the pitch of a note is equivalent to doubling
the frequency for each complete octave.

A	27.5	D	73.4	G	196.0	C	523.3	F	1396.9	B	3951.1
B	30.9	E	82.4	A	220.0	D	587.3	G	1568.0	C	4186.0
C	32.7	F	87.3	B	246.9	E	659.2	A	1760.0		
D	36.7	G	98.0	C	261.6	F	698.5	B	1975.5		
E	41.2	A	110.0	D	293.7	G	784.0	C	2093.0		
F	43.7	B	123.5	E	329.6	A	880.0	D	2344.3		
G	49.0	C	130.8	F	349.2	B	987.8	E	2637.0		
A	55.0	D	146.8	G	392.0	C	1046.5	F	2793.8		
B	61.7	E	164.8	A	440.0	D	1174.0	G	3136.0		
C	65.4	F	174.6	B	493.9	E	1318.5	A	3520.0		

26.31.6 Cesius–Fahreheit conversion table

C	F	C	F	C	F	C	F
0	32	265	509	530	986	795	1,463
5	41	270	518	535	995	800	1,472
10	50	275	527	540	1,004	805	1,481
15	59	280	536	545	1,013	810	1,490
20	68	285	545	550	1,022	815	1,499
25	77	290	554	555	1,031	820	1,508
30	86	295	563	560	1,040	825	1,517
35	93	300	572	565	1,049	830	1,526
40	104	305	581	570	1,058	835	1,535
45	113	310	590	575	1,067	840	1,544
50	122	315	599	580	1,076	845	1,553
55	131	320	608	585	1,085	850	1,562
60	140	325	617	590	1,094	855	1,571
65	149	330	626	595	1,103	860	1,580
70	158	335	635	600	1,112	865	1,589
75	167	340	644	605	1,121	870	1,598
80	176	345	653	610	1,130	875	1,607
85	185	350	662	615	1,139	880	1,616
90	194	355	671	620	1,148	885	1,625
95	203	360	680	625	1,157	890	1,634
100	212	365	689	630	1,166	895	1,643
105	221	370	698	635	1,175	900	1,652
110	230	375	707	640	1,184	905	1,661
115	239	380	716	645	1,193	910	1,670
120	248	385	725	650	1,202	915	1,679
125	257	390	734	655	1,211	920	1,688
130	266	395	743	660	1,220	925	1,697
135	275	400	752	665	1,229	930	1,706
140	284	405	761	670	1,238	935	1,715
145	293	410	770	675	1,247	940	1,724
150	302	415	779	680	1,256	945	1,733
155	311	420	788	685	1,265	950	1,742
160	320	425	797	690	1,274	955	1,751
165	329	430	806	695	1,283	960	1,760
170	338	435	815	700	1,292	965	1,769
175	347	440	824	705	1,301	970	1,778
180	356	445	833	710	1,310	975	1,787
185	365	450	842	715	1,319	980	1,796
190	374	455	851	720	1,328	985	1,805
195	383	460	860	725	1,337	990	1,814
200	392	465	869	730	1,346	995	1,823
205	401	470	877	735	1,355	1,000	1,832
210	410	475	887	740	1,364	1,005	1,841
215	419	480	896	745	1,373	1,010	1,850
220	428	485	905	750	1,382	1,015	1,859
225	437	490	914	755	1,391		
230	446	495	923	760	1,400		
235	455	500	932	765	1,409		
240	464	505	941	770	1,418		
245	473	510	950	775	1,427		
250	482	515	959	780	1,436		
255	491	520	968	785	1,445		
260	500	525	977	790	1,454		

26.31.7 Temperature conversion formulae

°F to °C	$°C = 5/9(°F - 32)$	°R to °F	$°F = 9/4(°R + 32)$
°C to °F	$°F = 9/5(°C + 32)$	°R to °C	$°C = 5/4(°R)$
°F to °R	$°R = 4/9(°F - 32)$	Absolute zero $= -273.14°C$.	

26.32 Paper sizes

ISO standards (BS 4000)
A series

Description	Size (mm)	Description	Size (mm)
4A0	$1,682 \times 2,378$	A5	148×210
2A0	$1,189 \times 1,682$	A6	105×148
A0	$841 \times 1,189$	A7	74×105
A1	594×841	A8	52×74
A2	420×594	A9	37×52
A3	297×420	A10	26×37
A4	210×297		

B series

Description	Size (mm)	Description	Size (mm)
B0	$1,000 \times 1,414$	B6	125×176
B1	$707 \times 1,000$	B7	88×125
B2	500×707	B8	62×88
B3	353×500	B9	44×62
B4	250×353	B10	31×44
B5	176×250		

26.33 Fuses

Fuses are sometimes coded with the use of coloured dots on the fuse body. Ratings of colour-coded fuses are as follows:

Colour	Rating		
Black	60 mA	Dark blue	1.0 A
Grey	100 mA	Light blue	1.5 A
Red	150 mA	Purple	2.0 A
Brown	250 mA	White	3.0 A
Yellow	500 mA	Black and	
Green	750 mA	White	5.0 A

Diameters of fuse wires for various amperage ratings and common materials is shown below:

	Copper			*Tin*			Lead	
Fusing current (A)	diameter (in)	s.w.g.		diameter (in)	s.w.g.		diameter (in)	s.w.g.
1	0.0021	47		0.0072	37		0.0081	35
2	0.0034	43		0.0113	31		0.0128	30
3	0.0044	41		0.0149	28		0.0168	27
4	0.0053	39		0.0181	26		0.0203	25
5	0.0062	38		0.0210	25		0.0236	23
10	0.0098	33		0.0334	21		0.0375	20
15	0.0129	30		0.0437	19		0.0491	18
20	0.0156	28		0.0529	17		0.0595	17

26.34 Statistical formulae

The arithmetic mean of a set of numbers $X_1, X_2 \ldots, X_N$ is their average. It is the sum of the numbers divided by the number of numbers and is denoted by $\overline{X}$

$$\overline{X} = \frac{X_1 + X_2 + X_3 \ldots X_N}{N} = \frac{\sum\limits_{i=1}^{N} X_1}{N}$$

The **standard deviation** is denoted by σ.

$$\sigma = \sqrt{\dfrac{\text{sum of squares of differences between numbers and mean}}{N}}$$

$$= \sqrt{\dfrac{\sum\limits_{i=1}^{N}(X_1 - \overline{X})^2}{N}}$$

26.35 Particles of modern physics

Stable particles are listed below.

	Name	Symbol	Rest energy M_0/MeV	Mean Lifetime τ/s	Common decay modes
Baryons	Proton	$p^{\pm}$	938.256(6)	stable	
	Neutron	n	939.550(5)	$9.32(14) \times 10^2$	$pe\nu$
	Lambda	Λ^0	1115.60(8)	$2.51(3) \times 10^{-10}$	$p\pi^-$ (65%)$n\pi^0$ (35%)
	Sigma	Σ^+	1189.4(2)	$8.02(7) \times 10^{-11}$	$p\pi^0$ (52%)$n\pi^+$ (48%)
		Σ^0	1192.46(12)	$< 10^{-14}$	$\Lambda\gamma$
		Σ^-	1197.32(11)	1.49×10^{-10}	$n\pi^-$
	Xi	Ξ^0	1314.7(7)	$3.03(18) \times 10^{-10}$	$\Lambda\pi^0$
		Ξ^-	1321.25(18)	$1.66(4) \times 10^{-10}$	$\Lambda\pi^-$
	Omega	Ω^-	1672.5(5)	$1.3(4) \times 10^{-10}$	$\Xi^0\pi^-$, $\Xi^-\pi^0$, ΛK^-(?)
Leptons	Photon	γ	0	stable	
	Neutrino	ν_e	0	stable	
		ν_μ	0	stable	
	Electorn	$e^{\pm}$	0.511004(2)	stable	
	Muon	$\mu^{\pm}$	105.659(2)	$2.1994(6) \times 10^{-6}$	$e\nu\overline{\nu}$
Mesons	Pion	$\pi^{\pm}$	139.576(11)	$2.602(2) \times 10^{-8}$	$\mu\nu$
		π^0	134.972(12)	$0.84(10) \times 10^{-6}$	$\gamma\gamma$(99%) γe^+e^- (1%)
	Kaon	$K^{\pm}$	493.82(11)	$1.235(4) \times 10^{-8}$	$\mu\nu$(64%)$\pi^{\pm}\pi^0$ (21%) 3π(5%)
		K^0	497.76(16)	50%K_1, 50%K_2	
		K_1		$8.62(6) \times 10^{-11}$	$\pi^+\pi^-$ (69%)$2\pi^0$ (31%)
		K_2		$5.38(19) \times 10^{-8}$	$\pi e\nu$(39%)$\pi\mu\nu$(27%) $3\pi^0$(21%)$\pi^+\pi^-\pi^0$ (13%)
	Eta	η^0	548.8(6)		$\gamma\gamma$(38%)$\pi\gamma\gamma$(2%)$3\pi^0$ (31%)$\pi^+\pi^-\pi^0$ (23%) $\pi^+\pi^-\gamma$(5%)

26.36 Calculus

Differentiation

The derivative of a function $y = f(t)$ is denoted by $\dfrac{dy}{dt}$ of $\dot{y}$ if t represents time

The second derivative of $y = f(t)$ is denoted by $\dfrac{d^2y}{dt^2}$ or $\ddot{y}$ if t is time.

Useful derivatives

function $y = f(t)$		derivative $\dfrac{dy}{dt}$
1		0
t		1
t^A	$(A \neq 0)$	At^{A-1}
$\sin \omega t$	$(\omega \neq 0)$	$\omega \cos \omega t$
$\cos \omega t$	$(\omega \neq 0)$	$-\omega \sin \omega t$
$\tan at$	$(a \neq 0)$	$a \sec^2 at$
$\exp at$	$(a \neq 0)$	$a \exp at$
$\log_e at$	$(a \neq 0)$	$\dfrac{1}{t}$

Standard integrals

function $f(t)$	standard integrals $\int f(t)dt$
1	t
t	$\dfrac{1}{2}t^2$
$t^N (N \neq -1)$	$\dfrac{1}{N+1}t^{N+1} \quad (N \neq -1)$
$\dfrac{1}{T}$	$\log_e T \quad (T > 0)$
$\sin \omega t$	$-\dfrac{1}{\omega} \cos \omega t \quad (\omega \neq 0)$
$\cos \omega t$	$\dfrac{1}{\omega} \sin \omega t \quad (\omega \neq 0)$
$\exp at \ (a \neq 0)$	$\dfrac{1}{a} \exp at \ (a \neq 0)$
$\dfrac{1}{a^2 - t^2}$	$\dfrac{1}{2a} \log_e \left(\dfrac{a+t}{a-t} \right) (-a < t < +a)$
$\log_e (at)$	$t[\log_e (at) - 1]$

26.37 Mensuration

A and a = area; b = base; C and c = circumference; D and d = diameter; h = height; $n°$ = number of degrees; p = perpendicular; R and r = radius; s = span or chord; v = versed sine.

Square: $a = \text{side}^2$; side = $\sqrt{a}$;
diagonal = side $\times \sqrt{2}$.

Rectangle or parallelogram: $a = bp$.

Trapezoid (two sides parallel): a = mean length parallel sides $\times$ distance between them.

Triangle: $a = \frac{1}{2}bp$

Irregular figure: a = weight of template $\div$ weight of square inch of similar material.

Side of square multiplied by 1.4142 equals diameter of its circumscribing circle.

A side multiplied by 4.443 equals circumference of its circumscribing circle.

A side multiplied by 1.128 equals diameter of a circle of equal area.

Circle: $a = \pi r^2 = d^2\pi/4 = 0.7854d^2 = 0.5cr$; $c = 2\pi r = dx = 3.1416d = 3.54a = (\text{approx.})\frac{22}{7}d$. Side of equal square = $0.8862d$; side of inscribed square = $0.7071d$; $d = 0.3183c$. A circle has the maximum area for a given perimeter.

Annulus of circle : $a = (D + d)(D - d)\dfrac{\pi}{4}$
$$= (D^2 - d^2)\dfrac{\pi}{4}$$

Segment of circle:

$$a = \text{area of sector} - \text{area of triangle}$$
$$= \frac{4v}{3}\sqrt{(0.625v)^2 + (\tfrac{1}{2}S)^2}.$$

Length of arc = $0.017453n°r$; length of

$$\text{arc} = \tfrac{1}{3}\left(8\sqrt{\tfrac{s^2}{4} + v^2} - s\right);$$

479

approx. length of arc = $\frac{1}{3}$ (8 times chord of $\frac{1}{2}$ arc − chord of whole arc).

$$d = \frac{\left(\frac{1}{2}\text{chord}\right)}{v} + v;$$

$$\text{radius of curve} = \frac{S^2}{8V} + \frac{V}{2}.$$

Sector of circle : $a = 0.5r \times$ length arc;

$$= n° \times \text{area circle} \div 360.$$

Ellipse: $a = \frac{\pi}{4}Dd = \pi Rr$; c (approx.)

$$= \sqrt{\frac{D^2 + d^2}{2}} \times \pi; c \text{ (approx.)} = \pi\frac{Da}{2}.$$

Parabola: $a = \frac{2}{3}bh$.

Cone or pyramid: surface

$$= \frac{\text{circ. of base} \times \text{slant length}}{2} + \text{base};$$

contents = area of base $\times \frac{1}{3}$ vertical height.

Frustrum of cone:

$$\text{surface} = (C + c) \times \frac{1}{2} \text{ slant height} + \text{ends};$$
$$\text{contents} = 0.2618h(D^2 + d^2 + Dd);$$
$$= \frac{1}{3}h(A + a + \sqrt{A \times a}).$$

Wedge: contents = $\frac{1}{6}$(length of edge + 2 length of back)bh.

Oblique prism: contents = area base $\times$ height.

Sphere: surface = $d^2\pi = 4\pi r^2$,

$$\text{contents} = d^3\frac{\pi}{6} = \frac{4}{3}\pi r^3.$$

Segment of sphere: r = rad. of base;

contents = $\frac{\pi}{6}h(3r^2 + h^2)$; r = rad. of sphere;

Spherical zone:

contents = $\frac{\pi}{2}h(\frac{1}{3}h^2 + R^2 + r^2)$; surface of convex part of segment or zone of sphere = πd(of sph.)$h = 2\pi rh$.

Mid. sph. zone: contents = $(r + \frac{2}{3}h^2)\frac{\pi}{4}$

Spheroid:

contents = revolving axis^2 $\times$ fixed axis $\times \frac{\pi}{6}$.

Cube or rectangular solid contents = lenght × breadth × thickness.

Prismoidal formula: contents

$$= \frac{\text{end areas} + 4 \times \text{mid. area} \times \text{length}}{6}$$

Solid revolution: contents = a of generating plane × c described by centroid of this plane during revolution. Areas of similar plane figures are as the squares of like sides. Contents of similar solids are as the cubes of like sides.

Rules relative to the circle, square, cylinder, etc.:

To find circumference of a circle:

Multiply diameter by 3.1416; or divide diameter by 0.3183.

To find diameter of a circle:

Multiply circumference by 0.3183; or divide circumference by 3.1416.

To find radius of a circle:

Multiply circumference by 0.15915; or divide circumference by 6.28318.

To find the side of an inscribed square:

Multiply diameter by 0.7071; or multiply circumference by 0.2251; or divide circumference by 4.4428.

To find side of an equal square:

Multiply diameter by 0.8862; or divide diameter by 1.1284; or multiply circumference by 0.2821; or divide circumference by 3.545.

To find area of circle:

Multiply circumference by $\frac{1}{4}$ of the diameter, or multiply the square of diameter by 0.7854; or multiply the square of circumference by 0.7958; or multiply the square of $\frac{1}{2}$ diameter by 3.1416.

To find the surface of a sphere or globe:

Multiply the diameter by the circumference; or multiply the square of diameter by 3.1416; or multiply 4 times the square of radius by 3.1416.

Cylinder.

To find the area of surface:

Multiply the diameter by $3\frac{1}{7}$ × length.

Capacity = $3\frac{1}{7}$ × radius^2 × height.

Values and Powers of:

$$\pi = 3.1415926536, \text{ or } 3.1416, \text{ or } \tfrac{22}{7} \text{ or } 3\tfrac{1}{7};$$
$$\pi^2 = 9.86965; \sqrt{\pi} = 1.772453;$$
$$\frac{1}{\pi} = 0.31831; \quad \frac{\pi}{2} = 1.570796;$$
$$\frac{\pi}{3} = 1.047197$$

Radian = 57.2958 degrees.

Table A

Fig. 1. Diagram
for Table A.

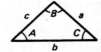

Parts given	Parts to be found	Formulae
abc	A	$\cos A = \dfrac{b^2 + c^2 - a^2}{2bc}$
abA	B	$\sin B = \dfrac{b \times \sin A}{a}$
abA	C	$C = 180° - (A + B)$
aAB	b	$b = \dfrac{a \times \sin B}{\sin A}$
aAB	c	$c = \dfrac{a \sin C}{\sin A} = \dfrac{a \sin (180° - A - B)}{\sin A}$
abC	B	$B = 180° - (A + C)$

482

Table B

Fig. 2. Diagram for Table B.

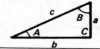

Parts given			
a&c	$\sin A = \dfrac{a}{c}$	$\cos B = \dfrac{a}{c}$	$b = \sqrt{c^2 - a^2}$
a&b	$\tan A = \dfrac{a}{b}$	$\cot B = \dfrac{a}{b}$	$c = \sqrt{a^2 + b^2}$
c&b	$\cos A = \dfrac{b}{c}$	$\sin B = \dfrac{b}{c}$	$a = \sqrt{c^2 - b^2}$
A&a		$B = 90° - A$	$b = a \times \cot A \quad c = \dfrac{a}{\sin A}$
A&b		$B = 90° - A$	$a = b \times \tan A \quad c = \dfrac{b}{\cos A}$
A&c		$B = 90° - A$	$a = c \times \sin A \quad b = c \times \cos A$

Fig. 3. In any right-angled triangle:

$$\tan A = \frac{BC}{AC}, \qquad \sin A = \frac{BC}{AB}$$
$$\cos A = \frac{AC}{AB}, \qquad \cot A = \frac{AC}{BC}$$
$$\sec A = \frac{AB}{AC}, \qquad \operatorname{cosec} A = \frac{AB}{BC}$$

Fig. 4. In any right-angled triangle:
$$a^2 = c^2 + b^2$$
$$c = \sqrt{a^2 - b^2}$$
$$b = \sqrt{a^2 - c^2}$$
$$a = \sqrt{b^2 + c^2}$$

Fig. 5. $c + d : a + b :: b - a : d - c.$
$$d = \frac{c + d}{2} + \frac{d - c}{2}$$
$$x = \sqrt{b^2 - d^2}$$

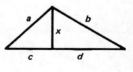

In Fig. 6. where the lengths of three sides only are known:

$$\text{area} = \sqrt{s(s-a)(s-b)(s-c)}$$

$$\text{where } s = \frac{a+b+c}{2}$$

Fig. 7. In this diagram:

$a : b :: b : c$ or $\dfrac{b^2}{a} = c$

Fig. 8. In an equilateral triangle $ab = 1$, then $cd = \sqrt{0.75} = 0.866$, and $ad = 0.5$; $ab = 2$, then $cd = \sqrt{3.0} = 1.732$, and $ad = 1$; $cd = 1$, then $ac = 1.155$ and $ad = 0.577$; $cd = 0.5$, then $ac = 0.577$ and $ad = 0.288$.

Fig. 9. In a right-angled triangle with two equal acute angles, $bc = ac$, $bc = 1$, then $ab = \sqrt{2} = 1.414$; $ab = 1$, then $bc = \sqrt{0.5} = 0.707$.

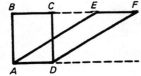

Fig. 10 shows that parallelograms on the same base and between the same parallels are equal: thus $ABCD = ADEF$.

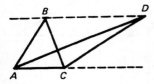

Fig. 11 demonstrates that triangles on the same base and between the same parallels are equal in area; thus $ABC = ADC$.

484

26.38 Trigonometrical relationships

$$\sin\left(\frac{\pi}{2} - \alpha\right) = \cos\alpha$$

$$\sin(-\alpha) = -\sin\alpha$$

$$\sin(\pi - \alpha) = \sin\alpha$$

$$\sin(\pi + \alpha) = -\sin\alpha$$

$$\sin(2\pi - \alpha) = \sin(-\alpha) = -\sin\alpha$$

$$\sin(2N\pi + \alpha) = \sin\alpha \ (N \text{ an integer})$$

$$\frac{\sin\alpha}{\cos\alpha} = \tan\alpha$$

$$\cos\left(\frac{\pi}{2} = \alpha\right) = \sin\alpha$$

$$\cos(-\alpha) = \cos\alpha$$

$$\cos(\pi - \alpha) = -\cos\alpha$$

$$\cos(\pi + \alpha) = -\cos\alpha$$

$$\cos(2\pi - \alpha) = \cos(-\alpha) = \cos\alpha$$

$$\cos(2\pi N + \alpha) = \cos\alpha \ (N \text{ and integer})$$

$$\tan\left(\frac{\pi}{2} - \alpha\right) = \frac{1}{\tan\alpha}$$

$$\tan(-\alpha) = -\tan\alpha$$

$$\tan(\alpha + N\pi) = \tan\alpha \ (N \text{ an integer})$$

$$\sin^2\alpha + \cos^2\alpha = 1$$

$$\sin^2\alpha = 1/2(1 - \cos 2\alpha)$$

$$\cos^2\alpha = 1/2(1 + \cos 2\alpha)$$

$$\tan^2\alpha + 1 = \sec^2\alpha$$

$$\left.\begin{array}{l}\sin(\alpha + \beta) = \sin\alpha\cos\beta + \sin\beta\cos\alpha \\ \cos(\alpha + \beta) = \cos\alpha\cos\beta - \sin\alpha\sin\beta \\ \tan(\alpha + \beta) = \dfrac{\tan\alpha + \tan\beta}{1 - \tan\alpha\tan\beta}\end{array}\right\} \begin{array}{l}(\alpha, \beta \text{ can be} \\ \text{positive or} \\ \text{negative})\end{array}$$

$$\sin 2\alpha = 2\sin\alpha\cos\alpha$$

$$\cos 2\alpha = \cos^2\alpha - \sin^2\alpha$$

$$= 2\cos^2\alpha - 1$$

$$= 1 - 2\sin^2\alpha$$

$$\tan 2\alpha = \frac{2\tan\alpha}{1 - \tan^2\alpha} \qquad (\tan\alpha \neq \pm 1)$$

26.39 Transistor circuits and characteristics

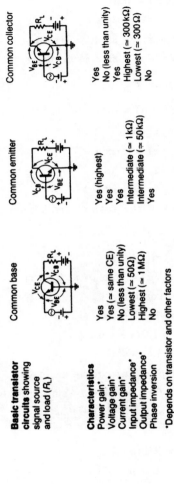

	Common base	Common emitter	Common collector
Basic transistor circuits showing signal source and load (R_L)			
Characteristics			
Power gain*	Yes	Yes (highest)	Yes
Voltage gain*	Yes (≈ same CE)	Yes	No (less than unity)
Current gain*	No (less than unity)	Yes	Yes
Input impedance*	Lowest (≈ 50Ω)	Intermediate (≈ 1kΩ)	Highest (≈ 300 kΩ)
Output impedance*	Highest (≈ 1MΩ)	Intermediate (≈ 50kΩ)	Lowest (≈ 300Ω)
Phase inversion	No	Yes	No

*Depends on transistor and other factors

26.40 Astronomical data

Distance of earth from sun
  (mean) $= 1.496 \times 10^{11}$ m
Distance of earth from sun
  (at aphelion) $= 1.521 \times 10^{11}$ m
Distance of earth from sun
  (at perihelion) $= 1.471 \times 10^{11}$ m
Distance of moon from earth
  (mean) $= 3.844 \times 10^{8}$ m
Escape velocity at surface of
  earth $= 11.2$ kms^{-1}
Escape velocity at surface of
  moon $= 2.38$ kms^{-1}
Escape velocity at surface of sun $= 618$ kms^{-1}
Gravity at surface of earth $= 9.80665$ ms^{-2}
Gravity at surface of moon $= 1.62$ ms^{-2}
Gravity at surface of sun $= 273$ ms^{-2}
Land area of earth $= 148.8 \times 10^{6}$ km^{2}
Light year (ly) $= 9.4605 \times 10^{15}$ m
Mass of earth $= 5.977 \times 10^{24}$ kg
Mass of moon $= 7.349 \times 10^{22}$ kg
Mass of sun $= 1.99 \times 10^{30}$ kg
Mean density of earth $= 5,517$ kgm^{-3}
Mean density of moon $= 3,340$ kgm^{-3}
Mean density of sun $= 1,409$ kgm^{-3}
Ocean area of earth $= 361.3 \times 10^{6}$ km^{2}
Parsec (pc) $= 3.0856 \times 10^{16}$ m
Period of moon about earth (sidereal) $= 27.32$
  means solar days
Period of sun's rotation
  (with respect to earth) $= 27.28$ days
Period of sun's rotation (sidereal) $= 25.38$ days
Radius of earth: (polar) $= 6,356.8$ km
          (equatorial) $= 6,378.2$ km
Radius of moon $= 1,738$ km
Radius of sun $= 6.960 \times 108$ m
Rotational velocity at equator of
  earth $= 465$ms^{-1}
Sidereal day $= 86,164.0906$ mean solar
  seconds
Sidereal year $= 365.256$ mean solar days
Solar second (mean) $= 1/86,400$ of a mean
  solar day

Surface area of earth $= 5.101 \times 10^{14}$ m^2
Surface area of moon $= 3.796 \times 10^{13}$ m^2
Surface area of sun $= 6.087 \times 10^{18}$ m^2
Synodical or lunar month (mean) $= 29.531$
   mean solar days
Tropical (civil) year $= 365.256$ mean solar
   days
Velocity of earth in orbit around sun
   (mean) $= 29.78$ kms^{-1}
Volume of earth $= 1.083 \times 10^{21}$ m^3
Volume of moon $= 2.199 \times 10^{19}$ m^3
Volume of sun $= 1.412 \times 10^{27}$ m^3

26.41 Resistivities of selected metals and alloys

Material	Form	Resistivity (ohm – m × 10⁻⁹)	Temperature (°C)	Temperature coefficient
Alumel	Solid	33.3	0	0.0012
Aluminium	Liquid	20.5	670	
	Solid	2.62	20	0.0039
Antimony	Liquid	123	800	
	Solid	39.2	20	0.0036
Arsenic	Solid	35	0	0.0042
Beryllium		4.57	20	
Bismuth	Liquid	128.9	300	
	Solid	115	20	0.004
Boron		1.8×10^{12}	0	
Brass (66 Cu 34 Zn)		3.9	20	0.002
Cadmium	Liquid	34	400	
	Solid	7.5	20	0.0038
Carbon	Diamond	5×10^{20}	15	
	Graphite	1400	20	−0.0005
Cerium		78	20	
Cesium	Liquid	36.6	30	
	Solid	20	20	
		18.83	0	
Chromax (15 Cr, 35 Ni, balance Fe)		100	20	0.00031
Chromel	Solid	70 – 110	0	0.00011 – 0.000054
Chromium		2.6	0	
Cobalt		9.7	20	0.0033
Constantan (55 Cu, 45 Ni)		44.2	20	+0.0002
Copper (commercial annealed)	Liquid	21.3	1083	
	Solid	1.7241	20	0.0039
Gallium	Liquid	27	30	
	Solid	53	0	
Germanium		45	20	
German silver (18% Ni)		33	20	0.0004
Gold	Liquid	30.8	1063	
	Solid	2.44	20	0.0034
		2.19	0	
Hafnium		32.1	20	
Indium	Liquid	29	157	
	Solid	9	20	0.00498
Iridium		5.3	20	0.0039
Iron		9.71	20	0.0052 – 0.0062
Kovar A (29 Ni, 17 Co, 0.3 Mn, balance Fe)		45 – 85	20	
Lead	Liquid	98	400	
	Solid	21.9	20	0.004
PbO_2		92		
Lithium	Liquid	45	230	0.003
	Solid	9.3	20	0.005

(continued overleaf)

Material	Form	Resistivity (ohm – m $\times 10^{-9}$)	Temperature (°C)	Temperature coefficient
Magnesium		4.46	20	0.004
Manganese		5	20	
Manganin (84 Cu, 12 Mn, 4 Ni)		44	20	±0.0002
Mercury	Liquid	95.8	20	0.00089
	Solid	21.3	−50	
Molybdenum		5.17	0	
		4.77	20	0.0033
MNO$_2$		6,000,000	20	
Monel metal (67 Ni, 30 Cu, 1.4 Fe, 1 Mn)	Solid	42	20	0.002
Neodymium	Solid	79	18	
Nichrome (65 Ni, 12 Cr, 23 Fe)	Solid	100	20	0.00017
Nickel	Solid	6.9	20	0.0047
Nickel-silver (64 Cu, 18 Zn, 18 Ni)	Solid	28	20	0.00026
Niobium		12.4	20	
Osmium		9	20	0.0042
Palladium		10.8	20	0.0033
Phosphor bronze (4 Sn, 0.5 P, balance Cu)		9.4	20	0.003
Platinum		10.5	20	0.003
Plutonium		150	20	
Potassium	Liquid	13	62	
	Solid	7	20	0.006
Praseodymium		68	25	
Rhenium		19.8	20	
Rhodium		5.1	20	0.0046
Rubidium		12.5	20	
Ruthenium		10	20	
Selenium	Solid	1.2	20	
Silicon		85×10^3	20	
Silver		1.62	20	0.0038
Sodium	Liquid	9.7	100	
	Solid	4.6	20	
Steel (0.4–0.5 C, balance Fe)		13–22	20	0.003
Steel, manganese (13 Mn, 1 C, 86 Fe)		70	20	0.001
Steel, stainless (0.1 C, 18 Cr, 8 Ni, balance Fe)		90	20	
Strontium		23	20	
Sulfur		2×10^{23}	20	
Tantalum		13.1	20	0.003
Thallium		18.1	20	0.004
Thorium		18	20	0.0021
Tin		11.4	20	0.0042

Material	Form	Resistivity (ohm − m × 10^{-9})	Temperature (°C)	Temperature coefficient
Titanium		47.8	25	
Tungsten		5.48	20	0.0045
Tophet A (80 Ni, 20 Cr)		108	20	0.00014
Uranium		29	0	0.0021
W_2O_5		450	20	
WO_3		2×10^{11}	20	
Zinc	Liquid	35.3	420	
	Solid	6	20	0.0037
Zirconium		40	20	0.0044

26.42 Electrical properties of elements

	Symbol	Atomic number Z	Mass number N + Z	Atomic weight	Atomic radii ×10⁻¹⁰ m	Gram atomic volume (cm³)	Electron-negativity, relative scale	First ionization potential (electron volts)	Electron work function			Electrochemical equivalent	
									Thermionic	Photoelectric	Contact	Valence* involved	Amp-hours per gram
Actinium	Ac	89	227	227			1.1	6.9				3	0.35
Aluminium	Al	13	27	26.98	1.25	10	1.5	5.98		4.08	3.38	3	2.98
Americium	Am	95	243	243				6.05					
Antimony	Sb	51	121 – 123	121.75	1.41	18	2.05	8.64		4.01	4.14	5	1.1
Argon	Ar or A	18	40	39.948	1.74	24	0	15.76				n	0.67

Element	Symbol	Z	Mass numbers	At. wt.									
Arsenic	As	33	75	74.92	1.21	16	2.0	9.81		5.11		5	1.79
Astatine	At	85	210	210			2.2					1	0.39
Barium	Ba	56	138	137.34	1.98	38	0.9	5.21	2.11	2.48	1.73	2	7.43
Boron	B	5	11	10.81	0.88	5	2.0	8.3		4.5		3	0.335
Bromine	Br	35	79–81	79.904	1.14	23	2.85	11.81				1	0.477
Cadmium	Cd	48	114–112	112.40	1.41	13	1.7	8.99				2	1.337
Calcium	Ca	20	40	40.08	1.74	26	1.0	6.11	2.24	4.07	4.0	2	
Californium	Cf	98	251	251						2.706	3.33	2	
Carbon	C	6	12	12.011	0.77	5	2.6	11.26				4	8.93
Cerium	Ce	58	140	140.12	1.65	21	1.1	5.6	4.34	4.81		3	0.574
Cesium	Cs	55	133	132.905	2.35	71	0.7	3.89	2.6	2.84		1	0.2
Chlorine	Cl	17	35	35.453	0.99	19	3.15	12.97	1.81	1.92	4.46	1	0.756
Chromium	Cr	24	52	51.996	1.17	7	1.6	6.76				3	1.546
Cobalt	Co	27	59	58.933	1.16	7	1.8	7.86	4.60	4.37	4.38	2	0.91
Copper	Cu	29	63	63.546	1.17	7	1.9	7.72	4.40	4.20	4.21	2	0.84
Curium	Cm	96	247	247					4.26	4.18	4.46		
Dysprosium	Dy	66	164-162-163	162.50	1.59	19	1.2	5.93				3	0.495
Einsteinium	Es or E	99	254	254									
Erbium	Er	68	166-168-167	167.26	1.57	18	1.2	6.10				3	0.48
Europium	Eu	63	153-151	151.96	1.85	29	1.1	5.67				3	0.53
Fermium	Fm	100	257	257									
Fluorine	F	9	19	18.998	0.64	15	3.9	17.42				1	1.41
Francium	Fr	87	223	223			0.65						
Gadolinium	Gd	64	158-160-156	157.25	1.61	20	1.1	6.16				3	0.513

(continued overleaf)

26.42 (continued)

Element	Symbol	Atomic number Z	Mass number $Z+N$	Atomic weight	Atomic radii $\times 10^{-10}$ m	Gram atomic volume (cm³)	Electron-negativity, relative scale	First ionization potential (electron volts)	Electron work function — Thermionic	Electron work function — Photoelectric	Electron work function — Contact	Valence* involved	Electrochemical equivalent: Amp-hours per gram
Gallium	Ga	31	69–71	69.72	1.25	12	1.6	5.99	4.12		3.80	3	1.15
Germanium	Ge	33	74-73-70	73.50	1.33	13	1.0	7.80		4.5	4.5	4	1.48
Gold	Au	79	197	196.967	1.34	10	2.4	9.22	4.32	4.82	4.46	3	0.41
Hafnium	Hf	72	180-178-177	178.49	1.44	13	1.3	7.0	3.53			4	0.600
Helium	He	2	4	4.003		32	0	24.59				n	6.698
Holmium	Ho	67	165	164.93	1.58	19	1.2	6.02				3	0.488
Indium	In	49	115	114.82	1.50	16	1.7	5.78				3	0.700
Iodine	I	53	127	126.904	1.33	26	2.65	10.45		6.8		1	0.211
Iridium	Ir	77	193-191	192.22	1.26	9	2.2	9.1	5.3		4.57	4	0.555
Iron	Fe	26	56	55.847	1.17	7	1.8	7.87	4.25	4.33	4.40	3	1.440
Krypton	Kr	36	84–86	83.80	1.89	33	0	13.99				n	0.32
Lanthanum	La	57	139	138.905	1.69	22	1.1	5.61	3.3			3	0.579

Lawrencium	Lw	103	257	257								4	0.517
Lead	Pb	82	208-206-207	207.2	1.54	18	1.8	7.42		4.05	3.94	1	3.862
Lithium	Li	3	7	6.940	1.23	13	1.0	5.39		2.35	2.49	3	0.46
Lutetium	Lu	71	175	174.97	1.56	18	1.2	6.15				4	1.952
Magnesium	Mg	12	24	24.305	1.36	14	1.2	7.64		3.76	4.14	4	1.952
Manganese	Mn	25	55	54.938	1.17	7	1.5	7.43		3.76	4.14		
Mendelevium	Md or Mv	101	256	256					3.83	3.83			
Mercury	Hg	80	202-200-199	200.59	1.44	14	1.9	10.43		4.53	4.50	2	0.267
Molybdenum	Mo	42	98-96-92-95	95.94	1.29	9	1.8	7.10	4.20	4.25	4.28	6	1.67
Neodymium	Nd	60	142-144-146	144.24	1.64	21	1.1	5.49	3.3			3	0.557
Neon	Ne	10	20	20.179	1.31	17	0	21.56				n	1.33
Neptunium	Np	93	237	237.048			1.3	5.8					
Nickel	Ni	28	58	58.71	1.15	6	1.8	7.63		5.01	4.96	2	0.913
Niobium	Nb	41	93	92.906	1.34	11	1.6	6.88	5.03	4.5	4.01		
Nitrogen	N	7	14	14.007	0.70	14	3.05	14.53				5	9.57
Nobelium	No	102	254	254									
Osmium	Os	76	192-190-189	190.2	1.26	9	2.2	8.7		4.55		4	0.56
Oxygen	O	8	16	15.999	0.66	11	3.5	13.62				2	3.35
Palladium	Pd	46	108-106-105	106.4	1.28	9	2.2	8.33	4.99	4.97	4.49	4	1.005
Phosphorus	P	15	31	30.974	1.10	17	2.15	10.48				5	4.33
Platinum	Pt	78	195-194-196	195.09	1.29	9	2.2	9.0	5.32	5.22	5.36	4	0.549
Plutonium	Pu	94	242	242				5.8					
Polonium	Po	84	209	210	1.53		2.0	8.43				6	0.766
Potassium	K	19	39	39.098	2.03	46	0.8	4.34	2.7	2.24	1.60	1	0.685
Praseodymium	Pr	59	141	140.907	1.65	21	1.1	5.42				3	0.571

(continued overleaf)

26.42 *(continued)*

Element	Symbol	Atomic number Z	Mass number Z + N	Atomic weight	Atomic radii ×10⁻¹⁰ m	Gram atomic volume (cm³)	Electron-negativity, relative scale	First ionization potential (electron volts)	Thermionic	Photoelectric	Contact	Valence* involved	Electrochemical equivalent Amp-hours per gram
Promethium	Pm	61	145	145			1.1	5.55				5	0.580
Protactinium	Pa	91	231	231.036			1.5					2	0.237
Radium	Ra	88	226	226.025		45	0.9	5.28				n	0.121
Radon	Rn	86	222	222	2.14	50	0	10.75				7	1.007
Rhenium	Re	75	187-185	186.2	1.28	9	1.9	7.87	5.1	5.0		4	1.042
Rhodium	Rh	45	103	102.905	1.25	8	2.2	7.46	4.80	4.57	4.52	1	0.314
Rubidium	Rb	37	85-87	85.468	2.16	56	0.8	4.18		2.09		4	1.054
Ruthenium	Ru	44	102-104-101	101.07	1.24	8	2.2	7.37			4.52	4	1.054
Samarium	Sm	62	152-154-147	150.35	1.66	20	1.1	5.63	3.2			3	0.535
Scandium	Sc	21	45	44.956	1.44	15	1.3	6.54				3	1.783
Selenium	Se	34	80-78	78.96	1.17	16	2.45	9.75		4.8	4.42	6	2.037

Element	Symbol		Isotopes	Weight									
Silicon	Si	14	28	28.086	1.17	12	8.15	1.9	3.59	4.52	4.2	4	3.821
Silver	Ag	47	107-109	107.868	1.34	10	7.57	1.9	3.56	4.73	4.44	1	0.248
Sodium	Na	11	23	22.99	1.57	24	5.14	0.9		2.28	1.9	1	1.166
Strontium	Sr	38	88	87.62	1.92	34	5.69	1.0		2.74		2	0.612
Sulfur	S	16	32	32.064	1.04	16	10.36	2.6				6	5.01
Tantalum	Ta	73	181	180.948	1.34	11	7.88	1.3	4.19	4.14	4.1	5	0.741
Technetium	Tc	43	99	98.906			7.28	1.9					
Tellurium	Te	52	130-128-126	127.60	1.37	21	9.01	2.3		4.76	4.70	6	1.260
Terbium	Tb	65	159	158.925	1.59	19	5.98	1.2				3	0.505
Thallium	Tl	81	205-203	204.37	1.55	17	6.11	1.8		3.68	3.84	3	0.393
Thorium	Th	90	232	232.038	1.65	20	6.95	1.3	3.35	3.47	3.46	4	0.462
Thulium	Tm	69	169	168.934	1.56	18	6.18	1.2				3	0.475
Tin	Sn	50	120-118	118.69	1.40	16	7.34	1.8		4.38	4.09	4	0.903
Titanium	Ti	22	48	47.90	1.32	11	6.82	1.5	3.95	4.06	4.14	4	2.238
Tungsten	W	74	184-186-182	183.85	1.30	10	7.98	1.7	4.52	4.49	4.38	6	0.874
Uranium	U	92	238	238.029	1.42	13	6.08	1.7	3.27	3.63	4.32	6	0.676
Vanadium	V	23	51	50.94	1.22	8	6.74	1.6	4.12	3.77	4.44	5	2.63
Xenon	Xe	54	132-129-131	131.30	2.09	43	12.13	0				n	0.204
Ytterbium	Yb	70	174-172-173	173.04	1.70	25	6.25	1.2				3	0.465
Yttrium	Y	39	89	88.906	1.62	21	6.38	1.3				3	0.904
Zinc	Zn	30	64-66-68	65.38	1.25	9	9.39	1.6		3.73	3.78	2	0.820
Zirconium	Zr	40	90-94-92	91.22	1.45	14	6.84	1.6	4.21	3.82	3.60	4	1.175

*n = nonvalent

26.43 Wire data and drill sizes

26.43.1 Standard wire gauge and standard drill sizes

Standard wire gauge	Standard drill size — in	Standard drill size — mm	Decimal inch equivalent	Nearest obsolete number drill
50			0.0010	
49			0.0012	
48			0.0016	
47			0.0020	
46			0.0024	
45			0.0028	
44			0.0032	
43			0.0036	
42			0.0040	
41			0.0044	
40			0.0048	
39			0.0052	
38			0.0060	
37			0.0068	
36			0.0076	
35			0.0084	
34			0.0092	
33			0.0100	
32			0.0108	
31			0.0116	
30			0.0124	
		0.32	0.0126	
29			0.0136	
		0.35	0.0138	80
28			0.0148	
		0.38	0.0150	79
	$\frac{1}{64}$		0.0156	
		0.40	0.0157	78
27			0.0164	
		0.42	0.0165	
		0.45	0.0177	77
26			0.0180	
		0.48	0.0189	76
		0.50	0.0197	
25			0.0200	
		0.52	0.0205	75
		0.55	0.0217	
24			0.0220	
		0.58	0.0228	74
		0.60	0.0236	73
23			0.0240	
		0.62	0.0244	
		0.65	0.0256	72, 71

Standard wire gauge	Standard drill size (in)	Standard drill size (mm)	Decimal inch equivalent	Nearest obsolete number drill
		0.68	0.0268	
		0.70	0.0276	70
22			0.0280	
		0.72	0.0280	
		0.75	0.0295	69
		0.78	0.0307	
	$\frac{1}{32}$		0.0312	68
		0.80	0.0315	
21			0.0320	
		0.82	0.0323	67
		0.85	0.0335	66
		0.88	0.0346	
		0.90	0.0354	65
20			0.0360	
		0.92	0.0362	64
		0.95	0.0374	63
		0.98	0.0386	62
		1.00	0.0394	61, 60
19			0.0400	
		1.05	0.0413	59, 58
		1.10	0.0433	57
		1.15	0.0453	
	$\frac{3}{64}$		0.0469	56
		1.20	0.0472	
18			0.0480	
		1.25	0.0492	
		1.30	0.0512	55
		1.35	0.0532	
		1.40	0.0551	54
17			0.0560	
		1.45	0.0571	
		1.50	0.0591	53
		1.55	0.0610	
	$\frac{1}{16}$		0.0625	
		1.60	0.0630	52
16			0.0640	
		1.65	0.0650	
		1.70	0.0669	51
		1.75	0.0689	
15		1.80	0.0720	50
		0.0709		
		1.85	0.0728	49
		1.90	0.0748	
		1.95	0.0768	48
	$\frac{5}{64}$		0.0781	
14		2.00	0.0787	47

(*continued overleaf*)

| Standard wire gauge | Standard drill size | | Decimal inch equivalent | Nearest obsolete number drill |
	in	mm		
			0.0800	
		2.05	0.0807	46
		2.10	0.0827	45
		2.15	0.0846	
		2.20	0.0866	44
		2.25	0.0886	43
		2.30	0.0906	
13			0.0920	
		2.35	0.0925	
	$\frac{3}{32}$		0.0938	42
		2.40	0.0945	
		2.45	0.0965	41
		2.50	0.0984	40
		2.55	0.1004	39
		2.60	0.1024	38
12			0.1040	
		2.65	0.1043	37
		2.70	0.1063	36
		2.75	0.1083	
	$\frac{7}{64}$		0.1094	
		2.80	0.1102	35, 34
		2.85	0.1122	33
		2.90	0.1142	
11			0.1160	
		2.95	0.1161	32
		3.00	0.1181	31
		3.10	0.1220	
	$\frac{1}{8}$		0.1250	
		3.20	0.1260	
10			0.1280	
		3.30	0.1299	30
		3.40	0.1339	
		3.50	0.1378	29
	$\frac{9}{64}$		0.1406	28
		3.60	0.1417	
9			0.1440	
		3.70	0.1457	27, 26
		3.80	0.1496	25
		3.90	0.1535	24, 23
	$\frac{5}{32}$		0.1562	
		4.00	0.1575	22, 21
8			0.1600	
		4.10	0.1614	20
		4.20	0.1654	19
		4.30	0.1693	18

| Standard wire gauge | Standard drill size | | Decimal inch equivalent | Nearest obsolete number drill |
	in	mm		
	$\frac{11}{64}$		0.1719	
7		4.40	0.1732	17
			0.1760	
		4.50	0.1772	16
		4.60	0.1811	15, 14
		4.70	0.1850	13
	$\frac{3}{16}$		0.1875	
6		4.80	0.1890	12
			0.1920	
		4.90	0.1929	11, 10
		5.00	0.1968	9
		5.10	0.2008	8, 7
	$\frac{13}{64}$		0.2031	
		5.20	0.2047	6, 5
5		5.30	0.2087	4
			0.2120	
		5.40	0.2126	3
		5.50	0.2165	
	$\frac{7}{32}$		0.2188	
		5.60	0.2205	2
		5.70	0.2244	
4		5.80	0.2283	1
			0.2320	
		5.90	0.2323	
	$\frac{15}{64}$		0.2344	A
		6.00	0.2362	B
		6.10	0.2402	C
		6.20	0.2441	D
		6.30	0.2480	
	$\frac{1}{4}$		0.2500	E
		6.40	0.2520	
3		6.50	0.2559	F
		6.60	0.2598	G
		6.70	0.2638	
	$\frac{17}{64}$		0.2656	H
		6.80	0.2677	
		6.90	0.2717	I
		7.00	0.2756	J
			0.2760	
		7.10	0.2795	
2	$\frac{9}{32}$		0.2812	K
		7.20	0.2835	
		7.30	0.2874	
		7.40	0.2913	L
		7.50	0.2953	M

(continued overleaf)

	$\frac{19}{64}$		0.2969	
		7.60	0.2992	
1			0.3000	
		7.70	0.3032	N
		7.80	0.3071	
		7.90	0.3110	
	$\frac{5}{16}$		0.3125	
		8.00	0.3150	O
0		8.10	0.3189	
		8.20	0.3228	P
			0.3240	
		8.30	0.3268	
	$\frac{21}{64}$		0.3281	
		8.40	0.3307	Q
		8.50	0.3346	
		8.60	0.3386	R
		8.70	0.3425	
00	$\frac{11}{32}$		0.3438	
		8.80	0.3465	S
			0.3480	
		8.90	0.3504	
		9.00	0.3543	
		9.10	0.3583	T
	$\frac{23}{64}$		0.3594	
		9.20	0.3622	
		9.30	0.3661	U
		9.40	0.3701	
3/0			0.3720	
		9.50	0.3740	
			0.3750	V
		9.60	0.3780	
		9.70	0.3819	
		9.80	0.3858	W
		9.90	0.3898	
	$\frac{25}{64}$		0.3906	
		10.00	0.3937	
		10.00	0.3976	X
4/0			0.4000	
		10.20	0.4016	
		10.30	0.4055	
	$\frac{1}{32}$		0.4062	Y
		10.40	0.4094	
		10.50	0.4134	Z
		10.60	0.4173	
		10.70	0.4213	
5/0	$\frac{27}{64}$		0.4219	
		10.80	0.4252	
		10.90	0.4291	
			0.4320	
		11.00	0.4331	
		11.10	0.4370	
	$\frac{7}{16}$		0.4409	

Drill sizes proceed thus;
$\frac{1}{2}$ to 2 inches in $\frac{1}{64}$ inch steps;
12.7 to 14 mm in 0.1 mm steps;
14 to 25 mm in 0.25 mm steps;
25 to 50.5 mm in 0.5 mm steps.

26.43.2 BSI standard metric sizes of copper winding wires

Conductor diameter			Sectional area mm²	Weight per km kg	Nominal resistance at 20°C		Current rating at 4.65 amps per mm²* amps
Nom. mm	Max. mm	Min. mm			Per metre ohms	Per kg ohms	
5.000	5.050	4.950	19.63	174.6	0.0008781	0.005029	91.30
4.750	4.798	4.702	17.72	157.5	0.0009730	0.006178	82.40
4.500	4.545	4.455	15.90	141.4	0.001084	0.007666	73.95
4.250	4.293	4.207	14.19	126.1	0.001215	0.009635	65.96
4.000	4.040	3.960	12.57	111.7	0.001372	0.01228	58.43
3.750	3.788	3.712	11.04	98.19	0.001561	0.01590	51.36
3.550	3.586	3.514	9.898	87.99	0.001742	0.01980	46.03

(continued overleaf)

26.43 (continued)

Conductor diameter			Sectional area mm²	Weight per km kg	Nominal resistance at 20°C		Current rating at 4.65 amps per mm²* amps
Nom. mm	Max. mm	Min. mm			Per metre ohms	Per kg ohms	
3.350	3.384	3.316	8.814	78.36	0.001956	0.02496	40.99
3.150	3.182	3.118	7.793	69.28	0.002212	0.03193	36.24
3.000	3.030	2.970	7.069	62.84	0.002439	0.03881	32.87
2.800	2.828	2.772	6.158	54.74	0.002800	0.05115	28.63
2.650	2.677	2.623	5.515	49.03	0.003126	0.06370	25.65
2.500	2.525	2.475	4.909	43.64	0.003512	0.08048	22.83
2.360	2.384	2.336	4.374	38.89	0.003941	0.1013	20.34
2.240	2.262	2.218	3.941	35.03	0.004375	0.1249	18.32

2.120	2.141	2.099	3.530	31.38	0.004884	0.1556	16.41
2.000	2.020	1.980	3.142	27.93	0.005488	0.1965	14.61
1.900	1.919	1.881	2.835	25.21	0.006081	0.2412	13.18
1.800	1.818	1.782	2.545	22.62	0.006775	0.2995	11.83
1.700	1.717	1.683	2.270	20.18	0.007596	0.3764	10.55
1.600	1.616	1.584	2.011	17.87	0.008575	0.4799	9.349
1.500	1.515	1.485	1.767	15.71	0.009757	0.6211	8.217
1.400	1.414	1.386	1.539	13.69	0.01120	0.8181	7.158
1.320	1.333	1.307	1.368	12.17	0.01260	1.035	6.364
1.250	1.263	1.237	1.227	10.91	0.01405	1.288	5.706
1.180	1.192	1.168	1.094	9.722	0.01577	1.622	5.085
1.120	1.131	1.109	0.9852	8.758	0.01750	1.998	4.581
1.060	1.071	1.049	0.8825	7.845	0.01954	2.491	4.103
1.000	1.010	0.990	0.7854	6.982	0.02195	3.144	3.652
0.950	0.960	0.940	0.7088	6.301	0.02432	3.860	3.296
0.900	0.909	0.891	0.6362	5.656	0.02710	4.791	2.958
0.850	0.859	0.841	0.5675	5.045	0.03038	6.022	2.639
0.800	0.808	0.792	0.5027	4.469	0.03430	7.675	2.337
0.750	0.758	0.742	0.4418	3.928	0.03903	9.936	2.054
0.710	0.717	0.703	0.3959	3.520	0.04355	12.37	1.841
0.670	0.677	0.663	0.3526	3.134	0.04890	15.60	1.639
0.630	0.636	0.624	0.3117	2.771	0.05531	19.96	1.449
0.600	0.606	0.594	0.2827	2.514	0.06098	24.26	1.315

(continued overleaf)

26.43 (continued)

Conductor diameter			Sectional area mm²	Weight per km kg	Nominal resistance at 20°C		Current rating at 4.65 amps per mm²* amps
Nom. mm	Max. mm	Min. mm			Per metre ohms	Per kg ohms	
0.560	0.566	0.554	0.2463	2.190	0.07000	31.96	1.145
0.530	0.536	0.524	0.2206	1.961	0.07814	39.85	1.026
0.500	0.505	0.495	0.1963	1.746	0.08781	50.29	0.9130
0.475	0.480	0.470	0.1772	1.575	0.09730	61.78	0.8240
0.450	0.455	0.405	0.1590	1.414	0.1084	76.66	0.7395
0.425	0.430	0.420	0.1419	1.261	0.1215	96.35	0.6596
0.400	0.405	0.395	0.1257	1.117	0.1372	122.8	0.5843
0.375	0.380	0.370	0.1104	0.9819	0.1561	159.0	0.5136
0.355	0.359	0.351	0.09898	0.8799	0.1742	198.0	0.4603
0.335	0.339	0.331	0.08814	0.7836	0.1956	249.6	0.4099
0.315	0.319	0.311	0.07793	0.6928	0.2212	319.3	0.3624

0.300	0.304	0.296	0.07069	0.6284	0.2439	388.1	0.3287
0.280	0.284	0.276	0.06158	0.5474	0.2800	511.5	0.2863
0.265	0.269	0.261	0.05515	0.4903	0.3126	637.6	0.2565
0.250	0.254	0.246	0.04909	0.4364	0.3512	804.8	0.2283
0.236	0.240	0.232	0.04374	0.3889	0.3941	1,013.0	0.2034
0.224	0.227	0.221	0.03941	0.3503	0.4375	1,249.0	0.1832
0.212	0.215	0.209	0.03530	0.3138	0.4884	1,556.0	0.1641
0.200	0.203	0.197	0.03142	0.2793	0.5488	1,965.0	0.1461
0.190	0.193	0.187	0.02835	0.2521	0.6081	2,412.0	0.1318
0.180	0.183	0.177	0.02545	0.2262	0.6775	2,995.0	0.1183
0.170	0.173	0.167	0.02270	0.2018	0.7596	3,764.0	0.1055
0.160	0.163	0.157	0.02011	0.1787	0.8575	4,799.0	0.0935
0.150	0.153	0.147	0.01767	0.1571	0.9757	6,211.0	0.0822
0.140	0.143	0.137	0.01539	0.1369	1.20	8,181.0	0.0716
0.132	0.135	0.129	0.01368	0.1217	1.260	10,353.0	0.0636
0.125	0.128	0.122	0.01227	0.1091	1.450	12,878.0	0.0571
0.112	0.155	0.109	0.009852	0.08758	1.750	19,982.0	0.0458
0.100	0.103	0.097	0.007854	0.06982	2.195	31,438.0	0.0365
0.090	0.093	0.087	0.006362	0.05656	2.710	47,914.0	0.0296
0.080	0.083	0.077	0.005027	0.04469	3.430	76,751.0	0.0234
0.071	0.074	0.068	0.003959	0.03520	4.355	123,722.0	0.0184
0.063	–	–	0.003117	0.02771	5.531	199,603.0	0.0145
0.060	–	–	0.002827	0.02514	6.098	242,562.0	0.0132
0.056	–	–	0.002463	0.02190	7.000	319,635.0	0.0115
0.050	–	–	0.001963	0.01746	8.781	502,921.0	0.0091

(continued overleaf)

26.43 (continued)

| Conductor diameter | | | Sectional area mm² | Weight per km kg | Nominal resistance at 20°C | | Current rating at 4.65 amps per mm²* amps |
Nom. mm	Max. mm	Min. mm			Per metre ohms	Per kg ohms	
0.045	—	—	0.001590	0.01414	10.84	766,620.0	0.0074
0.040	—	—	0.001257	0.01117	13.72	1,228,290.0	0.0058
0.036	—	—	0.001018	0.009049	16.94	1,872,030.0	0.0047
0.032	—	—	0.0008042	0.007150	21.44	2,998,601.0	0.0037
0.030	—	—	0.0007069	0.006284	24.39	3,881,286.0	0.0033
0.028	—	—	0.0006158	0.005474	28.00	5,115,090.0	0.0029
0.025	—	—	0.0004909	0.004364	35.12	8,047,663.0	0.0023

*4.65 amps per mm² is equivalent to 3000 amps per in².
Preferred sizes shown in bold type.

26.43.3 Metric wire sizes: turns per 10 mm

Nominal bare diameter mm	Turns per 10 mm min	Nominal bare diameter mm	Turns per 10 mm min
5.000	1.9	**0.500**	18.3
4.750	2.0	0.475	19.2
4.500	2.2	**0.450**	20.2
4.250	2.3	**0.400**	22.6
4.000	2.4	0.400	22.6
3.750	2.6	0.375	24.0
3.550	2.7	**0.355**	25.3
3.350	2.9	0.335	26.7
3.150	3.1	**0.315**	28.4
3.000	3.2	**0.300**	29.7
2.800	3.4	**0.280**	31.8
2.650	3.6	**0.265**	33.8
2.500	3.8	**0.250**	35.2
2.360	4.1	**0.236**	37.2
2.240	4.3	**0.224**	39.1
2.120	4.5	**0.212**	41.2
2.000	4.8	**0.200**	43.5
1.900	5.0	**0.190**	45.5
1.800	5.3	**1.180**	47.9
1.700	5.6	**0.170**	50.5
1.600	5.9	**0.160**	53.5
1.500	6.3	**0.150**	56.5
1.400	6.8	**0.140**	60.2
1.320	7.2	**0.132**	63.7
1.250	7.5	**0.125**	67.1
1.180	8.0	**0.112**	74.6
1.120	8.4	**0.100**	82.6
1.060	8.8	**0.090**	90.9
1.000	9.4	**0.080**	102.0
0.950	9.9	**0.071**	113.6
0.900	10.4	0.063	128.2
0.850	11.0	0.060	133.3
0.800	11.6	0.056	142.9
0.750	12.4	**0.050**	161.3
0.710	13.0	0.045	178.6
0.670	13.8	**0.040**	200.0
0.630	14.6	0.036	222.2
0.600	15.3	0.032	250.0
0.560	16.4	0.030	263.2
0.530	17.3	0.028	285.7
		0.025	322.6

Preferred sizes shown in bold type.

26.43.4 Copper wire data (SWG)

Standard wire gauge	Diameter in inches	Resistance in ohms per yard	Resistance in ohms per pound	Pounds per ohm	Weight in pounds per 1000 yards	Yards per pound	Turns per inch				
							Enamel covered	Single silk covered	Double silk covered	Single cotton covered	Double cotton covered
10	0.128	0.001868	0.0120	83.3	148.8	6.67		7.64	7.55	7.35	7.04
11	0.116	0.002275	0.0200	50.0	122.2	8.16		8.41	8.30	8.06	7.69
12	0.104	0.002831	0.0280	35.7	98.22	10.23		9.35	9.22	8.93	8.48
13	0.092	0.003617	0.0550	18.1	76.86	13.00	→	10.5	10.4	10.0	9.43
14	0.080	0.004784	0.0820	12.2	58.12	17.16		12.1	11.8	11.4	10.6
15	0.072	0.005904	0.1400	7.14	47.08	21.23	15.0	13.3	13.1	12.5	11.6
16	0.064	0.007478	0.2021	4.95	37.20	26.86	15.0	14.9	14.6	14.1	13.2
17	0.056	0.009762	0.3423	2.38	28.48	35.00	17.1	16.9	16.5	15.9	14.7
18	0.048	0.01328	0.6351	1.56	20.92	47.66	19.8	20.0	19.4	18.5	17.2
19	0.040	0.01913	1.315	0.757	14.53	68.66	23.7	23.8	23.0	21.7	20.0

20	0.036	0.02362	2.012	0.497	11.77	85.00	26.1	26.3	25.3	23.8	21.7
21	0.032	0.02990	3.221	0.309	9.299	107.6	29.4	29.4	28.2	26.3	23.8
22	0.028	0.03905	5.498	0.181	7.120	140.6	33.3	38.5	31.8	29.4	26.3
23	0.024	0.05313	10.14	0.098	5.231	191.6	38.8	28.5	36.4	33.3	29.4
24	0.022	0.06324	14.38	0.069	4.395	228.3	42.1	42.1	40.0	35.7	31.3
25	0.020	0.07653	21.08	0.0471	3.632	275.3	46.0	46.0	43.5	38.5	33.3
26	0.018	0.09448	32.21	0.0309	2.942	340.0	50.6	50.6	47.6	41.7	35.7
27	0.0164	0.11138	46.55	0.0215	2.442	410.0	55.9	55.1	51.6	44.6	37.9
28	0.0148	0.1398	70.12	0.0141	1.989	503.0	61.4	60.4	56.2	48.1	40.2
29	0.0136	0.1655	98.65	0.0101	1.680	596.6	66.2	65.2	60.2	51.0	42.4
30	0.0124	0.1991	142.75	0.0069	1.396	716.6	73.3	72.0	67.1	54.4	44.7
31	0.0116	0.2275	185.80	0.0054	1.222	820.0	77.8	76.3	70.9	56.8	46.3
32	0.0108	0.2625	248.20	0.0040	1.059	943.3	83.0	81.3	75.2	63.3	50.5
33	0.0100	0.3061	337.50	0.0029	0.9081	1,100	88.9	87.0	80.0	66.7	52.6
34	0.0092	0.3617	471.00	0.0023	0.7686	1,300	98.0	93.4	85.5	70.4	54.9
35	0.0084	0.4338	676.50	0.0014	0.6408	1,556	106	101	91.8	80.6	61.0
36	0.0076	0.5300	1,009	0.00098	0.5254	1,903	116	110	102	86.2	64.1
37	0.0068	0.6620	1,574	0.00064	0.4199	2,380	128	120	110	92.6	67.6
38	0.0060	0.8503	2,598	0.000385	0.3269	3,056	143	133	121	100	71.4
39	0.0052	0.132	4,645	0.000217	0.2456	4,066	168	149	134	109	75.8
40	0.0048	0.328	6,360	0.000156	0.2092	4,766	180	159	142	144	78.1
41	0.0044	0.581	9,020	0.000112	0.1758	5,700	194	169	150		
42	0.0040	0.913	13,150	0.000076	0.1453	6,866	211	191	167		
43	0.0036	0.362	20,120	0.000050	0.1177	7,500	230	206	179		
44	0.0032	0.989	32,210	0.000030	0.0929	10,766	253	225	192		
45	0.0028	0.904	54,980	0.000015	0.0712	14,066	282	247	208		

26.43.5 Wire gauges

Number of gauge	SWG in	SWG mm	AWG or B and S in	AWG or B and S mm	BWG in	BWG mm	Gold and silver (Birmingham) in	Gold and silver (Birmingham) mm	Lancashire steel pinion wire in	Lancashire steel pinion wire mm
7/0	0.500	12.70	–	–	–	–	–	–	–	–
6/0	0.464	11.78	–	–	–	–	–	–	–	–
5/0	0.432	10.97	–	–	–	–	–	–	–	–
4/0	0.400	10.16	0.46	11.68	0.454	11.53	–	–	–	–
3/0	0.372	9.44	0.409	10.388	0.425	10.787	–	–	–	–
2/0	0.348	8.83	0.364	9.24	0.380	9.65	–	–	–	–
1/0	0.324	8.23	0.324	8.23	0.340	8.63	–	–	–	–
1	0.300	7.62	0.289	7.338	0.300	7.62	0.004	0.101	0.227	5.757
2	0.276	7.06	0.257	6.527	0.284	7.21	0.005	0.127	0.219	5.558
3	0.252	6.40	0.229	5.808	0.259	6.578	0.008	0.203	0.212	5.380

4	0.232	5.89	0.204	5.18	0.238	6.04	0.010	0.254	5.257	0.207
5	0.212	5.38	0.181	4.59	0.220	5.58	0.012	0.304	5.181	0.204
6	0.192	4.88	0.162	4.11	0.203	5.156	0.013	0.330	5.105	0.201
7	0.176	4.46	0.144	3.66	0.180	4.57	0.015	0.381	5.048	0.199
8	0.160	4.06	0.128	3.24	0.165	4.187	0.016	0.406	4.997	0.197
9	0.144	3.66	0.114	2.89	0.144	3.753	0.019	0.482	4.921	0.194
10	0.128	3.24	0.101	2.565	0.134	3.40	0.024	0.61	4.845	0.191
11	0.116	2.94	0.090	2.28	0.120	3.04	0.029	0.736	4.777	0.188
12	0.104	2.642	0.080	2.03	0.109	2.768	0.034	0.863	4.697	0.185
13	0.092	2.336	0.071	1.79	0.095	2.413	0.036	0.914	4.620	0.182
14	0.080	2.03	0.064	1.625	0.083	2.108	0.041	1.041	4.57	0.180
15	0.072	1.828	0.057	1.447	0.072	1.828	0.047	1.143	4.513	0.178
16	0.064	1.625	0.050	1.27	0.065	1.65	0.051	1.295	4.437	0.175
17	0.056	1.422	0.045	1.14	0.058	1.473	0.057	1.447	4.360	0.172
18	0.048	1.219	0.040	1.016	0.049	1.244	0.061	1.549	4.263	0.168
19	0.040	1.016	0.035	0.889	0.042	1.066	0.064	1.625	4.161	0.164
20	0.036	0.914	0.031	0.787	0.035	0.889	0.067	1.701	4.085	0.161
21	0.032	0.812	0.028	0.711	0.032	0.812	0.072	1.828	3.988	0.157
22	0.028	0.711	0.025	0.635	0.028	0.711	0.074	1.879	3.937	0.155
23	0.024	0.61	0.022	0.558	0.025	0.635	0.077	1.955	3.886	0.153
24	0.022	0.558	0.020	0.508	0.022	0.558	0.082	2.082	3.835	0.151
25	0.020	0.508	0.017	0.431	0.020	0.508	0.095	2.413	3.753	0.148
26	0.018	0.457	0.015	0.381	0.018	0.457	0.103	2.616	3.702	0.146
27	0.016	0.406	0.0148	0.376	0.016	0.406	0.113	2.87	3.626	0.143
28	0.0148	0.376	0.012	0.304	0.0148	0.376	0.120	3.04	3.528	0.139

(continued overleaf)

26.43 (continued)

Number of gauge	SWG		AWG or B and S		BWG		Gold and silver (Birmingham)		Lancashire steel pinion wire	
	in	mm	in	mm	in	mm	in	mm	in	mm
29	0.0136	0.345	0.0116	0.29	0.0136	0.345	0.124	3.14	0.134	3.401
30	0.012	0.304	0.010	0.254	0.012	0.304	0.126	3.193	0.127	3.217
31	0.0116	0.29	0.008	0.203	0.010	0.254	0.133	3.376	0.120	3.04
32	0.0108	0.274	0.0079	0.199	0.009	0.228	0.143	3.626	0.115	2.917
33	0.010	0.254	0.007	0.177	0.008	0.203	0.145	3.677	0.112	2.840
34	0.009	0.228	0.006	0.152	0.0076	0.192	0.148	3.753	0.110	2.79
35	0.008	0.203	0.0056	0.142	0.005	0.127	0.158	4.013	0.108	2.743
36	0.0076	0.193	0.005	0.127	0.004	0.101	0.167	4.237	0.106	2.692

26.43.6 Metric sizes of insulated round winding wires

Nominal conductor diameter		
Preferred metric size mm	Non-preferred metric size mm	Approximate inch equivalent
5.000		0.1969
4.750		0.1870
4.500		0.1772
4.250		0.1673
4.000		0.1575
3.750		0.1476
3.550		0.1398
3.350		0.1319
3.150		0.1240
3.000		0.1181
2.800		0.1102
2.650		0.1043
2.500		0.0984
2.360		0.0929
2.240		0.0882
2.120		0.0835
2.000		0.0787
1.900		0.0748
1.800		0.0709
1.700		0.0669
1.600		0.0630
1.500		0.0591
1.400		0.0551
1.320		0.0520
1.250		0.0492
1.180		0.0465
1.120		0.0441
1.060		0.0417
1.000		0.0394
0.950		0.0374
0.900		0.0354
0.850		0.0335
0.800		0.0315
0.750		0.0295
0.710		0.0280
	0.670	0.0264
0.630		0.0248
	0.600	0.0236
0.560		0.0220
	0.530	0.0209

(continued overleaf)

26.43 (*continued*)

Nominal conductor diameter

Preferred metric size mm	Non-preferred metric size mm	Approximate inch equivalent
0.500		0.0197
	0.475	0.0187
0.450		0.0177
	0.425	0.0167
0.400		0.01575
	0.375	0.0148
0.355		0.0140
	0.335	0.0132
0.315		0.0124
	0.300	0.0118
0.280		0.0110
	0.265	0.0104
0.250		0.0098
	0.236	0.0093
0.224		0.0088
	0.212	0.00835
0.200		0.0079
	0.190	0.0075
0.180		0.0071
	0.170	0.0067
0.160		0.0063
	0.150	0.0059
0.140		0.0055
	0.132	0.0052
0.125		0.0049
0.112		0.00441
0.100		0.00394
0.090		0.00354
0.080		0.00315
0.071		0.00280
0.063		0.00248
	0.060	0.00236
	0.056	0.00220
0.050		0.00197
	0.045	0.00177
0.040		0.00157
	0.036	0.00142
0.032		0.00126
	0.030	0.00118
	0.028	0.00110
0.025		0.00098

GLOSSARY

Absorption coefficient The ratio of the sound energy absorbed by a surface, to the total sound energy incident on it.

Access time Time interval between a received instruction to read data stored in memory and the output of the data from memory.

Accumulator 1 A secondary cell, which produces a potential difference. 2 A register within the central processing unit of a computer.

a.c. Abbreviation for alternating current.

Acoustic feedback Unwanted feedback of sound waves from the output of an acoustic system to its input, causing unpleasant audible oscillations commonly known as howling.

Acoustic wave Synonym for sound wave.

Adder Circuit in a digital computer which performs addition.

Address Number that identifies a particular item of data in memory or input/output channel of a digital computer.

Admittance Reciprocal of impedance, symbol Y. The unit of admittance is the *Siemen*.

Aerial Construction, usually of metal, which radiates or receives radio waves. Synonym for antenna.

AFC Abbreviation for automatic frequency control.

Alphanumeric Alphabetical or numerical ordering.

Alternating current An electric current which periodically changes direction.

AM Abbreviation for amplitude modulation.

Ammeter Indicating meter used to measure current.

Ampere Unit used to measure current.

Amplify Make larger, electronically.

Amplifier Electronic circuit which increases some aspect of an applied signal.

Amplifier stage A single stage of a complete piece of electronic equipment to amplify an electronic signal.

Amplitude The peak value of an alternating current.

Amplitude modulation Type of modulation in which the amplitude of a carrier signal is varied above and below its nominal amplitude, by an amount proportional to the varying amplitude of a message signal.

Analogue Term used for a non-digital signal. Some part of the analogue signal varies as the analogue of a reference.

Analogue/digital converter A circuit which converts an analogue signal to a digital one.

AND gate Logic circuit whose output is high if all of its inputs are high.

Angular frequency The frequency of a periodic wave in radians s^{-1}. Symbol ω.

Anode Positive electrode of a system.

Antenna Aerial.

Antiphase Waveforms completely out of phase, i.e. differing by 180°.

Aspect ratio Ratio of the width of a television picture to its height. Typically 4:3.

Assembler Computer program which converts a program written in assembly code to a machine code program.

Astable multivibrator A multivibrator circuit which produces an output of two continuously alternating states, i.e. a square wave oscillator.

Asynchronous Untimed data transfer.

Attenuation Reduction in some aspect of a signal. Opposite of amplification.

Attenuator A circuit which attenuates an applied input signal.

Audio frequency Sound waves within the frequency range of the human ear, i.e. having a frequency between about 20 to 20,000 Hz.

Automatic frequency control Circuit to control the frequency of an applied signal.

Automatic gain control Circuit to control the amplitude of an applied signal.

Automatic volume control Synonym for automatic gain control.

Avalanche breakdown Phenomenon which occurs in a reverse biased semiconductor junction, in which free charge carriers within the junction multiply.

Background noise *See* Noise.

Balanced A transmission line with two conducting wires, each of which has the same resistance to ground, is said to be balanced.

Band 1 A coloured ring on an electronic component. 2 A specific range of communications frequencies.

Band-pass filter A filter which allows a specific range of frequencies to pass, while attenuating all other frequencies.

Band-stop filter A filter which attenuates a specific range of frequencies, while passing all other frequencies.

Bandwidth The band of frequencies a circuit passes, without the circuit's output amplitude falling by a specific fraction (usually one half) of the maximum amplitude.

Base One of the three terminals of a bipolar transistor.

BASIC A high-level computer programming language.

Bass An audio amplifier tone control which attenuates or amplifies bass (i.e. low) frequencies.

Batch processing A computing method used in large computing systems, in which a number of previously prepared programs are run in a single batch.

Battery A source of electricity, consisting of two or more cells connected together.

Band Unit of data modulation rate, corresponding to one transmitted signalling element per second. Often incorrectly confused with data signalling rate, measured in bits per second.

Beat A periodic signal produced when two signals of similar frequency are combined. The beat is caused by interference: the frequency of the beat is defined by the difference in frequency between the two interfering signals. Synonym for heterodyne.

Bel Unit used to express power ratios in electronics. *See* Decibel.

Beta The common emitter, forward current transfer ratio of a transistor. Symbol: β or h_{fe}.

Bias For a transistor to operate correctly the proper potentials have to be present at its emitter, base and collector. Normally the term bias refers to the voltage applied to the base to bring the operating point to a linear part of the amplification curve. For germanium transistors this is usually 0.3 V with respect to the emitter and for silicon transistors at least 0.6 V.

Bias voltage A standing voltage applied to an electronic component.

Binary code Numerical representation which has a base of two and, therefore, only two digits: 0 and 1.

Bipolar transistor A transistor in which both types of charge carriers (i.e. electrons and holes) are used in operation.

Bistable Abbreviation of bistable multivibrator: a circuit which has two stable states. Commonly known as flip-flop.

Bit Abbreviation of binary digit. One of the two digits (0 or 1) of binary code.

Black box Any self-contained circuit, or part of a system, which may be considered a separate entity. Because of this, a user or circuit designer does not need to understand the black box's internal operation – just its effect on external circuits.

Blocking capacitor A capacitor used in a circuit to prevent direct current flow between two parts of the circuit.

Bode diagram A graph in which gain and/or phase shift caused by a circuit, is plotted against frequency of applied signal.

Breadboard A plug-in method of temporarily assembling circuits, for design or test purposes.

Breakdown The sudden change from a high resistance to a low resistance which occurs when the breakdown voltage of a reverse-biased semiconductor junction is exceeded.

Bridge A network of components, generally arranged in a square formation.

Bridge rectifier A full-wave rectifier circuit, composed of four diodes in a bridge.

Brightness A surface's brightness is the property by which the surface appears to emit light in the direction of view. This is a subjective quantity.

Broadcast Radio or television transmission.

Bubble memory A type of computer memory device which, although solid-state, is not of semiconductor origin. Data is stored as tiny domains of magnetic polarisation.

Bucket-brigade *See* Charge coupled device.

Buffer Circuit interfacing two other circuits, used to prevent interference from one to the other.

Bug A computer program fault.

Bus 1 A conductor between two or more parts of a circuit, generally of high current carrying capacity. 2 A set of conductors between parts of a computer system.

Byte A group of bits, treated as a single unit of data in a computer system. Generally, though not necessarily, a byte is taken to be a group of eight bits.

Cable A set of conductors, insulated from each other but enclosed in a common outer sheath.

Capacitance The property of two isolated conductors whereby they hold an electrical charge. Symbol: C. Unit: F.

Capacitor An electronic component, which has two isolated conductive plates. A capacitor may therefore hold an electrical charge.

Carrier 1 A signal which is modulated by a message signal to allow communications, e.g. amplitude modulation. 2 A hole, or electron in a semiconductor device, which carries charge.

Cathode Negative electrode of a system.

Cathode ray A beam of electrons, generated in a cathode ray tube.

Cathode ray oscilloscope An electronic test instrument which allows a signal to be displayed on its screen, as a graph of voltage against time. Abbreviated: CRO.

Cathode ray tube Glass evacuated tube allowing a beam of electrons to be generated, focused and positioned onto its face (screen). Cathode ray tubes form the display device in common TVs and cathode ray oscilloscopes. Abbreviated: CRT.

CCITT Abbreviations for International Telegraph and Telephone Consultative Committee. A body which recommends standards concerning voice and data communications systems.

Ceefax *See* Teletext.

Cell Device which produces a potential difference by chemical means. Two or more cells in combination form a battery.

Central processing unit The part of a digital computer that controls the computer operation. Abbreviation: CPU.

Channel 1 A communication path between a transmission source and receiver. 2 The region between the source and drain of a field effect transistor.

Charge carrier A hole or electron in a semiconductor.

Charge coupled device A semiconductor memory device comprising a number of memory cells, each of which may hold a charge. Each charge is passed along from cell to cell, earning the device the nickname bucket-brigade device. Abbreviation: CCD.

Chip 1 A small piece of semiconductor material containing a single electronic component, or an electronic circuit. A chip is found within every transistor or integrated circuit. 2 Nickname for an integrated circuit.

Clock A circuit or device which generates a periodic signal (generally a square wave) to synchronise operations of a digital system.

CMOS Abbreviation for complementary metal oxide semiconductor.

Coax Abbreviation for coaxial cable.

Coaxial cable A cable with an inner conductor comprising one or more strands of wire, and an outer conduction sheath. The conductors are insulated from each other and the whole arrangement is covered in an outer layer of insulating material.

Coil Conductor(s) wound in a number of turns.

Collector One of the three terminals of a bipolar transistor.

Colour code Method of marking an electronic component with information regarding its value, tolerance and any other aspects which may be of interest to its user.

Complementary pair Most modern transistor audio amplifiers make use of a pair of transistors, one npn and the other pnp, with similar characteristics and closely matched gains in the driver or output stage: they are referred to as a complementary pair.

524

Computer An automatic system, which processes information according to instructions contained in a stored program.

Conductor A material with low resistance to the flow of electric current.

CPU Abbreviation for central processing unit.

CRO Abbreviation for cathode ray oscilloscope.

Crosstalk Interference between signals of two adjacent communication channels.

CRT Abbreviation for cathode ray tube.

Current Rate of flow of electricity. Symbol: I. Unit: ampere (abbreviated: amp: A).

Cut-off frequency Frequency at which a circuit output falls to a specified fraction (usually one half) of the maximum.

Cycle Complete set of changes in a regularly repeating wave.

Darlington pair A combination of two transistors which operate as if they are a single transistor, with a gain given by the product of the individual transistors gains.

dB *See* decibel.

d.c. Abbreviation for direct current.

d.c. voltage Common term to mean direct voltage.

Debug The action of finding and correcting computer program faults.

Decibel Dimensionless unit expressing the ratio of two powers. Under certain conditions it may also be used to express the ratio of two voltages or currents.

Demodulation *See* Modulation.

Demodulator A circuit which demodulates a received, modulated signal in a communications system. Synonymous with detector.

Demultiplexer *See* Multiplexer.

Detector *See* Demodulator.

Device An electronic component or system which contains at least one active element.

Diac Bi-directional voltage breakdown diode; passes current above a certain breakdown voltage. Normally employed with a triac in an a.c. control circuit.

Dielectric A material which is an insulator and can sustain an electric field. The layer of insulating material between the conducting plates of a capacitor is a dielectric.

Dielectric constant The ratio of the capacitance of a capacitor with a dielectric, to the capacitance of the capacitor with the dielectric replaced by a vacuum. Synonymous with relative permittivity. Symbol: μ_t. *See* Permittivity.

Differential amplifier An amplifier which produces an output signal which is a function of the difference between its two inputs. Principle of the operational amplifier.

Differentiator A circuit which produces an output signal which is a function of the differential of its input signal.

Digital A circuit or system responding to, operating on, and producing fixed, discrete voltages. Where only two levels are used, the circuit or system is said to be binary digital.

Digital computer *See* Computer.

Digital multimeter A multimeter which is capable of measuring and displaying a number of electrical quantities as a decimal value.

Digital voltmeter A voltmeter which displays a measured voltage as a decimal value.

DIL Abbreviation for dual-in-line.

Diode An active electronic component with two electrodes, which allows current flow in one direction but not in the other. Many derivative types of diode exist.

Diode transistor logic Family of logic integrated circuits built using diodes and transistors. Abbreviation: DTL.

Dioptre The unit of measure of lens power; the reciprocal of the focal length, expressed in metre.

Dipole aerial Simplest type of aerial, in which a standing wave of current is symmetrical about its midpoint.

Direct current A unidirectional, constant current. Abbreviation: d.c.

Direct voltage A unidirectional, more or less constant voltage.

Distortion Extra unwanted components in the output of a system, which have been added by the system itself. There are many types of distortion.

Doping The addition of impurities to a pure semiconductor material in order to affect the numbers and types of charge carriers present. Donor impurities are added to form an n-type semiconductor. Acceptor impurities are added to form a p-type semiconductor.

Double pole switch A switch with two electrically independent switching mechanisms.

Drain One of the three connections of a field/effect transistor.

Drift A variation of an electrical property with time.

Dry battery A battery of two or more dry cells.

Dry cell A cell whose contents are in non-liquid form.

Dry joint A faulty soldered joint.

DTL Abbreviation for diode transistor logic.

Dual-in-line Standard package for integrated circuits, in which the connection pins are in two parallel rows, either side of the body.

Duplex Simultaneous operation of both channels of a communications link.

DMM Abbreviation for digital multimeter.

DVM Abbreviation for digital voltmeter.

EAROM Abbreviation for electrically alterable read only memory.

Earphone Small loudspeaker which fits into the ear.

Earth The arbitrary zero point in electrical potential.

Earth current Current which flows to earth.

Earth fault Fault occurring in a circuit or system, when a conductor is connected to earth or a low resistance occurs between the conductor and earth. Causes an unacceptable earth current.

ECL Abbreviation for emitter coupled logic.

Edge connector A connector which is pushed onto the edge of a printed circuit board. Tracks on the printed circuit board are taken to the edge forming connections.

EEROM, E²ROM Abbreviation for electrically erasable read only memory.

EHT Abbreviation for extra high tension.

Electrically alterable read only memory *See* Read only memory. Abbreviation: EAROM.

Electrically erasable read only memory *See* Read only memory. Abbreviation: EEROM, E^2ROM.

Electrode Part of a component or system which gives out or takes in charge carriers.

Electrolysis Chemical change caused by an applied current through an electrolyte.

Electrolyte Material which allows conduction due to its dissociation into ions.

Electrolytic capacitor A capacitor in which the dielectric is formed by electrolytic action.

Electromagnet A component which becomes a magnet only when a current flows through it.

Electromagnetic spectrum Complete frequency range of electromagnetic energy.

Electromotive force Potential difference produced by an electrical energy source. Abbreviation: EMF. Symbol: E. Unit: volt.

Electron Atomic particle which possesses a negative charge, of magnitude 1.602×10^{-9} coulomb and a mass of 9.109×10^{-3} kg. Movement of electrons in one direction is equal to current flow in the opposite direction.

Electron beam Beam of electrons given off from an electron gun, typically in a cathode ray tube. Synonymous with cathode ray.

Electric gun Arrangement which is used to generate an electron beam in a cathode ray tube or similar.

Electronvolt The energy gained by one electron when passing across a potential difference of one volt. Symbol: eV.

Element Substance consisting of atoms of only one type.

EMF Abbreviation for electromotive force.

Emitter One of the three terminals of a bipolar transistor.

Emitter follower A single transistor amplifier whose output is between emitter and earth.

Enable Activate a circuit or device.

Encoder Circuit, system, or device producing an output which is a coded version of the input.

Energy bands Theoretical levels of energy which electrons of an atom possess.

Equalisation Process whereby the distortion produced by a system may be compensated for.

Equaliser Circuit or device which causes equalisation.

Equivalent circuit A circuit comprising simple (generally passive) elements, used to model the action of a complete circuit under specified conditions.

Erase Remove stored information.

Error Difference between the correct value of something and its actual value.

Exclusive OR gate Logic circuit with two or more inputs, whose output is high if and only if one input is high. Abbreviation: XOR.

Facsimile A picture transmission system by which pictorial images can be transmitted using an ordinary communications link. Abbreviation: fax.

Failure Ceasing of a component system's ability to function correctly.

Failure rate The number of failures which may be assumed by a component or system, in a given time. The failure rate is given by:

$$f = \frac{1}{\text{MTBF}}$$

where MTBF is the mean time between failures.

Fall-time The time taken by a logic device or circuit to change output state from high to low.

Fan-out The maximum number of circuits which may be driven by the output of a similar circuit.

Farad Unit of capacitance. Symbol: F.

Feedback The return of some part of a circuit or system's output, to its input in such a way as to control the function of the circuit or system.

FET (field effect transistor) The f.e.t. makes use of the electric field established in a p- or n-type channel of semiconductor material to control the flow of current through the channel. The field is established by the bias applied to the gate connections and the f.e.t. is thus a voltage-controlled device. This means that it has a much higher input impedance than ordinary transistors. The main connections are the source, drain and gate but some f.e.t.s have additional connections.

Fibre optics *See* Optical fibre.

Field 1 Region affected by some phenomena. 2 Set of bits forming a unit of data with a specific purpose. 3 Set of lines of a displayed television picture.

Field effect transistor A unipolar transistor (i.e. with only one p-n junction). Abbreviation: FET.

531

Filter Circuit which passes some applied frequencies of signals while restricting others.

Fleming's left-hand rule When the thumb, first finger and middle finger of the left hand are held naturally at right angles, the thumb represents the direction of motion, the first finger represents the magnetic field, and the middle finger represents the current, I, in an electric motor.

Fleming's right-hand rule Similar to Fleming's left-hand rule, but representing a dynamo, with the right hand.

Flicker The eye's perception of fluctuation of brightness when the fluctuations occur more rapidly than the persistance of vision.

Flicker noise *See* Noise.

Flip-flop Nickname for a bistable multivibrator.

Floating Term describing a part of a circuit which is not connected.

Floating-point representation Means of expressing a number with the use of a mantissa and an exponent.

Floppy disk Magnetic memory medium used by computers as auxiliary memory.

FM Abbreviation for frequency modulation.

Forward bias When a voltage is applied across a semiconductor junction, the junction is said to be forward biased when the current through the junction is the greater of the two ways. Thus, a diode is forward biased when it conducts and reverse biased when it does not conduct.

Frame One complete television picture.

Frequency Number of complete oscillations or cycles of a periodic signal in one second. Unit: hertz (Hz).

Frequency is related to the wavelength (λ) of the signal by the signal velocity (v), where

$$v = f\lambda.$$

Frequency division multiplex A system in which a number of message signals are combined into one. Each message signal is modulated onto a different carrier wave frequency so that a number of frequency band channels exist. Abbreviation: FDM.

Frequency modulation Type of modulation in which the carrier frequency is varied up and down by the message signal. The carrier amplitude remains constant. Abbreviation: FM.

Frequency range Range of frequencies a circuit will operate on.

Frequency response Variation with frequency of the gain of a circuit. Drawn as a graph, usually of gain in decibels against a logarithmic scale of frequency.

Frequency spectrum A graph, chart, or table showing the frequencies of all electromagnetic waves, related to types, e.g. X-rays, radio waves, audio waves, etc.

FSD Abbreviation for full scale deflection.

Full scale deflection Maximum value displayed by measuring equipment. Abbreviation: FSD.

Full wave rectifier A circuit which rectifies both positive and negative half cycles of an applied a.c. wave.

Function generator A circuit or piece of test equipment which generates a variety of waves, e.g. sine, sawtooth, square, for use in testing other circuits.

Fundamental frequency Generally the lowest sine wave frequency present in a complex periodic waveform.

Fuse Device which is intended to cause an open circuit when the current taken by a circuit goes above a specified level. Generally, a fuse is formed using a short length of fuse-wire which, at a specified frequency and voltage, will melt, i.e. 'blow' at the fuse's current rating, thus breaking the power supply connection.

Fusible link memory Type of read only memory device consisting of a matrix of fusible links. Data may be programmed into the device by 'blowing' selected links.

Gain Measure of a circuit's effect on the amplitude of an applied signal. Can be stated in terms of the ratio between output and input signals as a decimal number, or in decibels.

Gain control A control which may be used to vary the gain of a circuit.

Ganged Term used to describe variable components which are mechanically coupled so that they all vary simultaneously when a single control is varied.

Gate 1 A circuit having two or more inputs and one or more outputs. The output(s) varies as a direct result of the states of the inputs. 2 One of the terminals of a field effect transistor.

Gating signal A signal which, when applied to a gate is used to control the gate's output such as the output may be on (and produce an output signal which is some function of another input) or off (producing no output).

Geostationary orbit A satellite orbit in which the satellite lies about 36,000 km above the earth's equatorial plane, such that the satellite appears stationary to an observer on earth.

Geosynchronous orbit Similar to a geostationary orbit, but the satellite traces a figure-of-eight orbit thus appearing to move up and down in one-day cycles to an observer on earth.

Germanium Semiconductor element used in the majority of early transistors and diodes.

Giga- Unit prefix which means a multiplication factor of 10^{12}. Abbreviation: G.

Graphics Display of graphical symbols and scenes, generated by a computer.

Ground Synonym for earth. Abbreviation: Gnd.

Guard hand Range of frequencies between two ranges of transmission frequencies, left unoccupied to minimise interference.

Half-adder Elementary digital circuit composed of logic gates. *See* Adder.

Half-duplex A pair of transmission channels over which two-way communications may take place, although only one channel is operational at any one time, is said to allow half-duplex communications.

Half-wave rectifier A circuit which rectifies only one half of each cycle of an applied a.c. wave.

Half effect An electromagnetic phenomenon which occurs when a current carrying conductor is placed in a magnetic field, the direction of which is perpendicular to the directions of both the current and its own magnetic field.

Ham Colloquial term denoting an amateur radio transmitting/receiving enthusiast.

Hardware Physical parts of a computer system, e.g. printer, keyboard, VDU, etc.

Harmonic A signal present in a complex periodic waveform, which is a multiple of the fundamental frequency. The second harmonic is three times etc.

Head Transducer of a magnetic recording system which allows electrical signals to be changed into a magnetic field to write data onto the medium, or converts magnetic data into electrical signals.

Headset Pair of earphones.

Heatsink Metal attachment mechanically connected to a heat producing element in a circuit (e.g. a power transistor) to ensure heat is dissipated away from the element, preventing damage by excessive heat.

Henry Unit of magnetic inductance. Symbol: H.

Hertz Unit of frequency. Equivalent to one cycle of a periodic wave which occurs in one second. Symbol: Hz.

Heterodyne Production of beats by combination of two signals which interfere. Used in a superheterodyne radio receiver to produce an intermediate frequency.

HF Abbreviation for high frequency.

h_{fe} and h_{FE} *See* Beta.

Hifi Acronym for high fidelity.

High fidelity Commonly used term denoting audio reproduction equipment of good quality.

High frequency Bands of radio transmissions around 10 MHz. Abbreviation: HF.

High level programming language A computer programming language which is more like human language or mathematical notation than the machine code used by the central processing unit of the computer.

High logic level Term denoting a logic 1 level (in positive logic).

High pass filter A filter which allows signal frequencies above a specific corner frequency to pass without attenuation. Signal frequencies below the corner frequencies are attenuated.

High tension Voltages in the range between about 50 V to 250 V. Abbreviation: HT.

Holding current The value of current which must be maintained to hold a thyristor in its on state. If the current through the thyristor falls below the holding current, the thyristor turns off and ceases conduction.

Hole An empty space in a semiconductor material due to a 'missing' electron. As electrons are negatively charged, holes are positive. Holes, like electrons, may be thought of as charge carriers, moving through the semiconductor material thus forming a current.

Hole current The current through a semiconductor due to the movement of holes under an applied voltage.

Howl Colloquial term for the sound caused by acoustic feedback.

HT Abbreviation for high tension.

Hum Capacitive or magnetic interference between a mains powered device such as a power supply, and local equipment such as an amplifier. Often heard in audio frequency systems as a low drone of mains supply frequency, or a harmonic of that frequency.

Hunting A system's oscillation about its desired point, caused by over-correction.

Hybrid integrated circuit An integrated circuit comprising a number of discrete components attached to a substrate and interconnected to form a circuit. *See* Integrated circuit.

Hybrid-π A type of equivalent circuit used to show transistor operation.

Hysteresis Phenomenon occurring in some circuits or systems, in which the output lags behind a changing input. A hysteresis loop if formed – a graph of output against input – which shows that the value of output depends on whether the input is increasing or decreasing in value.

Hz Abbreviation for hertz.

IC Abbreviation for integrated circuit.

IEC Abbreviation for International Electrotechnical Commission.

IEE Abbreviation for Institute of Electrical Engineers.

IEEE Abbreviation for Institute of Electrical and Electronic Engineers Inc.

IF Abbreviation for intermediate frequency.

IGFET Abbreviation for insulated gate field effect transistor.

I^{2}L Abbreviation for integrated injection logic.

Illuminance Luminous flux perpendicularly reaching a surface per unit area. The unit of illuminance is the lumen m^{-2} or lux.

Image frequency Unwanted input frequency to a radio, causing a spurious output. Synonym for second channel frequency.

Image interference Interference caused by an image frequency.

Impedance The opposition of a circuit to alternating current flow.

Impedance matching The matching of impedances between two circuits, to ensure maximum transfer of power from one to the other.

Impulse noise Noise in an electronic system caused by a single disturbance. *See* Noise.

Impulse voltage A single, rapidly occurring, pulse of voltage. Impulses are generally unwanted, as they tend to cause impulse noise throughout the system.

Impurities Atoms in a semiconductor material, not of the semiconductor element itself. Impurities may occur naturally or may be deliberately introduced. *See* Doping.

Incandescence When a material gives off visible light due to its high temperature.

Incandescent lamp Lamp which emits light when its filament is heated by an electric current. The filament often reaches temperatures of 2,500°C and over.

Index error An error occurring in a measuring instrument such that when no measurand is present (i.e. with zero input) a non-zero reading is obtained. Synonymous with zero error.

Indirect wave Radio wave which is reflected by the ionosphere, i.e. it does not travel directly from transmitter to receiver.

Induce To cause a change in electrical or magnetic conditions in a system, by changing the electrical or magnetic conditions of another, local, system.

Inductance A constant occurring when a circuit is magnetically linked with the current flowing through it. Unit: henry.

Induction Abbreviation of electromagnetic induction or electrostatic induction.

Inductor A component which has inductance. Generally, inductors are constructed of some form of coil.

Information technology The study of the combined effects of electronics, computing and communications. Abbreviation: IT.

Infrared radiation Invisible electromagnetic radiation with wavelengths of about 730 nm to about 1 mm.

Inject To introduce carriers into a semiconductor junction area.

Input 1 The signal applied to a system. 2 The terminal at which the input signal is applied.

Input impedance The impedance which a circuit presents to an input signal.

Input/output Term applying to operations performed by, or devices connected to a computer, which allow the computer to receive and send out data. Abbreviation: I/O.

Instantaneous value The value of any measurand which varied with time, e.g. instantaneous voltage, instantaneous current.

Institute of Electrical and Electronics Engineers Inc A standardisation body in the USA. Abbreviated: IEEE.

Institute of Electrical Engineers A UK standardisation body. Abbreviation: IEE.

Instruction set The complete list of operations which may be performed by the central processing unit of a computer.

Insulate To prevent unwanted current flow by sheathing a conductor with non-conductive material.

Insulated gate field effect transistor Type of construction of field effect transistor, used in MOSFETs. Abbreviation: IGFET.

Insulator A non-conductor. A material with a very high resistance to flow of electric current. Current flow is assumed to be negligible.

Integrated circuit A device which contains a complete circuit. One of two main methods are used in manufacture. A *hybrid integrated circuit* is manufactured from discrete components, attached to a substrate and interconnected by layers of metallization. A monolithic integrated circuit is made by building up all components within the circuit onto a single chip of silicon. In recent usage, integrated circuit, chip, microchip, have all become synonymous. Abbreviation: IC.

Integrated injection logic A type of monolithic integrated circuit construction. Abbreviation: I^2L.

Integrator Circuit which performs the equivalent of a mathematical integration on an applied input signal.

Intelligent Any system with processing and capabilities, whose actions may be controlled by stored instructions, is said to be intelligent.

Intelsat Acronym of International Telecommunications Satellite Consortium.

Intensity Denoting magnetic or electric field strength.

Interactive Term used to denote computer operation where user and computer communicate in a continuous manner. Generally refers to an on-line situation.

Interference A disturbance to any signal in a system, causing additional, unwanted signals. Interference may be natural or manmade. *See* Hum, Crosstalk, Image frequency.

Interlaced scanning Scanning method used in a television system in which lines are scanned in two separate scans, even lines and odd lines.

Intermediate frequency Signal generated in a heterodyne-based radio receiver, where the received radio signal is combined with a local oscillation signal. Abbreviation: IF. *See* Heterodyne, Beats.

Internal resistance The small resistance inherent in any source of electricity. The internal resistance limits the voltage which may be produced by the electricity source under load conditions.

International Electrotechnical Commission An international standardisation body. Abbreviation: IEC.

International Radio Consultative Committee An international standardisation committee: part of the International Telecommunications Union. Abbreviation: CCIR.

International Telecommunications Union An international standardisation body. Abbreviation: ITU.

International Telegraph and Telephone Consultative Committee An international standardisation committee: part of the International Telecommunications Union. Abbreviation: CCITT.

Inversion The production of an opposite polarity in a semiconductor device, due to an applied electric field.

Inverter 1 A circuit which produces an output which is the inverse of an applied input signal. A digital inverter produces an output which is the opposite logic state to that of the input. 2 A circuit which converts a direct current to an alternating current.

I/O Abbreviation for input/output.

Ion A particle of material (an atom, molecule, group of atoms, or group of molecules) with an electric charge. Negative ions are called anions: positive ions are called cations.

ISO Abbreviation for International Standards Organisation.

Isolate To disconnect two parts of a system, ensuring that no electrical connection exists.

ITU Abbreviation for International Telecommunications Union.

Jack A connector pair (plug and socket) allowing quick and easy input or output connections, to or from a circuit or system. Many sizes and types are available.

Jam To cause interference in radio-type transmissions, rendering correct reception impossible.

JEDEC Abbreviation for Joint Electron Device Engineering Council.

JFET Abbreviation for junction field effect transistor.

JK flip flop A type of flip-flop circuit.

Johnson noise A type of noise generated by the random movement of electrons in resistive components, due to thermal activity. Synonymous with thermal noise.

Josephson junction The junction between a thin layer of insulating material and a superconducting material. A superconducting current can flow across the junction even without an applied voltage.

JUGFET Abbreviation for junction field effect transistor.

Junction 1 The boundary between two layers of material in an electronic device. 2 An electrical connection.

Junction box An enclosed container, in which wires or leads from circuits may be joined, by screw terminals or other means.

Junction capacitance Capacitance between pn junctions in a semi-conductor device. Also called barrier, depletion layer or transition capacitances. *See* Neutralisation.

Junction field effect transistor A type of field effect transistor. Abbreviation: JFET; JUGFET.

Junction transistor Abbreviation for bipolar junction transistor.

Keyboard Part of a computer, with a typewriter-style appearance, allowing input of instructions and data to the computer.

Kilo- 1 A prefix denoting a multiple of 10^3. Abbreviation: k. 2 A prefix to a computing term, denoting a multiple of 2^{10}, i.e. 1024. Abbreviation: K.

Kirchhoff's laws Two basic laws of electricity. The first states that: the algebraic sum of all currents into and out of a point in a circuit is zero. The second states that: the algebraic sum of the products of current and resistance in each part of circuit is equal to the algebraic sum of the voltages.

Klystron An electron gun device, used as an amplifier or oscillator at high radio frequencies.

Lag The delay between one waveform and another, measured in time or as an angle. *See* Lead.

Lamination Thin sheet material, used to make the laminated core of a wound component, e.g. a transformer, a relay.

Land A contact on a printed circuit board.

Language Short for programming language.

Large scale integration A level of integration used in the manufacture of integrated circuits. Abbreviated: LSI.

Laser An acronym for light amplification by stimulated emission of radiation. A laser device is a source of coherent, monochromatic light. The light may not be visible but may be of ultra-violet or infrared origin.

Latch Common name for a bistable multivibrator or flip-flop.

LCD Abbreviation for liquid crystal display.

LC network A tuned circuit containing inductance and capacitance.

Lead 1 The amount by which one waveform is in front of another, measured as a time interval or as an angle.

See Lag. 2 An electrical conductor, used to make an electrical connection between two parts of a circuit or system.

Lead-acid cell A secondary cell or accumulator comprising lead metal cathodes and lead dioxide anodes, with a dilute sulphuric acid electrolyte. Lead-acid batteries, formed by a number of series connected cells are commonly used in cars.

Leading edge The portion of a pulse which signals the commencement of the pulse.

Leakage Current flow through a circuit or component due to faulty insulation.

Leakage current That current which flows due to leakage.

Leclanché cell A primary cell comprising a carbon-rod anode and a zinc cathode, with an ammonium chloride solution electrolyte. Leclanché cells with a paste-based ammonium chloride electrolyte are said to be dry and form the basis of many available cells used in common battery-powered appliances.

LED Abbreviation for light emitting diode.

Left-hand rule *See* Fleming's rules.

LF Abbreviation for low frequency.

Light emitting diode A semiconductor diode which emits light as the result of an electroluminescent effect. As electron and hole combine near the junction of the diode, sufficient energy is released to form light. The emitted light is of a particular frequency and so of a particular colour. Light emitting diodes of red, yellow, orange, green and blue are available, as well as infrared varieties. Abbreviation: LED.

Light-pen A device used with a computer to produce data on a cathode ray tube screen by writing on the screen in a similar manner to writing with a conventional pen on paper. The computer locates the light-pen's position on the screen and produces an image at that point. As the light-pen is moved across the screen many images combine to form, say, a line.

Light sensitive devices Light and heat both affect the conductivity of a pn junction. Devices are available in which a pn junction is exposed to light so as to make use of this property. Light falling on the junction liberates current carriers and allows the device to conduct.

Linear Any circuit or system which produces an output directly proportional to the input at any time, is said to be a linear circuit or system.

Lines The physical paths followed by the electron beam of a television receiver's cathode ray tube across the screen. Standard UK television pictures are composed of 625 lines.

Line of flux Imaginary line in a magnetic field. The direction at any point along a line of flux is that of the magnetic flux density.

Line of force Imaginary line in an electric field, the direction of which at any point represents the field's direction at the point.

Liquid crystal display A display comprising a thin layer of liquid crystal material between two electrodes. Application of a potential difference across the electrodes causes the liquid crystal material to change in respect of light transmission. Abbreviation: LCD.

Load 1 A circuit or system which absorbs power from any other circuit or system. 2 The output power provided by a circuit or system.

Load characteristic A characteristic curve, typically for a transistor, in which the relationship between variables is plotted.

Load impedance The impedance presented to a circuit or system by its load.

Load line A line drawn on the load characteristics of a component which shows the relationship between voltage and current in the circuit.

Local oscillator An oscillator within a radio receiver operating on the superheterodyne principle.

Logic circuit A circuit which performs a logical operation such as AND, OR, NOT, NAND, NOR, EXOR.

Logic diagram A diagram showing the logic elements of a logic circuit.

Logic symbol Graphical symbol representing a logic element.

Long-tailed pair A circuit containing two transistors coupled together so that their emitters are joined with a common emitter bias resistor which provides a constant current. The long-tailed pair forms the basis of a differential amplifier.

Long wave A radio frequency wave with a wavelength between 1 to 10 km.

Loss A dissipation of power due to the resistance of current flow.

Lossless A circuit or system which theoretically loses no power due to resistance.

Lossy A circuit or system which loses a great deal of power due to resistance.

Loudness Subjective measure of sound intensity. Although dependent on intensity it also varies with frequency and timbre of the sound.

Loudness level A comparison of a sound's loudness with a standard sound loudness. The standard sound is a sinusoidal note which has a frequency of 1,000 Hz. The unit of loudness level is the phon.

Loudspeaker A transducer which converts electrical energy into sound energy. Typically, loudspeakers are electromagnetic devices which rely on the applied electrical signal to move a coil of wire in a magnetic field. Attached to the coil is a cone of material which thus also moves with the electrical signal. The cone causes a movement of air, which the ear detects as sound.

Low frequency Radio signals in the frequency band of 30 kHz to 300 kHz, having wavelengths between 1 and 10 km. Abbreviation: LF.

Low level programming language A computer programming language which comprises instructions to the computer in machine code, i.e. binary codes which the computer can directly understand.

Low logic level Term denoting a logic 0 level (in positive logic).

Low-pass filter A filter which allows signal frequencies below a specific corner frequency to pass without attenuation. Signal frequencies above this corner frequency are attenuated.

LSI Abbreviation for large scale integration. *See* Integrated circuit.

Luminance A surface's luminance is the objective measure of the light emitted per unit projected area of surface, the plane of projection being perpendicular to the direction of view. The unit of luminance is the candela m^{-2}.

Luminous intensity The unit of luminous intensity is the candela (cd).

Machine code The binary codes understood by the central processing unit of a computer.

Magnet Term applied to a substance which generates a magnetic field. A magnet can be temporary or permanent.

Magnetic bubble memory *See* Bubble memory.

Magnetic circuit A closed path of lines of magnetic flux.

Magnetic field Space surrounding a magnet which contains a magnetic flux. A magnetic field may be represented by lines of force.

Magnetic field strength Magnetising force. Symbol: H. Unit: ampere per metre ($A\,m^{-1}$).

Magnetic flux The flux through an area in the space surrounding a magnet. Symbol: Φ. Unit: weber.

Magnetic flux density Magnetic induction. The magnetic analogue of the electric field. Symbol: B. Unit: tesla (T), or weber per square metre ($w\,m^{-2}$).

Magnetism The collection of properties exhibited by a magnet or magnets.

Mains Domestic electricity supply distributed through the National Grid system. A voltage of 240 VAC, at a frequency of 50 Hz is obtained from all domestic outlets.

Mains hum *See* Hum.

Make To close a circuit by means of a switch or similar component make and break: a type of switch which is automatically opened and closed, thus making and breaking the circuit, by the circuit which it forms part of.

Man-made noise *See* Noise.

Mark-space ratio The ratio between a pulse's duration and the time between successive pulses.

Maser An acronym for *microwave amplification by stimulated emission of radiation*. Similar to laser except radiations are part of the microwave frequency band, not light.

Mask Photographic reproduction of the circuit to be integrated into an integrated circuit chip by photographic or other means.

Matched termination A load attached to a circuit or system such that it absorbs all the power available from the circuit or system.

Mean life The mean time to failure of a component, circuit or system.

Measurand The quantity to be measured by measuring equipment.

Medium frequency Radio signals in the frequency band of 300 kHz to 3 MHz, having wavelengths between 100 m to 1 km. Abbreviation: MF.

Medium wave A radio frequency wave with a wavelength placing it in the medium frequency band.

Mega- 1 A prefix to a number, denoting a multiple of 10^6. 2 A prefix used in computing, to denote a multiple of 2^{20} (i.e. 1,048,576). Symbol: M.

Megger A portable insulation testing equipment.

Memory Any device associated with a digital circuit (particularly a computer) which is capable of storing information in digital form. Synonymous with store.

Memory location A storage element with a unique address.

Meter Any measuring equipment.

Meter movement The part of an analogue meter which indicates the measured value, typically constructed of a finely balanced moving-coil in a magnetic field. The coil rotates when a current flows through it, the amount of rotation is proportional to the value of current.

Meter resistance The internal resistance of a meter.

MF Abbreviation for medium frequency.

Micro- A prefix, denoting a multiple of 10^{-6}. Symbol: μ.

Microcomputer 1 A single integrated circuit which contains all the parts which can be combined to function as a computer, i.e. central processing unit, memory, timing and control circuits. 2 A computer which comprises an integrated circuit microprocessor. 3 A home computer.

Microphone A transducer which converts sound energy into electrical energy.

Microprocessor An integrated circuit which contains the central processing unit of a computer.

Microwave An electromagnetic wave with a frequency between infra-red and radio waves in the electromagnetic spectrum. Microwave wavelengths range from about 3 mm to 1.3 m.

Mike Abbreviation for microphone.

Milli- A prefix, denoting a multiple of 10^{-3}. Symbol: m.

Mismatch When a circuit's load does not have the same impedance as the load itself.

Mixer 1 An audio circuit to combine two or more signals. The output signal is merely the addition of the input signals. 2 A radio circuit which combines two or more signals to produce an output signal of a different frequency to the inputs.

Modem Acronym for modulator-demodulator. Any appliance which converts signals from one circuit or system to signals of another circuit or system. Typically modems are used to connect two computers via telephone circuits.

Modulation The alteration of a signal's parameter by another parameter. For instance, a carrier wave's amplitude may be modulated by a music signal. Other parameters which may be modulated include: phase, frequency, or a combination of more than one.

Monochromatic light Light of a single colour, i.e. it has only one frequency.

Monochromatic television Black and white television.

Monostable multivibrator A circuit which has one stable state. On application of a triggering pulse, the output of the monostable multi-vibrator assumes a second state for a defined period of time, after which it returns to the stable state. Synonymous with one-shot.

Morse code Internationally agreed code for the transmission of alphanumeric symbols, in which each symbol is transmitted as a combination of short and long pulses (dots and dashes).

Morse telegraphy Electric telegraphy transmitting alphanumeric symbols as Morse code.

MOSFET Abbreviation of metal oxide semiconductor or field effect transistor. A type of field effect semiconductor.

MOST Type of f.e.t. with oxide insulating layer between the metal gate and semiconductor channel. It has a higher input impedance than the junction type f.e.t.

Motorboating Term used to describe an oscillation arising in low or audio frequency amplifiers, resembling a motorboat engine.

Moving coil A device which relies on its motion due to current through a coil in a magnetic field.

Multiplex Combination of two or more signals, such that a single signal is obtained which may be transmitted and later demultiplexed back into the original signals.

Multiplexer A circuit which allows the multiplex process to take place.

Multivibrator A circuit which contains two inverters coupled so that the output of one forms the input of the other. Resistive coupling of the two inverters produces a bistable multivibrator, or flip-flop. Resistive/capacitive coupling produces a monostable multivibrator. Capacitive coupling produces an astable multivibrator.

NAND gate Logic circuit whose output is high if one or more of its inputs are low, and low if all its inputs are high.

Nano- A prefix denoting a multiple of 10^{-9}. Symbol: n.

Natural frequency The frequency at which free oscillation occurs in a resonant electrical, electronic, or mechanical system.

N-channel The conducting channel of a field-effect transistor of N-type semiconductor material. The term is also used to refer to the transistor, i.e. N-channel field-effect transistor.

Negative bias A voltage applied to an electrode of some electronic component, which is negative with respect to a fixed reference potential.

Negative feedback Type of control procedure in which all or some part of a system's output signal is fed back to the system's input terminal. Generally, by changing the amount of negative feedback the system's gain is changed. The gain may thus be controlled by choosing the required amount of negative feedback.

Negative modulation Type of modulation procedure followed in the transmission of television signals, such that a black display results from a more positive signal and a white display results from a negative signal. This principle is followed to ensure that any noise which a television receiver picks up produces a darker image and is thus less noticeable than it would be if positive modulation were used.

Neon indicator Type of indicator, relying on the gas-discharge properties of the inert gas, neon. A voltage of about 80 V is required to illuminate such indicators, and so they are typically used as indicators to display the presence of mains voltages.

Neper A dimensionless unit used to express the ratio of two signal powers. One neper equals 8.686 decibels. Symbol: Np.

Network Alternative term describing a circuit.

Neutral 1 One of the three lines of the domestic mains electric supply. 2 Descriptive term implying no overall positive or negative charge.

Neutralisation In radio frequency transistors there is a tendency for self-oscillation to occur due to the collector-base capacitance. In modern r.f. transistors this capacitance can be made very small. To overcome the effect in early r.f. transistors it was usual to use a small amount of capacitive negative feedback in each stage, this being known as neutralisation.

Nicad Abbreviation for nickel cadmium.

Nickel-cadmium cell A secondary cell, with a nickel-based anode, a cadmium cathode and a potassium hydroxide electrolyte. Abbreviation: nicad cell.

Node Any point on a transmission line, where standing wave is of zero value.

Noise Unwanted signals occurring in an electronic system, causing spurious output signals. Noise can be the result of man-made causes, or natural causes. Many different types of noise exist, named after their basic nature, e.g. thermal noise, atmospheric noise, white noise, impulse noise.

Noise factor The ratio of a device's or circuit's input signal-to-noise, to its output signal-to-noise. Synonymous with noise figure.

Noise figure *See* Noise factor.

Non-linear Any circuit or system which produces an output which is not directly proportional to its input at all times is said to be non-linear.

Nonvolatile memory Type of memory in which data is maintained even when power is disconnected.

NOR gate Logic circuit whose output is high if all the inputs are low. If one or more inputs are high, the output is low.

NOT gate Logic circuit whose output is always the inverse of the input. Synonymous with inverter.

NPN transistor A bipolar transistor formed by three layers of semi-conductor material – the outside two layers being of N-type material, the middle layer of P-type material.

NTSC Abbreviation for National Television System Committee. An American committee, responsible for television standards. The initials NTSC are often comically described as never the same colour – referring to the constant colour changes inherent in the system.

N-type semiconductor Semiconductor material containing a higher concentration of negative charge carriers, i.e. electrons, than positive charge carriers, i.e. holes.

Numerical control Type of automatic control system in which a number generated by the controlling device is compared with a number generated by another device. The difference between the two numbers is detected by the controlling device and used to generate a control signal.

Nyquist diagram Graph of a system's performance, which may be used to determine the system's stability under untested criteria.

OCR Abbreviation for optical character reader.

Octave A difference or interval between two sounds, whereby one sound is twice the frequency of the other.

Off-line A computer peripheral which is unconnected to the computer is said to be off-line.

Ohm The unit of resistance, reactance and impedance. One ohm is the resistance between two points when a constant current of one amp flows as the result of an applied voltage of volt between the points.

Ohmic A material which follows Ohm's law is said to be ohmic.

Ohmmeter An instrument which measures resistance.

Ohmmetre The unit of electric resistivity. Symbol: Ωm.

Ohm's law Law which defines the linear relationship between the voltage applied across a material, the current produced through the material, and the resistance of the material. Ohm's law can be written:

$$V = IR$$

One shot Synonym for a monostable multivibrator.

On-line A computer peripheral which is connected to and receiving or transmitting data from or to a computer, is said to be on-line.

Opamp Abbreviation for operational amplifier.

Open circuit Term applying to a circuit or system whose output is not connected to any following circuit or system. The output is therefore unloaded. Measurements of electrical parameters at this time are said to be under open circuit or no lead conditions.

Operating point The point on a semiconductor device's characteristic curve, representing electrical parameters when defined conditions are applied to the device.

Operational amplifier An amplifier, generally in integrated circuit form, which is usable with only a few components and power supply connections.

Optical character reader A computer input peripheral which is capable of converting symbols printed on paper into digital signals.

Oracle The Independent Broadcasting authority's version of broadcast teletext.

OR gate Logic circuit whose output is high if one or more of its inputs are low.

Oscillation A periodic variation of an electrical parameter.

Oscillator A circuit system which produces on oscillation.

Oscilloscope Test equipment which is able to produce a visual display of one or more oscillations of voltage. Generally the device used to display the voltages is a cathode ray tube and such oscilloscopes are often referred to as cathode ray oscilloscopes (shortened to CRO).

Output 1 The part of a circuit or system which produces an output signal. 2 The signal produced by a circuit or system.

Output impedance The impedance of the output of a circuit or system.

Overall efficiency The ratio of the power absorbed by a circuit or system to the power supplied by a source.

Overdamping Damping applied to a period oscillation which prevents the oscillation from completing one cycle before stopping.

Overdriven Term, generally applied to a linear system such as an amplifier, which refers to the state when the size of input signal is such that the system's output is non-linearly related. In the case of an overdriven amplifier the output sounds harsh and is known as distorted.

Oxidation A process in the manufacture of semiconductor devices when the semiconductor base material undergoes a reaction with oxygen, to form a semiconductor oxide.

PA Abbreviation for public address system.

PABX Abbreviation for private automatic branch exchange.

Packing density The number of transistors or gates in unit area on an integrated circuit chip.

Pair Two similar conductors, insulated from each other but running in parallel, forming a transmission line. Generally, the pair is in the form of wire, e.g. twisted wire pair, coaxial cable.

PAL Abbreviation for phase alternation by line.

PAM Abbreviation for pulse amplitude modulation.

Parallel Components are said to be in parallel if current from a single source divides and flows through them then later reunites.

Parallel circuit A circuit containing two or more components connected in parallel.

Parallel plate capacitor A capacitor formed from two parallel conductive plates, between which is the dielectric.

Parallel resonant circuit A circuit containing a capacitance in parallel with an inductance, which exhibits resonance.

Parameter A criterion of an electronic component, circuit, or system. Typical parameters are voltage, current, resistance, capacitance, etc.

Parametric amplifier 1 A microwave frequency amplifier, whose reactance is varied in a regular manner. 2 An audio frequency amplifier which can amplify or attenuate specific frequency signals, while passing other signals unaltered. It is thus a bandpass/band reject filter, whose centre frequency may be adjusted.

Pascal A high level programming language.

Passive Any component which does not introduce gain is known as a passive device.

P-channel The conducting channel of a field-effect transistor of P-type material. The term is also used to refer to the transistor, i.e. P-channel field-effect transistor.

PCM Abbreviation for pulse code modulation.

PD Abbreviation for potential difference.

Peak-to-peak amplitude The difference between extreme values of a periodic oscillation.

Peak value The extreme value of a periodic oscillation.

Period The time to complete a single cycle of an oscillation. Symbol: T.

Periodic Term used to describe any variable which exhibits a regularly occurring form.

Peripheral devices Devices which connect to a computer.

Permanent memory Non-volatile memory, i.e. memory, the contents of which remain intact without a supply of power.

PFM Abbreviation for pulse frequency modulation.

Phase The amount by which a periodic variable has progressed from a reference point. Phase can be measured as an angle or in radians. Two periodic variables with the same frequency and waveform which reach corresponding stages simultaneously are said to be in phase. If this does not occur, they are said to be out of phase.

Phase alternation by line A colour television system, variations of which have been adopted throughout Europe, in which the colour signal (known as the chrominance signal) is resolved into two components and transmitted separately. The phase difference of these two components is reversed on alternate lines, a procedure which helps to reduce errors due to received phase variations. Abbreviation: PAL.

Phase difference Difference in phase between two sine waves of the same frequency.

Phase modulation Modulation in which the phase of a carrier wave is varied by an amount proportional to the amplitude of the message signal. Abbreviation: PM.

Phase shift keying Alternative name for simple phase modulation of a digital signal. Abbreviation: PSK.

Photocell A transducer which converts light to some parameter of electricity.

Photodiode A semiconductor diode device, which conducts electric current by an amount proportional to the quantity of light falling on it.

Photoresist Photosensitive material which changes in molecular ways upon exposure to light. Photoresists are used in the manufacture of semiconductors and integrated circuits and printed circuit boards.

Pick-up A transducer which converts recorded signals into electrical signals.

Pico- Prefix denoting a multiple of 10^{12}. Symbol: p.

Picture element The smallest portion of a graphic or pictorial display system which can be resolved by the system. Often shortened to pixel.

Piezoelectric crystal A crystal which displays the piezoelectric effect.

Piezoelectric effect An effect observed in certain materials when a voltage is generated across the faces of the material as a mechanical stress is applied.

p-i-n diode A diode which contains a layer of intrinsic, i.e. pure, semiconductor between the P and N layers.

PM Abbreviation for phase modulation.

PN junction The junction between two layers of semiconductors of P-type and N-type origin.

PNP transistor A bipolar transistor formed by three layers of semiconductor materials – the outside two layers being of P-type material, the middle layer of N-type material.

Point contact device One in which the pn junction is formed at the contact between a metal 'cats-whisker' and the semiconductor material. Point contact diodes have advantages in some applications.

Polarised Term used to describe any component or device which must be inserted into a circuit a particular way round.

Positive feedback Type of control procedure in which part of the output signal of a circuit is fed back to the input terminal in such a way that the circuit regenerates the signal, resulting in greater amounts of signal fed back, resulting in further regenerates. Generally, the result of positive feedback is to form an oscillation.

Pot Abbreviation for potentiometer.

Potential Abbreviation for potential difference.

Potential difference The voltage across two points.

Potential divider A circuit consisting of a number of series components. Tapping at one of the junctions between components allows a fraction of the total applied voltage to be obtained. Synonymous with voltage divider.

Potentiometer A form of variable resistor with three contacts. A voltage is applied across the outer two (across the total resistance) and the third contact (the wiper) may be varied along the length of the resistance forming a variable voltage divider. Abbreviation: pot.

Power Rate at which energy is used up or work is done. The electrical unit of power is the watt. Abbreviation: W. Symbol: P.

Power ratio The unit of acoustical or electrical power measurement in comparison with a standard level is the bel. In practical terms, power ratios are usually expressed in decibels (dB).

Power supply A source of electrical power for electronic circuits. Usually the power supply is integral to the equipment. Abbreviation: PSU (for power supply unit).

Power transistor A transistor which operates at high values of power.

PPM Abbreviation for pulse position modulation.

Preamp Abbreviation for preamplifier.

Preamplifier Part of an amplifying system which amplifies small applied input signals, generally amplifying in terms of voltage amplitude only.

Preferred values Predetermined component values. Their use makes component manufacture relatively simple, as only a selected few values need be manufactured, not every possible value.

Prestel *See* Videotex.

Primary cell A cell whose structure does not allow it to be recharged.

Printed circuit board Method of manufacturing electronic products, in which all or most of the circuit is constructed on a thin board (the printed circuit board). Connections between components are formed with thin strips of copper. Abbreviation: PCB.

Printer Computer peripheral which prints characters or symbols onto paper.

Program The complete set of instructions which can control the operation of a computer.

Programmable read only memory Computer memory which may be programmed, i.e. have data written into it, once. After this it may only be read from. Abbreviation: PROM.

Programming language Any language which may be understood by computers and humans. Computers ultimately require instructions in machine code, so this is the simplest programming language. It *can* be understood by humans but not easily. Low level programming languages resemble machine code and are thus still difficult in terms of human use. High level programming languages resemble human languages and are thus easier for humans to use.

PROM Abbreviation for programmable read only memory.

P-type semiconductor Semiconductor material containing a higher concentration of positive charge carriers, i.e. holes, than negative charge carriers, i.e. electrons. In effect holes are simply a depletion of electrons, but nevertheless can be viewed as small objects which carry a charge through a semiconductor.

Public address system Sound reproduction system used to amplify sound and thus allow it to be relayed to many people over a large area. Abbreviation: PA.

Pulse A single variation in voltage or current from a zero value, to a maximum and back to zero.

Pulse amplitude modulation Pulse modulation system in which the amplitude of a pulse is modulated with respect to the amplitude of a message signal. Abbreviation: PAM.

Pulse code modulation Pulse modulation system in which pulses are produced corresponding to the message signal. Abbreviation: PCM.

Pulse modulation Any modulation system in which a train of pulses is used as the carrier. Abbreviation: PM.

Pulse position modulation Pulse modulation system in which the position of each pulse is related to the message signal. Abbreviation: PPM.

Pulse width modulation Pulse modulation system in which the width of each pulse is modulated with respect to the message signal. Abbreviation: PWM.

Push-pull Circuit operation in which two devices operate totally out of phase.

Q-factor Abbreviation for quality factor.

Quadrophonic Referring to a sound reproduction system with four separate sound channels.

Quadrature Two sine waves of the same frequency but 90° out of phase are referred to as being in quadrature.

Quality factor A variable which describes the selectivity of a circuit. It is typically used in conjunction with resonant circuits. The quality factor may be calculated from the expression:

$$Q = \frac{B}{f}$$

where Q is the quality factor, B is the bandwidth of the circuit, and f is the centre frequency of the circuit. Abbreviation: Q-factor.

Quantisation The production of a number of quantised, i.e. discrete, values which may be used to describe a continuous waveform. The best example of the use of quantisation is in the process of pulse modulation, where the sampled values are used to define some aspect of a pulse train.

Quartz A type of crystal which exhibits the phenomenon of piezoelectricity.

Quartz-crystal oscillator An oscillator which relies on the principle that crystal will vibrate at a fixed natural frequency.

Quiescent current Current which flows through any component or part of a circuit under normal conditions, when no signal is applied.

Quiescent point Point on a semiconductor's characteristic curve representing the parameters of the device when in a quiescent state.

Radar An acronym of radio direction and ranging. A system capable of locating distant objects using reflected radio waves.

Radiation Any form of energy transmitted as electromagnetic waves, or as streams of particles.

Radio The use of electromagnetic radiation within the frequency range of about 3 kHz to 300 GHz to transmit information without connecting wires.

Radiowave Any electromagnetic radiation with a frequency within the radio frequency range of about 3 kHz to 300 GHz.

RAM Abbreviation for random access memory.

Random access memory Type of computer memory which may be accessed randomly, i.e. directly (nonsequentially).

Raster Term describing the pattern of lines on a television-type display screen, which occurs at all times.

Ratings Specification sheets for transistors cover many facets of the device's operation but most parameters are needed only by the designer. The ratings which need to be known for replacement purposes are $V_{CF(max)}$, the maximum collector to emitter voltage; I_C the collector current; h_{fc}, the gain and f_1 the cut-off frequency. The output power also needs to be observed.

RC Abbreviation for resistor-capacitor.

RC network Abbreviation for resistor-capacitor network. Any circuit or network which consists primarily of a resistor and a capacitor.

Reactance The part of the total impedance of a circuit, which is due to capacitance or inductance, and not to resistance. Reactance causes the current and voltage to become out of phase (in a circuit of pure resistance, current and voltage are in phase). Symbol: X. Unit: ohm.

Reactive load A load with reactance, which thus causes the applied current and voltage to be out of phase.

Reactor A component with reactance, i.e. a capacitor or inductor.

Read To retrieve information previously stored in a computer-type memory device.

Read only memory A computer-type memory device, from which information can only be read from, and not stored into. The information held in a read only memory is generally stored at the manufacture stage and is specific to the operation of the computer. Abbreviation: ROM.

Read-write bend Device used to record and retrieve information to and from a magnetic memory.

Real time operation Use of a computer during the actual time a process is occurring, to monitor and control the process.

Receiver The part of a communication system which receives encoded information from a transmitter, and decodes it to the form required.

Record Any permanent or semi-permanent storage of electrical information.

Rectifier Any device which passes current in only one direction. A rectifier is thus an a.c.-to-d.c. convertor.

Redundancy 1 The use of extra components in a circuit or system to ensure that breakdown of one component does not affect operation of the circuit or system.

Redundancy is a method of increasing reliability. 2 Inclusion of extra information in a transmitted signal which may be eliminated without loss of essential information.

Refractive index The ratio (n) of the velocity of light in free space to that in the material.

Refresh The restoration of information stored so that the information is not lost, typically in a dynamic memory device, or in devices with a destructive read operation.

Regeneration Synonym for positive feedback.

Register One of the temporary storage locations within the central processor of a computer, used to store the results of operations and calculations performed.

Regulator A circuit or device which manitains a constant output voltage or current, regardless of input voltage or output current requirements.

Rejection band The band of frequencies which are not passed through a filter.

Relative permittivity The ratio of the difference between the permittivity of a capacitor dielectric and the permittivity of free space.

Relaxation oscillator An oscillator which relies for its operation on an increasing and decreasing current or voltage within each period of oscillation.

Relay An electrical component in which an applied voltage or current electromagnetically operates a switching mechanism. The contacts of the switch can be isolated from the electromagnet providing a means whereby separate circuits may be interfaced without the need for electrical contact. Modern relays, although providing the same function, are often of a solid state form.

Reliability The ability of a component, circuit, or system to perform its functions for a given period of time.

Reluctance The magnetic equivalent of resistance.

Repeater A device or circuit which amplifies, regenerates, or restores to its original condition a signal in a telecommunications system, which has deteriorated due to transmission over a distance.

Resistance The ability of a material to resist the flow of electric current and to convert electrical energy into heat. A material's resistance is given by the ratio of applied voltage across it to the current flow through it caused by this voltage. Symbol: R. Unit: ohm.

Resistivity The ability of a material to have a resistance dependent on the material's cross-sectional area and its length. The resistivity of a material is given by

$$\rho = \frac{RA}{L}$$

where ρ is the resistivity, R is the resistance, A the cross-sectional area, and L the length.

Resistor An electronic component which possesses resistance. A pure resistor possesses only resistance, no capacitance or inductance, but all practical resistors possess some small amount of capacitance or inductance. Usually these are sufficiently small to be negligible.

Resonance Phenomenon arising when a circuit or system is excited by an applied signal, so that a small input signal produces a relatively large output signal, at the system's resonant frequency.

Resonant frequency The frequency at which a resonant circuit naturally resonates. Symbol: ω.

Reverberation The persistance of sound inside an enclosure, due to multiple reflections from the inside surfaces of the enclosure.

569

Reverberation time The time required from the cessation of a sound, for the intensity to fall by 60 dB (that is, one millionth of the original value). The unit of reverberation time is the second.

Reverse bias Voltage applied to a PN junction, such that the P-type layer of semiconductor is negative with respect to the N-type layer. Synonymous with reverse voltage.

Reverse Voltage Synonym for reverse bias.

Rewrite Synonym for refresh.

RF Abbreviation for radio frequency.

Rheostat A variable resistor used specifically to alter the current flow in a circuit.

Right-hand rule *See* Fleming's right-hand rule.

Ringing The delay which a system exhibits in returning to its quiescent state after a sharp pulse input, due to inherent resonance within the system. Generally a period of oscillation occurs, gradually dying away. Damping the system will reduce this period.

Ripple A small a.c. signal superimposed on a d.c. voltage or current, typically found on the output of a d.c. power supply, where the frequency of the ripple is mains frequency, i.e. 50 Hz or sometimes twice this frequency.

Rise time The time taken for a pulse's leading edge to rise from 10% to 90% of its final value.

RMS Abbreviation for root mean square.

ROM Abbreviation for read only memory.

Root mean square Term used to describe the effective value of an a.c. waveform. It is the square root of the mean value of the squares of the instantaneous values of the waveform. In the specific case of a sinewave, the root mean square value is equal to the peak value divided by $\sqrt{2}$. Abbreviation: RMS.

RS flip-flop *See* Bistable.

Rumble Unwanted noise heard in a hi-fi system, caused by mechanical vibrations in the record playing deck, of low frequency.

Sampling The extraction of portions of an electrical analogue signal, used to produce a series of discrete values.

Satellite Artificial body in orbit around the earth for purposes of communications, either one-way from the satellite to the earth, or two-way from earth to satellite and back.

Saturation When the output current of an electronic device is constant and independent of input.

Sawtooth oscillator A relaxation oscillator which produces a sawtooth shaped waveform.

Scanning Process of controlling the electron beam horizontally across and vertically down the face of a cathode ray tube device.

Schematic Circuit diagram.

Schmitt trigger Bistable circuit in which the binary output is determined by the magnitude of the input signal in such a way that the circuit exhibits hysteresis – the output changes when the input exceeds a predetermined level, and changes back when the input falls below a lower predetermined level.

Scramble Process of rendering a communications signal unintelligible at the receiver unless a descrambling circuit is used.

Screen 1 Surface of a cathode ray. 2 Shield to prevent electromagnetic interference.

SCS Acronym for silicon controlled switch.

SECAM Acronym for sequential couleur à memoire; a line-sequential colour television standard.

Secondary cell Rechargeable cell.

Secondary emission Emission of electrons from a material as the result of a bombardment by high-velocity electrons or positive ions.

Secondary voltage Voltage developed across the secondary windings of a transformer.

Selectivity Ability of a radio receiver to discriminate against carrier frequencies different to that selected.

Semiconductor device Device whose operation is based on the use of semiconductor material. In addition to transistors and diodes there is a wide range of components which make use of semiconductor effects.

Semiconductor material Material whose conductive properties depend on the addition of minute quantities of impurity atoms. Unlike normal conductors, semiconductors increase in conductivity with an increase in temperature.

Sensitivity 1 The change in output of a device per unit change in input. 2 Ability of a radio receiver to respond to weak input signals.

Sensor Transducer.

Serial transmission Communication method in which characters are transmitted in turn along a single line.

Series Components in series have one current flowing through each.

Shift register Digital store of information, in which the information is displaced one place in either direction on application of a shift pulse.

Short circuit Unwanted electrical connection between two points in a circuit.

Short wave Radiowave in the wavelength range from 10 to 100 metres.

Shunt Parallel connection.

Siemens Symbol: S. The SI unit of electrical conductance.

Signal Variable electrical parameter.

Signal generator Device which can generate a controlled signal.

Signal-to-noise ratio The ratio of the value of signal at a point in a system, compared with the value of noise at the same point. Usually expressed in decibels.

Silicon Semiconductor element, most widely used element to form semiconductor devices.

Silicon controlled rectifier Abbreviation SCR. Thyristor.

Simplex Communications channel operating in one direction only.

Sinusoidal Waveform identical in shape to a sine function.

Slew rate The rate at which the output of a circuit can be driven from one limit to the other.

Smoothing circuit Circuit designed to reduce ripple in a direct current or voltage.

Solid state circuit A circuit in which the current flows through solid material instead of through a gas or vacuum.

Super alpha pair *See* Darlington pair.

Synchronous Clocked.

Telecommunications The transfer of information by any electromagnetic means.

Telemetry Measurement at a distance using electromagnetic means.

Telephone Communication of speech and/or other sounds via electromagnetic means.

Television Communication of video and audio information by electromagnetic means.

Thermal runaway Semiconductor materials are very sensitive to heat – germanium much more so than silicon. Circuit design has to take account of this and many components have to be included to prevent increased current flow due to heat. Without such protection heat induced current will raise the temperature leading to a further increase in current and so on, a process known as thermal runaway which can destroy a semiconductor. *See* Heat sink.

Thermionic emission Electron emission from the surface of a body, due to the temperature of the body.

Thermistor A semiconductor whose resistance varies with temperature. Some have a negative temperature coefficient, that is resistance falls with an increase in temperature, others have a positive temperature coefficient. Typical applications are to provide compensation for the effects of heat on circuit operation.

Thévenin's theorem A theorem used to simplify the analysis of resistance networks.

Threshold of hearing The sound level or intensity which is just audible, for an average listener. For a pure sinusoidal tone of 1,000 Hz, it corresponds approximately to a root mean square pressure of 2×10^{-5} Pa.

Thyristor Three junction, four layer semiconductor rectifier which conducts when either the voltage across it reaches a breakdown point or when triggered by a pulse at its gate electrode. Once triggered it remains conducting until the voltage across it becomes zero.

Transducer Any device that converts one parameter into another, where one of the parameters is an electrical signal.

Transformer Device which transforms electrical energy at its input to electrical output at its output. Usually the voltages of the electrical energies differ.

Transistor Semiconductor device in which the current flowing between two electrodes may be modulated by the voltage or current applied to other electrodes.

Triac Bi-directional thyristor.

Tunnel diode A heavily doped semiconductor diode which exhibits a negative-resistance characteristic, i.e. over part of its characteristic increased forward bias leads to a reduction in the current flowing.

Type numbers The numbers in a transistor designation rarely describe anything about its characteristics. In the 2N series adjacent type numbers are frequently widely coded with the first letter A (germanium) or B (silicon) followed by a second letter which indicates the type:

A	Diode	P	Photo type
C	a.f. (low power)	S	Switching (low power)
D	a.f. (power)	U	Switching (power)
E	Tunnel diode	Y	Diode (power)
F	h.f. (low power)	Z	Zener diode
L	h.f. (power)		

UHF Ultra high frequency.

Ultrasonics Sound frequencies above the limits of human ears, generally classed as above 20 KHz.

Ultraviolet radiation Electromagnetic radiation of wavelengths between visible light and X-rays.

Unijunction transistor Three terminal transistor comprising an n-type silicon bar with a base contact at each end (base 1 and base 2) and a p-type emitter region. Current flow from one base to the other is controlled by the emitter current; when the emitter voltage reaches a certain level the emitter-base 1 junction virtually short circuits. With a suitable charging circuit at the emitter the device operates as a relaxation oscillator.

Valency The ability of atoms to unite with other atoms due to the electrons that exist in the outer orbit, or valency band, being able to form a shared orbit with other atoms.

Varicap diode Varactor. When reverse biased, all pn junctions exhibit capacitance, as the depletion layer at the junction forms an insulator between the conductive regions. This property is used in the varicap, in purposes such as automatic tuning and AFC in radio receivers.

VDU Abbreviation for visual display unit.

VHF Abbreviation for very high frequency.

Voltage-dependent resistor Resistor using semiconductor material whose resistance varies with applied voltage.

Voltage drop Voltage between any two points of a circuit, due to the current flow between them.

Watt Symbol: W. SI unit of power.

Wave A periodic motion, through a medium (which may be space) in which the propagation from a point is a function of time and/or position.

White noise Noise with a wide, flat frequency response.

Word A string of bits corresponding to a unit of information in a digital circuit.

Wow Low frequency (below 10 Hz) periodic variations in the pitch of the sound output of a sound reproduction system.

Write To enter information into a storage element.

Yagi aerial Directional aerial array – most television aerials are based on the Yagi aerial.

Zener diode Voltage regulating diode. A pn junction diode which has a defined reverse breakdown voltage. Once in the breakdown region large increases in current produce negligible variation in the voltage across it.

INDEX

A/D 37
Abbreviations 359
Abnormal propagation 26
Absolute delay 40
AC voltage gradient 4
Activity 113
ADC 37
Adjacent channel selectivity 170
AGC 163
Ageing 111
Air density 23
Air-articulated dielectric cable 58
Alternating current 4
AM 124
AM broadcast station classes (USA) 348
AM splash 124
AM transmission 157
AM transmitter 157, 159
Amateur abbreviations 446
Amateur radio bands (UK) 431
Amateur radio emission designation 440
American military nomenclature 114
Ampere's rule 460
Ampere's theorem 460
Ampere-hour 262, 397
Amplitude modulation 124–125, 133, 141

Amplitude modulation transmitters 155
Amplitude shift keying 136
Analog-to-digital converter 37
Analogue modulation 124
Analogue signals 121
Angle modulated transmitters 157, 159
Angle modulation 129
Angular velocity 273
Anhydrous nitrogen 57
Antenna 13
Antenna characteristics 61
Antenna gain 61
Antenna placement 24
Antenna radiation angle 24
Antenna types 66
Antennas 61
Aperture 65
Aperture ratio 80
Arc length 479
ASCII control characters 459
ASK 136
Assembly of connectors 288ff
Astronomical data 487
AT-cut crystal 109, 111
Atmospheric attenuation 6

Atmospheric conditions 20
Atmospheric losses 7
Atmospheric noise 12
Atmospheric oxygen attenuation 7
Attenuation 116–117
Attenuation loss 56
Audible frequency range 472
Audible intensity 473
Audio connectors 284
Audio frequency response 170
Audio output 170
Auroral propagation 26
Australia TV 345
Automatic calling 298
Automatic gain control 163
Axial mode helix 81
Azimuthal 21

Back scatter 21
Balanced line hybrids 41
Balanced modulator 127
BALUN 52
Band-IV 343
Bandpass circuit 94
Bands 2
Band-V 344
Bandwidth 12, 61, 125, 130, 134, 137, 147, 168
Bandwidth BW_1 116
Bandwidth BW_2 117
Bandwidth requirements 121
Base band 121
Base band lines 39
Base band widths 121
Base bandwidth 137

BASK 136
Batteries 262, 264
Baud rate 123
Baur's constant 461
BAW 101
Baying antennas 76
BBC AM radio stations 310
BBC VHF broadcasting 311
BBC VHF FM radio stations 311
BBC VHF text tone transmissions 317
BBC weather forecasts 414
BC-cut crystal 109
Beamwidth 61
Beat frequency oscillator 136
Beaufort scale 416
Bessel functions 131
BFO 136
Binary amplitude shift keying 136
Binary coded information 123
Binary decibel values
Binary FFSK 137
Binary phase shift keying 138
Binary signal rate 141
Bit error rate 41
Bit rate 123
Blocking 170
Boolean algebra 402
Boundaries of sea areas 415
BPSK 138
British Broadcasting Corporation 318
British Telecom 39
Broadcast transmitters 56

Broadcasting 307
Broadside array 68–68
BS-6552 285
BSI standard metric wires 503
BT weather forecasts 414
BT-cut crystal 109, 111
Bulk acoustic wave 101
Bulk delay 40

Caesium frequency standard 107
Calculus 478
Camera connectors 285
Candela 397
Capacitance 5, 402, 406
Capacitance, coaxial cable 59
Capacitive reactance 78
Capacitor colour code 407
Capacitor letter code 409
Capacity, cell 262
Capture area 65
CCITT 40
CCITT frequencies 137
CCITT recommendations 464
CCITT V-24 295
CCITT-2 code 122
CDMA 142
Celcius-Fahrenheit conversion table 474
Cell case temperature 272
Cell characteristics 262
Cells 272
Centre frequency 117
Centronics interface 301

Ceramic filter 116
Channel capacity 123
Characteristic impedance 5, 42, 53, 70, 402
Characteristics of UHF TV 340
Circle 479
Circuit condition 116
Circular polarization 64
Class-A stations 348
Class-B stations 349
Class-C stations 349
Class-D stations 349
Classes of radio station 152
Coaxial cable 52, 56, 60
Coaxial cable capacitance 59
Coaxial cable cut-off frequencies 59
Coaxial cable, types of 56
Coaxial connector 286
Code conversion tables 453
Code of Practice 138
Collinear antennas 74
Colour codes 407
Colpitts oscillators 98
Communications by satellite link 275
Communications receiver 164
Component symbols 375
Conical antenna 76
Connections 295
Connectors 284, 295
Constant current automatic charging 272
Continuous wave 136, 141

Conversion factors 447
Coordinated universal time 418
Copper wire data (SWG) 510
Coulomb 397
Coulomb's law 461
Coupled bandpass circuits 93
Critical coupling 96
Critical frequency 18
Cross modulation 170
Cross polarization 64
Crystal filter 116
Crystal oscillator 133
CTCSS decoding 172
CTCSS encoding 172
Current distribution 5
Curvature of earth 9
Cut-off frequencies 59
CW 141
Cylinder 481

D2-MAC 282
Data circuit terminating equipment 295
Data processing 134
Data terminating equipment 295
dB 28, 397
dB to any ratio conversion 37
$dB\mu V$ 34
dBa 34
dBa0 34
dBd 34, 61
dBi 34
dBm 35
dBm to watts conversion 37
dBm0 35
dBm0p 36
dBmp 36
dBr 36

dBrn 36
dBrnc 36
dBrnc0 36
DBS 276
DBS television channels 277
DBS transmission 282
dBu 36
dBV 36
dBW 36
DCE 295
DDS 37
Decibel 397
Decibel figures 29–34
Decibel glossary 34
Decibel scale 28
Decibels referred to absolute values 28
Decimal multipliers 399
De-emphasis 133
Density inversion 9
Department of Trade and Industry 145
Depth of modulation (AM) 124
Derivatives 478
Desensitization 170
Designation of radio emissions 146
Deviation FM 133
Deviation ratio 130
Dielectric 57, 59
Dielectric constant 403
Dielectric losses 92
Dielectric properties of air 22
Differential delay 118
Differential phase 118
Differential phase shift keying 139
Differentially ionized 21
Differentiation 478

Diffraction 10–11, 15
Digital modulation 134
Digital signals 121
Digital-to-analogue converter 37
DIN standards 284
Dipole 65, 68
Dipole antenna 5
Dipole lengths (amateur bands) 440
Direct Broadcasting Satellite 276
Direct component 18
Direct digital synthesis 37, 106
Direct wave 16
Direction of propagation 55
Directional array 68
Directivity 61, 70, 73
Discone antenna 76–77
Disposable batteries 264
Distance 7
Distortion 130
Distortion with feedback 405
D-layer 26
D-MAC 282
Domestic class 348
Domestic Data Bus 286
Doppler effect 13
Doppler frequency shift 14
Double sideband amplitude modulation 126
Double sideband suppressed carrier 126, 127
Double-shielded coaxial cable 58
Doublet antenna 5

Down lead 68
DPSK 139–140
Drill sizes 498
Drive level 116
Drive level linearity 120
Drive level stability 118
DSB 124
DSBSC 126, 127, 128
DTE 295
Dual modulus pre-scaler 105
Duct 24
Ducting 9, 23–24
Duplex separation 170
Dynamic resistance 403

E.m.f. 108
Earth conductivity 9
Earth currents 8
Earth orbits 273
Effect of aging 112
Effective capture 164
Effective height 62, 89
Effective length 62
Effective parallel resistance 113
Effective radiated power 63
Effective series resistance 113
Efficiency 63
EIA 299
Electric current 54
Electric field 4, 6, 85
Electric quantities 368
Electrical properties of elements 492
Electrical relationships 395
Electromagnetic interference 58

Electromagnetic wave 1, 4, 54–55, 403
Electronic multiple conversion 400
Electrostatic discharge 59
Ellipse 480
Elliptical orbit 275
Elliptical waveguide 55
EMI 58, 85
End-fire array 68–69
Energy quanta 463
Engineering information (broadcast) 318
E-plane 6, 64, 69
Equatorial orbit 273
Equivalent circuit (crystal) 109
Equivalent noise temperature 166
ERP 63
ESR 113
Euroconnector 285

Fading 12
Farad 398
Faraday effect 461
Faraday's law 461
Fast code time and date BCD 309
Fast frequency shift keying 137
FFSK 137–139
Field strength 15
Figure of Merit 166
Filter 128
Filtering 134
Filters 116
Finishing charge 264
Flat loss 116
Fleming's rules 461
Flexible coaxial cable 56–57

Flexible microwave coaxial cable 58
Float charge 264, 270
FM 25, 128–129, 133
FM broadcast band 22
FM broadcast frequencies (USA) 349
FM receiver 164
FM transmission 130
FM transmitter 133, 159
Foam cable 56
Forward gain 61
Forward scatter 21
Four-level Gray code 140
Fractional bandwidth 117
Fractions inch metric equivalents 452
Free space loss 7
Free space power loss 6
Frequency 1, 7, 114, 403
Frequency band table 3
Frequency designations 147
Frequency modulation 104, 129, 133, 159
Frequency modulation deviation 133
Frequency planning 144
Frequency selective fading 12
Frequency shift 14
Frequency shift keying 136
Frequency stability 102, 111, 114
Frequency synthesizer 101, 155

Frequency tolerance 114
Friis noise equation 167
Front-to-back ratio 63
Frustrum of cone 480
FSK 136–139, 142
FT-cut crystal 109
Fundamental constants 395
Fundamental frequency 110
Fundamental units 396
Fuses 475

Gain 28
Gas-filled line 57
Gaussian filter 134
Gaussian minimum phase shift keying 137
General frequency allocations 148
Geostationary orbits 273
German Industrial Standard Board 284
Global positioning system 282
Glossary 517
GMSK 137
GPS 282
Gray code 140
Gray coding 135–136
Great circle 355
Great Circle Bearings 353
Great circle path 25, 355
Greek alphabet 397
Greenwich mean time 355
Grey line propagation 26

Ground plane antenna 79
Ground stations 273, 275
Ground wave 15
Group delay 40, 118
Grover equation 89

Half-power beam width 61–62, 81
Hall effect 461
Hand-portable antenna 79
Hardline 57
Hartley oscillators 98
Hartley-Shannon theorem 123
H_{eff} 89
Helical antenna 80
Helical line 57
Henry 398
HEO 273
Hertz 398
HF 17, 21, 42, 67–68, 70, 148
High earth orbit 273
High frequency 16
High stability oscillators 112
Holder style 114
Home Office 145
Homodyne 164
Horizontal (E) plane 68
HPBW 61, 81
H-plane 6, 64, 69, 76
Human voices 473
Huygen's principle 11

Ideal receiver 165
IF 161, 163
IF amplifier 163
IF bandwidth 168
Image frequency 163

Impedance 40, 63, 66, 118, 404
Impedance at resonance 93
Impedance matching 38
Impedance of free space 5
Impedance, Characteristic 5
Inband intermodulation distortion 120
Independent Television Commission 319
Inductance 5, 404, 406
Initial charge rate 264
Insertion loss 40, 43, 116
Insertion phase 118
Integrals 478
Interfaces 284, 205
Interfering waves 11
Intermediate frequency 161, 163
Internal resistance 263
International "Q" codes 442
International abbreviations 446
International allocation of call signs 426
International Morse code 444
International planning 144
International Telecommunications Union 144
Intersymbol interference 137–138
Inverted-L 67
Ionized clouds 21
Ionized layers 11
Ionosphere 16–17, 26

Ionospheric scatter 21
Ionospheric skip 21
Irregular figure 479
ISI 137
ISO standard (BS-4000) 475
Isotropic radiator 2, 6
ITU 144

J.M.E. Baudot 122
Joule 398
Joule's law 461

K (dielectric constant) 403
Kerr effect 462
KeyLine 40
K-factor 9, 23
Kilovolt-ampere 398
Kilowatt 398

Lambert's cosine law 462
Latitude 354
Laws of electricity 460
Layered air masses 23
Lead acid batteries 269
Length of arc 479
Lenz's law 462
LEO 275
Letter symbols by unit name 363
LF 42, 67, 148
Light, velocity 399
Linear noise factor 56
Lithium battery 264, 269
LNA 167
Load capacitance 113
Load impedance 118
Lobe angle 69
Log-periodic antenna 70
Long path 25
Longitude 354

Longitudinal lines 26
Loop antennas 84, 87
Loop filter 104
Loop inductance 89
Loss 28
Loss frequency response 40
Lossy electrical device 56
Loudspeaker 164
Low earth orbit 275
Low noise amplifier 167
Low-pass (Gaussian) filter 134
Lumen 398
Luminescence 26

MAC format 282
Magnetic field 4, 55, 85
Manchester encoding 135
Matched line loss 56
Matching circuits 52
Maximum bit rate 41
Maximum frequency error 41
Maximum fundamental frequency 123
Maximum input level 118
Maximum usable frequency 18
Maxwell's law 462
Maxwell's rule 462
MCW 136
Mechanical axis 108
Mechanical deformation 59
Mechanical movement 58
Medium frequency 15–16

Medium wave 20
Memory effect (NiCad batteries) 270
Mensuration 478
Meteor ion trail 22
Meter conversion 404
Meters to kilohertz 459
Methods of coupling 93
Metric sizes of wire 515
Metric wire sizes (turns/10-mm) 509
MF 42, 67, 148
MFSK 138
Mho 398
Microwave antennas 80
Microwave band designation 441
Microwave communications 6
Microwave frequencies 23
Miller circuit 462
Miller effect 462
Minimum frequency shift keying 138
Mismatch 78
m-level data symbol 135
Mobile antennas 76, 78
Mobile data transmission 12
Mobile environment 12
Mobile radio 13
MODEM connector 297
Modulated continuous wave 136
Modulation 121, 124
Modulation depth 169
Modulation index 125, 129, 132–133
Morse code 136, 444

Motorcycle antenna 78
Mounting 114
MPT-1317 138
MPT-1326 100
MPT-1362 76
MSF Rugby 309
MSK 137
MUF 18
Multiple scatter 22
Musical notes
  (frequencies) 473
Mutual inductance 94
MW 15

Narrow band FM 141
National planning 144
Natural quartz crystal
  109
Negative feedback 404
New Zealand TV 345
Newton 398
Nicad 262, 264, 270
Nickel cadmium
  batteries 264, 270
Noise 12, 40, 164
Noise factor 56, 165
Noise figure 165 – 166
Noise in cascade
  amplifiers 167
Noise temperature
  165 – 166
Noise voltage 12
Noise, shot 13
Noisy ground plane 58
Non-chargeable primary
  batteries 264
Non-reciprocal direction
  26
Northern lights 26
Number code 444
Numerically controlled
  oscillator 106

Oblique prism 480
Ohm 398

Ohm's law 405
Omnidirectional 74
Omnidirectional normal
  mode helix 81
Omnidirectional
  radiation pattern 71
On/off keying 136
Optical axis 108
Oscillator frequency
  133
Oscillator requirements
  97
Oscillators 97
Out-of-band
  intermodulation
  distortion 120
Overmodulation 157
Overtone crystals 110
Overtone frequency
  101
Overtone oscillator 101

PAL signals 282
Paper sizes 475
Parabolic reflector 65
Parallel resonance 93
Parallel resonant circuits
  91
Parallelogram 479
Particles of physics 477
Pascal 398
Passband 116
Passband performance
  116
Path 25
Path length 16
Patterson equation 89
Peak carrier voltage
  125
Peak envelope power
  129
PEP 129
Peritelevision 285
Permeability 5

Permittivity 5
Phase linearity 118
Phase locked loops 102
Phase modulated
    transmitters 159
Phase modulation 133
Phase shift keying 138
Phonetic alphabet 445
Physical properties 396
Piezoelectric devices
    108
Piezoelectric effect 108
Piezoelectricity 59
Pilot carrier 129
Planck's constant 463
PLL 102
PM 159
Polar diagram 68, 76
Polarization 64, 71,
    311
Polarized 6
Polyethylene 56
Potential difference 88
Power 263, 405
Power in unmodulated
    carrier 126
Power output 157
Power ratio 405
Power relationships AM
    wave 126
Power: Volume ratio
    263
Power: Weight ratio
    263
Powers of numbers 465
Powers of sixteen 469
Powers of ten 467
Powers of two 465
Pre-emphasis 133, 159
Primary batteries 264
Prime meridian 354
Prismoidal 481
Programmable
    equipment 170

Programmable read only
    memory 171
PROM 171
Propagation 20
Propagation velocity 13
Propagation, direction of
    55
Propagation, methods of
    14
Pseudo-noise generation
    141–142
PSK 138–139, 142
Pullability 113

Q (figure of merit) 405
Q 113
Q codes 442
Q factor 93
QPSK 140
QRK code 443
QSA code 443
Quadrature phase shift
    keying 140
Quanta 463
Quarter wave section
    53
Quartz crystal
    108–109
Quartz crystal
    characteristics 111
Quartz crystal oscillator
    99, 157, 155
Quaternary phase shift
    keying 140

Radar systems 6
Radian 482
Radiation 6
Radiation lobe 61
Radiation pattern 64,
    67, 69
Radiation resistance
    63–64
Radiator 6

Radio astronomy allocations 447
Radio Authority 319
Radio communication 275
Radio Communications agency 100, 138, 145–146, 159, 170
Radio communications channels 2
Radio directional finding 84
Radio engineering 28
Radio equipment 155
Radio frequency carrier 121
Radio frequency filter 116
Radio frequency lines 42
Radio frequency oscillators 97
Radio frequency spectrum 2
Radio frequency wave 124
Radio horizon 10, 20–21, 24, 403
Radio Investigation Service 146
Radio obstacles 20
Radio spectrum 2
Radio station classes 152
Radio waves 17
Radio waves, Behavior 7
Radio waves, formation 4
Radio-3 317
Radiological Protection Board 80
Radius of curve 480
Radome 81

Rating for telephony 416
Rayleigh fading 20
RC time constants 410
RDF 84
Reactance 405
Reactance 92
Reactance of capacitors 414
Reactance of inductors 414
Receiver aperture 65
Receiver functions 160
Receiver specifications 169
Receiver, ideal 165
Receiver, types of 161
Receivers 160
Rechargeable batteries 264, 269
Recharging conditions 264
Rectangle 479
Rectangular waveguide 54
Reflected component 18
Reflected power 51
Reflected wave 16
Reflection 11
Reflection coefficient 51
Refraction 9, 11, 23
Refraction modes 22
Refractive index 9
Region-2 Class 348
Regional planning 144
Republic of Ireland TV 344
Resistance 405
Resistance, dynamic 403
Resistivity 489

Resistor colour code 407
Resistor digit code 409
Resistor letter code 409
Resonance 92–93, 406
Resonant circuits 91
Resonant frequency 111
Response of coupled circuits 95
Return loss 51
RF amplifier 157, 165
RF cable 50
RF line 42
RF PA 157
RF transformer 52
Rhombic antenna 69–70
Ripple 117
RL time constants 411
RMS 126
RS-232C 299
RS-449 299
RST code 444
Rubidium frequency standard 107

S/N 124, 168
Sabine's relation 463
Safety 80
Satellite communications 273
Satellite television 276
Satellite television formats 282
SCART (BS-6552) 285
Scatter 21
Scatter propagation modes 26
Second channel frequency 163
Segment of circle 479
Segment of sphere 480
SEI 8UA 146

Selectivity 170
Semi-flexible coaxial cable 57
Sensitivity 169
Series resonant (crystal) 99
Series resonant circuits 91
Shape factor 117
Short path 25
Shot noise 13
Shunt resistance 404
Side scatter 21
Sideband power 129
Siemens 398
Signal rating codes 416
Signal voltage 88
Signal-to-noise ratio 165, 168–170
Simple refraction 23
SINAD 161, 168–169
Single sideband AM wave 128
Single sideband suppressed carrier 127
SINPFEMO codes 417
SINPO code 418
SIO code 418
Skip distance 18
Sky wave propagation 16
Slow code time and date information 309
Small loop antenna 84
Small loop antenna patterns 87
Small loop geometry 85
S_N 165
Snell's law 463
SNR 165
Solid revolution 481
Sound 1, 471

Sound velocity 399, 471
Sounds and sound levels 471
Source impedance 118
South Africa TV 345
Southern lights 26
Space wave 16
Space wave propagation 18
Specification 159
Specifications, receiver 169
Spectrum 125
Speed of light 4
Sphere 480
Spherical zone 480
Spheroid 480
Sporadic E-layer reflections 18
Sporadic-E propagation 21
Spread spectrum 142
Spread spectrum transmission 140
Spreading bandwidth 141
Spurious attenuation 118
Spurious radiation 78
Spurious response attenuation 170
Spurious responses 113
Square 479
Squelching oscillator 101
SSB 127–128
SSBSC 127
Stacking antennas 76
Standard deviation 477
Standard frequency formats 308
Standard frequency transmissions 307

Standard integrals 478
Standard pitch 473
Standard time transmissions 307
Standard units 397
Standard wire gauge 498
Standing waves 51
Statistical formulae 476
Stereo pick-up colour code 409
Stereo services 311
Stopband 117
Stopband performance 117
Subcarrier 136
Subharmonic frequency 155
Subrefraction 24
Super refraction 23
Superheterodyne 161, 163
Superheterodyne, double 164
Surface wave 15
Switching bandwidth 170
Symbols 359
Synchronization of clocks 135

Teflon 57
Telegraphy 122
Television channels (USA) 352
Television connectors 285
TEM 55, 59–60, 85
TEM mode 60
TEM mode frequency 60
Temperature coefficient 111

Temperature
  compensation 99
Temperature conversion
  formulas 475
Temperature range 114
Terrestrial TV aerial
  dimensions 347
Terrestrial TV channels
  343
Tesla 398
Thermal noise 12, 58
Thevenin's theorem 463
Time constant 406
Timing 445
Total line loss 56
Transformer ratios 406
Transistor circuits 486
Transistor letter symbols
  (bipolar) 369
Transistor letter symbols
  (field effect) 372
Transistor letter symbols
  (unijunction) 372
Transition band 117
Transmission line
  considerations 56
Transmission line filters
  52
Transmission line noise
  58
Transmission lines 38
Transmitter 155
Transmitter
  specifications 159
Transmitter, FM 159
Transmitters 157
Transmitters, angle
  modulated 157, 159
Transmitters, phase
  modulated 159
Transverse electric 55
Transverse
  electromagnetic wave
  55, 59

Transverse magnetic 55
Trapezoid 479
TRF 161
Trickle charge current
  270
Trickle charging 264
Trigonometric relations
  485
Troposcatter systems 12
Tropospheric
  propagation 24
Tropospheric refraction
  22
Tropospheric scatter 20
Tunable oscillators 97
Tuned radio frequency
  161
Tuned resonant circuits
  91
Turns ratio 407
TV 25
TV channels (Australia)
  345
TV channels (New
  Zealand) 345
TV channels (Republic
  of Ireland) 344
TV channels (South
  Africa) 345
TV channels (USA) 346
Twilight zone 26

UHF 23, 52, 64, 67,
  71, 74, 81, 151
UK 625-line TV system
  342
UK broadcasting band
  310
UK television
  transmitters 320
Unbalanced cable 53
USA TV 346
US-Canadian AM
  Agreement 348

Varicap diode   102
VCO   102, 159
Velocity of sound   471
Vertical (H) plane   68
Very low frequency   15
VHF   21–22, 24–25,
   52, 64, 67, 71, 73–74,
   79, 81, 151
VHF slot antenna   74
Video connectors   284
Video recorder
   connectors   285
Virtual height   16
VLF   15, 148
Volt   399
Voltage controlled
   oscillators   102
Voltage distribution   5
Voltage standing wave
   ratio   50–51, 64
Volt-ampere   399
VSWR   50–51, 54, 61,
   64, 78, 80

Water vapor attenuation
   7
Watt   399

Wattage rating   407
Wavefront   1
Waveguide   54
Wavelength   1, 6, 403
Wavelength of tuned
   circuit   407
Wavelength-frequency
   conversion table   459
Weber   399
Wedge   480
Wind loading   81
Wire data   498
Wire gauges   512
World Administrative
   Radio Conference
   144
World allocation DBS
   satellite positions
   279
World time   418

X-cut crystal   108

Z-cut crystal   108
Zero-IF receiver   164
Zulu time   355